Artificial Intelligence in Hydrology

In Focus – Special Book Series

Artificial Intelligence in Hydrology

Editors
Elena Volpi, Jong Suk Kim, Shaleen Jain and Sangam Shrestha

IWA PUBLISHING

Published by IWA Publishing
Unit 104-105, Export Building
1 Clove Crescent
London E14 2BA, UK
Telephone: +44 (0)20 7654 5500
Fax: +44 (0)20 7654 5555
Email: publications@iwap.co.uk
Web: www.iwapublishing.com

First published 2024

British Library Cataloguing in Publication Data
A CIP catalogue record for this book is available from the British Library

ISBN: 9781789064858

Contents

doi: 10.2166/nh.2023.102

Editorial: artificial intelligence in hydrology

Nowadays, hydrological systems are becoming increasingly complex owing to the growing interaction between nature and humans at the local scale of river sections, lakes, reservoirs, catchments, etc., to the global scale. There is great demand for the development of models to evaluate, predict, and optimize the performance of complex hydrological systems whose behavior is characterized by a strong nonlinearity. However, traditional approaches can hardly handle this nonlinear behavior; moreover, the analysis of hydrological systems at the large scale, even global, requires dealing with large-volume and real-time data. In recent years, artificial intelligence (AI), especially deep learning, has shown great potential to process massive data and solve large-scale nonlinear problems. AI has been successfully applied to computer vision, machine translation, bioinformatics, drug design, and climate science. AI models have produced results comparable to and even better than expert human performance. It is expected that AI can significantly contribute to hydrology research as well as development.

This book is dedicated to presenting to the hydrology community some of the latest advances in the field of AI in hydrology. After unbiased and rigorous peer reviews, eight articles were selected to be published in this book. Both theoretical and experimental articles are included, covering new and emerging AI methods and models from various challenging problems in hydrology. We hope this collection will attract more attention from academia and industry to this exciting area.

Roushangar *et al.* (2021) focused on short-term and long-term drought prediction. A hybrid model was developed integrating data preprocessing, permutation entropy, and AI for point and interval predictions of the standardized precipitation index. The proposed approach was studied in northwest Iran and general for application in all areas. The performance of the hybrid model was extensively evaluated with the comparison with single models. The experimental results indicated the high ability of the proposed hybrid model in drought indices modeling.

Xu *et al.* (2021) presented the adaptation and application of the temporal convolutional network in simulating the hourly rainfall–runoff relationship. The optimal hyperparameter combination of the network was investigated at different periods of flood forecast. Two datasets in Jingle and Kuye watersheds were utilized to evaluate the performance of an adapted temporal convolutional network compared with excess infiltration and excess storage model, artificial neural network, and long short-term memory (LSTM). The experimental results showed that the adapted temporal convolutional network outperforms the other approaches with a faster convergence rate and a higher flood forecasting accuracy.

Lu *et al.* (2021) studied the application of red edge band in water body remote sensing extraction. In the study, the GF-6 WFV dataset was considered owing to its high spatial resolution, remarkable number of bands, and wide width. A decision tree model was proposed based on the combined operations of the green band, red band, near-infrared band, red edge 1 band, and red edge 2 band. The extensive experimental results showed the superiority of the proposed decision tree model in terms of extraction effect.

The application of stochastic artificial neural networks in hydrology was studied by Wu *et al.* (2021). An adapted genetic algorithm was employed to calibrate the weights in the model. The developed approach was applied for accurate and real-time hydrological forecasts. A large-scale training dataset was simulated to train the proposed model via a nonnormal multivariate Monte Carlo approach. Finally, a real-time correction method is used to adjust the hydrological forecast.

Baghanam *et al.* (2022) studied the problem of selecting dominant large-scale climate variables, or predictors, in the statistical downscaling of climate models. In their article, a wavelet coherence transform was employed to assess the high common powers and the multi-scale correlation between two time series, i.e., predictand and predictor, as a function of time and frequency. Continuous wavelet transform was employed to identify the potent periodicity in the time series of predictands. Finally, an artificial neural network implements statistical downscaling.

de Moura *et al.* (2022) evaluated the performance of the LSTM network for discharge prediction under changing climate conditions. The LSTM approach was benchmarked with the Hydrologiska Byråns Vattenbalansavdelning (HBV) bucket-type model with two parameterizations. Two insights arose from the article. On the one hand, LSTM had good predictive accuracy in both the calibration and validation periods. However, LSTM was always less robust than the HBV model. On the other hand, when using the maximum number of years in calibration, the LSTM was robust enough in its application under changing conditions when applied in a condition drier than the one used in calibration.

Nourani (2022) points out that identifying various pixels with similar patterns from satellite images was a reliable method to give an appropriate view of the soil moisture condition of a region. To this end, an AI-based self-organizing map method was employed to classify homogenous pixels using parameters extracted from satellite images. The results revealed the practicability of the self-organizing map clustering method to identify the specific points by which the soil moisture can represent the soil moisture condition.

Finally, Khouakhi *et al.* (2022) studied the AI methods for flood applications. The authors found that flood-related image datasets from social media, smartphones, cameras, etc., were valuable for managing flood risk, e.g., flood extent detection and flood depth estimation. The article discussed the increasing role of convolutional neural networks in flood research and pointed out the lack of open-labeled flood image datasets. The significance of such datasets was also discussed.

We would like to take this opportunity to thank all the anonymous reviewers for their effort and constructive comments, without which this book would not have been possible. The same gratitude is extended to the Editors-in-Chief of *Hydrology Research* (Prof. Bjørn Kløve and Prof. Nevil Wyndham Quinn) for their integrity and patient guidance. Finally, we thank all the authors for submitting their precious research work to this book.

Guest Editors
Elena Volpi
Università Roma Tre, Italy

Jong Suk KIM
Wuhan University, China

Shaleen Jain
University of Maine, USA

Sangam Shrestha
Asian Institute of Technology (AIT), Thailand

REFERENCES

Baghanam, A. H., Norouzi, E. & Nourani, V. 2022 Wavelet-based predictor screening for statistical downscaling of precipitation and temperature using the artificial neural network method. *Hydrology Research* **53** (3), 385–406. https://doi.org/10.2166/nh.2022.094.

de Moura, C. N., Seibert, J. & Detzel, D. H. M. 2022 Evaluating the long short-term memory (LSTM) network for discharge prediction under changing climate conditions. *Hydrology Research* **53** (5), 657–667. https://doi.org/10.2166/nh.2022.044.

Khouakhi, A., Zawadzka, J. & Truckell, I. 2022 The need for training and benchmark datasets for convolutional neural networks in flood applications. *Hydrology Research* **53** (6), 795–806. https://doi.org/10.2166/nh.2022.093.

Lu, Z., Wang, D., Deng, Z., Shi, Y., Ding, Z., Ning, H., Zhao, H., Zhao, J., Xu, H. & Zhao, X. 2021 Application of red edge band in remote sensing extraction of surface water body: a case study based on GF-6 WFV data in arid area. *Hydrology Research* **52** (6), 1526–1541. https://doi.org/10.2166/nh.2021.050.

Nourani, V. 2022 Application of the artificial intelligence approach and remotely sensed imagery for soil moisture evaluation. *Hydrology Research* **53** (5), 684–699. https://doi.org/10.2166/nh.2022.111.

Roushangar, K., Ghasempour, R., Kirca, V. S. O. & Demirel, M. C. 2021 Hybrid point and interval prediction approaches for drought modeling using ground-based and remote sensing data. *Hydrology Research* **52** (6), 1469–1489. https://doi.org/10.2166/nh.2021.028.

Wu, S.-J., Hsu, C.-T. & Chang, C.-H. 2021 Stochastic modeling of artificial neural networks for real-time hydrological forecasts based on uncertainties in transfer functions and ANN weights. *Hydrology Research* **52** (6), 1490–1525. https://doi.org/10.2166/nh.2021.030.

Xu, Y., Hu, C., Wu, Q., Li, Z., Jian, S. & Chen, Y. 2021 Application of temporal convolutional network for flood forecasting. *Hydrology Research* **52** (6), 1455–1468. https://doi.org/10.2166/nh.2021.021.

doi: 10.2166/nh.2022.094

Wavelet-based predictor screening for statistical downscaling of precipitation and temperature using the artificial neural network method

Aida Hosseini Baghanam [a,*], Ehsan Norouzi [a] and Vahid Nourani [a,b]

[a] Department of Water Resources Engineering, Faculty of Civil Engineering, University of Tabriz, Tabriz, Iran
[b] Faculty of Civil and Environmental Engineering, Near East University, Nicosia, North Cyprus, Turkey
*Corresponding author. E-mail: hosseinibaghanam@gmail.com

AHB, 0000-0003-4130-7941

ABSTRACT

One of the challenging issues in statistical downscaling of climate models is to select dominant large-scale climate variables (predictors). Correlation-based methods have been revealed to be efficacious to select the predictors; however, traditional correlation analysis has shown limited ability due to the nonstationary and nonlinear nature of climatic time series. Hence, in this study, Wavelet Coherence Transform (WTC) was employed to assess the high common powers and the multi-scale correlation between two time series (i.e., predictand and predictor) as a function of time and frequency. To this end, a coefficient correlation (CC) and a wavelet-based method were used for predictor screening and the results were compared in statistical downscaling. To apply the wavelet-based method, Continuous Wavelet Transform (CWT) was utilized to identify the potent periodicity in the time series of predictands. WTC was applied to determine the coherence between predictors and predictands in the potent periodicities, and Scale Average (SA) wavelet coherency was applied to rank them. In order to implement statistical downscaling, the ANN model was developed. In this study, three climate models including BNU-ESM Can-ESM5, and INM-CM5 have been used. The projection of the future climate based on the ANN downscaling revealed that precipitation will undergo a 7.1–28.92% downward trend, while the temperature will experience a 2.25–4.21 °C increase.

Key words: artificial neural network (ANN), climate models, statistical downscaling, wavelet coherence transform (WTC), wavelet transform

HIGHLIGHTS

- Statistical downscaling via ANN.
- Predictor screening via continuous wavelet and wavelet coherence.
- Predictor screening via linear and nonlinear relation of predictor and predictand.
- Predictor screening via correlation coefficient.
- Application of CMIP5 and CMIP6 for downscaling precipitation and temperature.

1. INTRODUCTION

Throughout the last decades, industrial activities and the atmospheric concentration of greenhouse gases (GHGs) such as carbon dioxide (CO_2) have led to human-induced climate change. The resulting impacts of climate change including changes in the climate system and climate-related variables, such as precipitation and temperature, have become one of the major areas of concern for researchers. Therefore, examination of their future patterns will assist decision-makers to carry out effective measures to deal with the predicament such as droughts, flash floods, temperature extremes, and melting of ice shields, to name but a few. In this context, climate models have been developed, and they have the ability to predict the future variations in temperature and precipitation as a consequence of the increasing concentration of GHGs. Climate models employ physically based equations to simulate the processes governing the exchange of energy and mass between the atmosphere, lands, and oceans.

One of the main limitations of climate models is the coarse spatial resolutions of the outputs, inasmuch as many geographical features could be neglected or not represented in the models. To acquire the desirable local or regional scales of climate models data, several downscaling methods have been employed (Wilby & Wigley 1997). Downscaling approaches are divided into two categories. Dynamical downscaling is a method in which the outputs of climate models, as the boundary condition of regional climate models (RCMs), are applied to derive higher resolution. Demanding compute-intensive processes, dynamical downscaling methods are less economical and, therefore, challenging to apply. Statistical downscaling is a method to establish a statistical relationship between local-scale observed data (predictands) and large-scale climate variables (predictors). This is a more straightforward approach and can be run more rapidly, allowing researchers to simulate multiple models in a brief period. Furthermore, this method can be used for different regions (Timbal *et al.* 2003). Previous studies applied statistical downscaling methods such as statistical downscale models (SDSM; Wilby *et al.* 2002), K-nearest neighbors (Gangopadhyay *et al.* 2005), support vector machine (SVM; Tripathi *et al.* 2006), rule induction and tree algorithms (Goyal *et al.* 2012), and artificial neural network (ANN; Moghim & Bras 2017) to statistically downscale the climate variables. Over the last few years, machine learning methods such as ANN have been employed to investigate the nonlinear relationship between hydro-climatic variables as well as downscaling GCMs data. Studies conducted by Li & Zheng (2003), Dibike & Coulibaly (2006) evaluated the performance of ANN and regression models to downscale daily and monthly precipitation and temperature and concluded that the final results are similar. Another study conducted by Harpham & Wilby (2005) examined the performance of ANN and SDSM for downscaling the daily precipitation, and the results showed that ANN has the edge over SDSM. In a study by Tisseuil *et al.* (2010), the competitiveness of ANN and linear models such as the generalized linear model (GLM) was investigated. It was concluded that ANN performs better than linear models. Chadwick *et al.* (2011) proposed an improved ANN simulation to downscale temperature and rainfall. Su *et al.* (2015) used the ANN model to forecast the fluctuation of streamflow and precipitation during 2011–2050. Given the results of downscaling with machine learning techniques (e.g., ANN), there are some opposing conclusions. While some studies indicated the effectiveness of ANN application in downscaling GCMs data, others commented that ANN can lead to undesirable performance. Khan *et al.* (2006) in a comparative study analyzed the performance of ANN and SDSM, and the results demonstrated that SDSM performed more accurately in downscaling local climate features. In a study conducted by Abdellatif *et al.* (2013), the advantage of using the GLM-ANN model in downscaling was investigated and concluded that the proposed method has superiority over traditional ANN.

These contrary results appear to be rooted in the property and quantity of the datasets. It has been observed that large datasets can decrease the efficacy of ANN-based modeling and it would suffer from a data redundancy (Bowden *et al.* 2005). Then, it stands to reason that a data pre-processing technique should be applied to big datasets like the outputs of climate models to select the potential predictors among a large number of data (Nourani *et al.* 2017).

Among recent studies, Ahmadi & Han (2012) employed a synthesis method using Gamma test (GT) and coefficient correlation analysis to forecast the precipitation. Correlation analysis, as a common approach, was applied by Sachindra *et al.* (2013) to determine the dominant outputs of general circulation models (GCMs). Okkan (2015) utilized the mutual information (MI) and correlation method to select the potential predictors to downscale GCMs data. In a joint study, Devak & Dhanya (2016) investigated the performance of the principal component analysis (PCA) to decline the dimensionality of GCM variables. The correlation-based analysis is a conventional method to screen and derive the features of the large datasets. Nevertheless, one of the limitations of this method is its low ability to spot nonlinear relationships between time series. Therefore, in light of the fact that hydro-climatic variables have nonstationary and nonlinear features, traditional correlation methods would not lead to desirable results. To address this challenge, wavelet analysis has been properly employed by different studies to investigate nonstationary signals in hydro-climatological processes (Partal & Kişi 2007; Belayneh *et al.* 2014; Rashid *et al.* 2016; Tan *et al.* 2016; Xu *et al.* 2019).

In the field of statistical downscaling, Rana & Moradkhani (2016) employed continuous wavelet transform (CWT) in the projection step to analyze the possible variability in future precipitation and temperature. More recently, Baghanam *et al.* (2018) and Nourani *et al.* (2018a) utilized wavelet entropy (WE) to downscale the precipitation. Based on the preceding studies, CWT and the discrete wavelet transform (DWT) have been applied for feature extraction and denoising of a single signal, however, the application of the two methods has not been investigated for predictor screening purposes.

Cross-wavelet transform (XWT) is used as a technique to investigate and detect the coherence and phase relations among signals in hydro-climatic studies conducted by Adamowski (2008), Holman *et al.* (2011), Soon *et al.* (2014), and Yu & Lin (2015). Maraun & Kurths (2004) applied the XWT and wavelet coherence transform (WTC) to evaluate the relations between ENSO and NAO. Jevrejeva *et al.* (2003) described a WTC method to study the impacts of Arctic Oscillation (AO) and NAO on the state of the ice in the Baltic Sea. Grinsted *et al.* (2004) tested WTC to investigate the correlation between geophysical time series, namely the Arctic Oscillation (AO) and the Baltic maximum sea ice (BMI). Tamaddun *et al.* (2017) used CWT and wavelet correlation spectrum to examine the influence of oceanic-atmospheric indices on the U.S. snow water. Earlier work by Nourani *et al.* (2018b) evaluated the influence of hydro-climatological variables on the fluctuations of Urmia Lake by applying WTC. Among the vast growing body of literature written on the application of WTC, it has not been used as a method combined with a machine learning skill to downscale the climate models data. Therefore, in this study, WTC, as a data screening method, was employed to select the dominant variables affecting the precipitation and temperature of Tabriz synoptic station. To this end, the coherence between the output of climate models and predictands was investigated. The Coupled Model Intercomparison Project (CMIP) provides climate models which are modern tools to investigate the climate condition in the future (Wilby & Wigley 1997). A host of studies have examined the performance of the CMIP5 models in projecting the climate variables (Kumar *et al.* 2013; Pendergrass & Hartmann 2014). The CMIP5 offers models under different representative concentration pathways (RCPs; Taylor *et al.* 2012; Stouffer *et al.* 2017). The new cutting-edge CMIP6 has recently been provided to compensate for the downside of the CMIP5. CMIP6 models are constantly being updated and comprise higher spatial resolutions than CMIP5 (Eyring *et al.* 2016). CMIP6 provides a new set of emission scenarios incorporating various shared socioeconomic pathways (SSPs; Riahi *et al.* 2017). Furthermore, CMIP6 is equipped with more simulations for various scenarios to portray inner inconsistency (Eyring *et al.* 2016; Stouffer *et al.* 2017). Numerous research works have investigated the performance of CMIP6 models in simulating the precipitation and temperature in various areas (Akinsanola *et al.* 2020; Jiang *et al.* 2020; Rivera & Arnould 2020; Wang *et al.* 2020; Ngoma *et al.* 2021).

Ultimately, the objectives of the present study can be listed as follows:

i. To provide a robust wavelet-based predictor screening tool and compare the results with the traditional classic coefficient correlation (CC) method. In this regard, several climate models such as BNU-ESM from CMIP5, and Can-ESM5 and INM-CM5 from CMIP6 were applied.
ii. To feed the screened dominant predictors into the ANN in order to implement downscaling.
iii. To project the future precipitation and temperature based on the ANN model.

2. MATERIAL AND METHODS

2.1. Study area and datasets

Tabriz city (latitude 38.07°N, longitude 46.14°) is the capital city of East Azerbaijan province located in northwestern Iran (see Figure 1). It lies in the valley of the Quru River. Tabriz is surrounded by mountains (e.g., Sahand at 3,707 m) and scattered hills in three directions, namely Pachkin, Eynaly, and Payan. The west side of the city lies within the Tabriz plain, which slopes gently toward the Urmia Lake (60 km to the west). The elevation of the city ranges between 1,350 and 1,600 meters above sea level. The climate of Tabriz is categorized as a continental climate with regular seasons. This gives the city dry and semi-hot summers and cold, wet winters. The city used to receive 360.2 mm of precipitation on average, per year in the past and during the recent decade, it has decreased to 280 mm. The average minimum and maximum temperatures of the city are −2 and 19 °C in January and July, respectively. The average relative humidity varies between 73% in December and 48% in June.

Climate change has led to many predicaments in various parts of the world specifically in arid regions. The city of Tabriz is located in an arid and semi-arid area near the Urmia Lake, the largest salt lake in the region, where the negative impacts of the climate change along with the water resources management policies have dried up the lake. Therefore, an assessment of the future precipitation and temperature variations of the city is required (Zarghami *et al.* 2011; Baghanam *et al.* 2020; Sadeqi & Dinpashoh 2020). This can help decision-makers to reach better solutions to climate-related challenges.

In this study, two predictands (i.e., precipitation and temperature) and the outputs of three climate models, as predictors, were applied. Monthly precipitation and temperature of Tabriz synoptic station for the period of 1951–2018, which have been provided by the Tabriz Meteorological Organization, were used in this study. The distribution of monthly precipitation and temperature of Tabriz station are depicted in Supplementary Appendix Figure 1. As shown in the figure, the maximum precipitation fluctuation occurs within April, October, and December indicating the wet season. Accordingly, the driest month takes place during July. In a similar way, the highest and the lowest temperature occur in July and January, respectively.

In order to implement the proposed downscaling method, the predictors from two climate models under the AR6-CMIP6 (Assessment Report 6 and Coupled Model Intercomparison Project datasets) and the predictors under the AR5-CMIP5 (Assessment Report 5 and Coupled Model Intercomparison Project datasets) from the IPCC data distribution center during January 1951 to December 2005 were extracted. Furthermore, future climate variables for CMIP5 models under

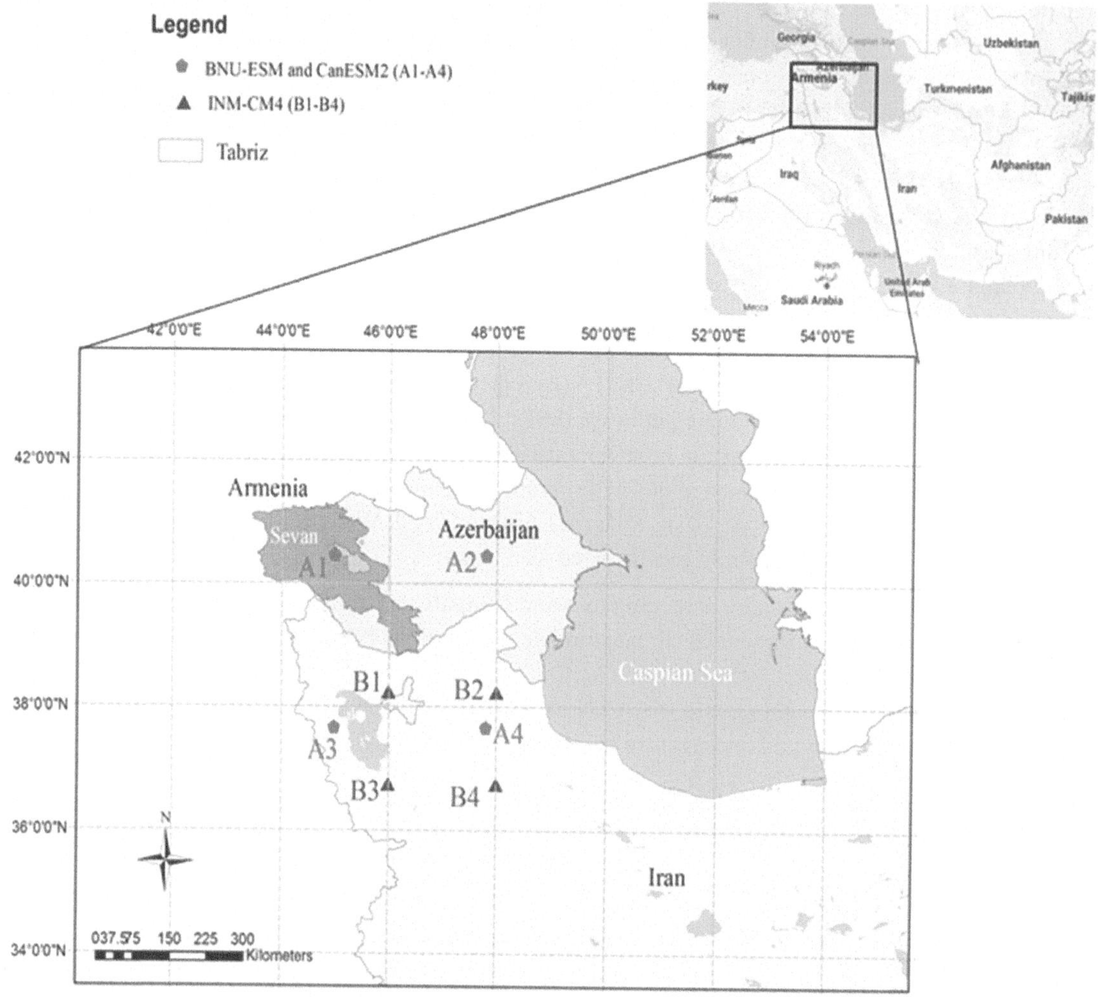

Figure 1 | Geographical location of the study area.

the RCPs (i.e., RCP4.5 and RCP8.5) and climate variables for CMIP6 models under SSPs (i.e., SSP2 and SSP5) for the future simulation were used in this study. The historical and future data for the period of 1951–2060 were retrieved from https://esgf-node.llnl.gov/. Climate models including BNU-ESM, Can-ESM5, and INM-CM5, which respectively have been developed by research centers in China, Canada, and Russia, were used in this study (Table 1), owing to the preceding research works proving that these three climate models appear to be more effective for investigating the climate of the region (Baghanam *et al.* 2018; Nourani *et al.* 2019a, 2019b). Since, in this study, two Earth system models (ESMs) and one GCM were used, henceforth the term climate models (CMs) will be used to refer to both models. Since multiple studies have recommended applying several grid points around the study station will lead to more thoroughgoing results (Frost *et al.* 2011; Guo *et al.* 2012; Beecham *et al.* 2014), in the present study, based on the resolution of CMs, four grid points for each CM were considered nearby the study area (Figure 1). Yet according to the proposed methodology, dominant predictors from potential grid points were selected later on. As depicted in Figure 1, A1, A2, A3, and A4 represent the considered four grid points for BNU-ESM and Can-ESM5, which share the same resolution and their grid points fully overlap. Similarly, B1, B2, B3, and B4 indicate grid points for INM-CM5. Table 1 shows the attributes of CMs and potential variables considered in this study.

2.2. Proposed methodology

The objective of this paper is to carry out a wavelet-based predictor screening method to develop an ANN-based statistical downscaling technique and project the future precipitation and temperature of Tabriz city. Figure 2 demonstrates the blueprint of the study containing three steps. The first step is to determine dominant predictors from the selected CMs by

Table 1 | Applied GCMs

Center	Center Acronym	Model	Grid Size	RCP (W/m^2)	Applied climate variables for three CMs
Beijing Normal University, China Canadian Centre for Climate Modelling and Analysis, Canada Russian Academy of Sciences, Institute of Numerical Mathematics, Russia	BNU CCCMA INM	BNU-ESM Can-ESM5	2.81°×2.81° 2.81°×2.81° 1.5°×2°	RCP4.5 RCP8.5 SSP2-4.5 SSP5-8.5 SSP2-4.5 SSP5-8.5	^{P}ua: eastward wind; ^{P}va: northward wind; ^{P}zg: geopotential height; ^{P}hur: relative humidity; ^{P}hus: specific humidity; tas: air temperature; uas: eastward near surface wind; vas: northward near surface wind; psl: air pressure at sea level; hfls: surface upward latent heat flux; prc: convective precipitation flux; pr: precipitation flux; hurs: near surface relative humidity; huss: near surface specific humidity; evspsbl: water evaporation flux

P: 100 pa, 200 pa, 300 pa, 500 pa, 700 pa, 1,000 pa, 2,000 pa, 3,000 pa, 5,000 pa, 7,000 pa, 10,000 pa, 15,000 pa, 20,000 pa, 30,000 pa, 40,000 pa, 50,000 pa, 60,000 pa, 70,000 pa, 85,000 pa, 92,500 pa, 100,000 pa.

means of feature extraction technique (i.e., WTC and CC). At the second step, dominant predictors screened from the first step were fed into the ANN model to downscale precipitation and temperature. Eventually, at the third step, based on the potential inputs and proper ANN model, which developed during the second step, the projection for the future precipitation and temperature of Tabriz synoptic station under the SSP2 and SSP5 for Can-ESM5 and INM-CM5 and RCP4.5 and RCP8.5 for BNU-ESM was executed. A full description of the proposed methodology is delineated hereunder.

2.3. First step

Throughout this step, firstly, dominant periodicities for observed precipitation, temperature, and predictors were separately obtained by CWT. This could bring detailed information that would not have been identified by the original time series. Thereafter, the correlation between predictors and predictands was estimated by WTC and CC. In order to examine the influence of each predictor scale average coherency for each predictor was calculated by extracting the matrix of significant scale (at the 95% confidence level) WTC.

It should be noted that, in the absence of the data pre-processing techniques for selection of the dominant predictors, there would have been 95 large-scale climate variables for BNU-ESM and INM-CM5 and 120 variables for Can-ESM5, meaning that in the aggregate, 1,240 variables should be constituted in the downscaling procedure. If the trial and error method for each parameter had been used, then 2^{1240}–1 trial and errors would have been required for determining the dominant predictors, which is a time-intensive progress.

WTC and CC used in this step are based on the nonlinear and linear correlation-based methods. This would be a contributing factor in investigating the most influential variables which have a linear or nonlinear relationship with hydrological parameters (i.e., precipitation and temperature). Afterward, selected predictors based on the WTC and CC methods were separately used in the ANN modeling to downscale the climate variables, and the performances of the two methods were evaluated.

2.4. Second step

In this stage, according to the dominant predictors obtained from three CMs in the first step, an ANN-based downscaling method for statistical downscaling was developed.

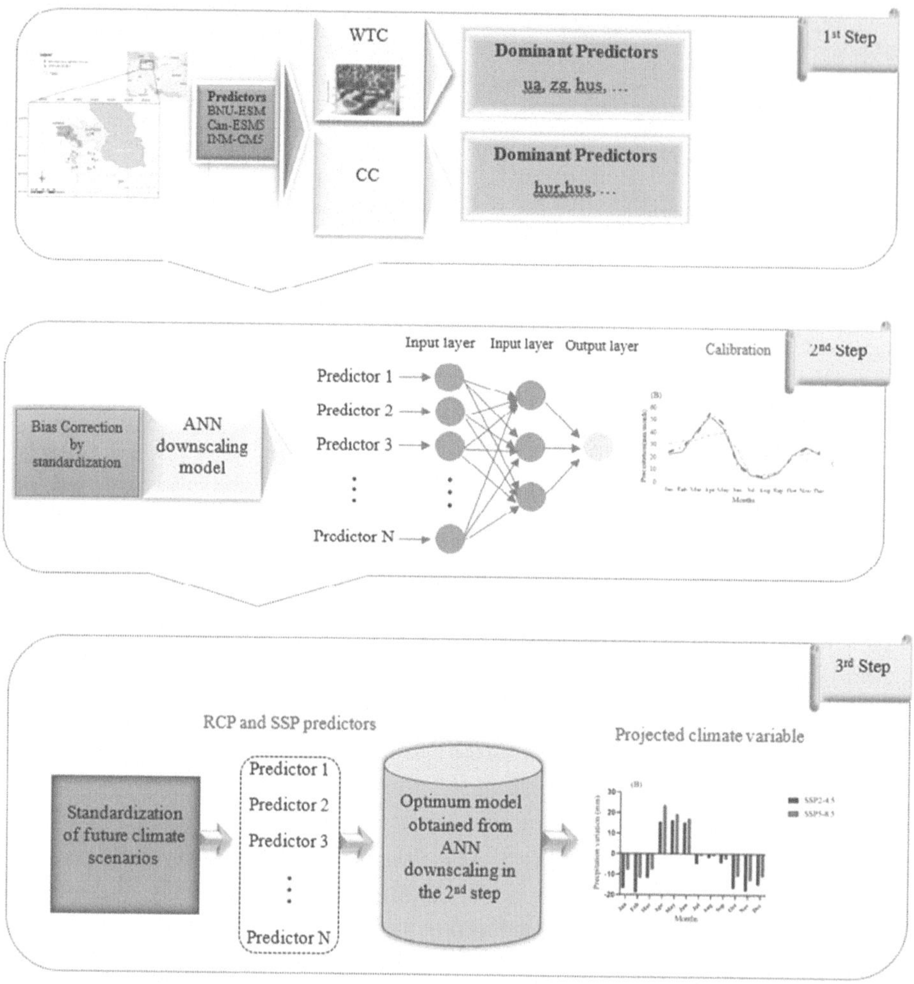

Figure 2 | Schematic of the proposed methodology.

CMs often expose substantial deviations from the locally observed data, known as biases. Therefore, to perform downscaling, these biases should be corrected. Standardization is one of the common techniques to eliminate CMs biases, which has been applied to data in several studies (Tripathi *et al.* 2006; Anandhi *et al.* 2009). Standardization is implemented by subtracting the mean of observed data and then dividing by the standard deviation.

2.5. Third step

Ultimately, based on the model developed in the previous step, simulation of the future precipitation and temperature under SSP2 and SSP5 from CMIP6 and RCP4.5 and RCP8.5 from CMIP5 during the period of 2021–2060 was conducted. It should be mentioned that due to the diverse possible perspectives on human-origin effects that influence the environment, such as the rate of population growth, greenhouse gases emission, and economic condition, each RCP and SSP represents a specific radiative forcing pathway. In this respect, SSP2 and RCP4.5 are related to the moderate while SSP5 and RCP8.5 are associated with high emission scenarios.

In order to implement the proposed methodology, the applied mathematical tools were explained briefly in the following sections.

2.6. Wavelet analysis

Throughout the recent decades, researchers sought to use a wide range of methods for extracting the features of time series. Among those, Fourier analysis can be mentioned. Nevertheless, Fourier analysis has limitations to be applied to nonstationary time series and can be used for investigating signals which have consistent amplitude and statistical features over time (Kaiser 2011). Thus, Windowed Fourier transform (WFT) was offered to fill this gap (Kaiser 2011). WFT, however, is limited to properly analyzing the time-frequency localization and yielded imprecise and ineffective results. The imprecision stems from its limited capability to identify low and high frequencies, and the ineffectiveness is because of the fact that it has to be done several times according to the scale of the windows to select the best dominant frequencies (Torrence & Compo 1998). Therefore, wavelet analysis as a scale-independent method can be deployed.

Being undoubtedly one of the most renowned mathematical transforms, wavelet transform (WT) has been widely used in signal and image processing. Because of the fact that many climatic phenomena have an intermittent nature in temporal and/or spatial domain, and some of their features and characteristics cannot be detected in the temporal domain, then by virtue of transferring them to other domains (e.g., wavelet transform, Fourier Transform, etc.), those features can be identified. Thus, wavelet transform affords an improved signal analysis method to investigate the time series in time and frequency space. In this study, the CWT was adopted to find meaningful periodicities in time series. CWT of a continuous signal *X(t)* is defined as (Mallat 1999):

$$T(a, b) = \frac{1}{\sqrt{a}} \int_{+\infty}^{-\infty} x(t)\Psi^{*}\left(\frac{t-b}{a}\right)dt; \; a, b \in R \tag{1}$$

where Ψ is the mother wavelet or wavelet function, * is the complex conjugate of the mother wavelet, *a* is the dilation factor, and *b* is the temporal translation value. The primary objective of the mother wavelet is to present a family of continuously translated and dilated wavelets. In this study, Morlet wavelet was adopted due to the fact that it is the most effective choice for time and frequency localization (Grinsted *et al.* 2004). CWT can also be adopted to detect the correlation between two distinctive hydrological processes (Labat 2010). In this respect, the wavelet spectrum of a continuous time series *X(t)* can be formulated as the following (Torrence & Compo 1998):

$$W_X(a, \; b) = \; C_{\Psi}^{X}(a, \; b)C_{\Psi}^{*X}(a, b) = |C_X(a, \; b)|^2 \tag{2}$$

where $C_{\Psi}^{X}(a, \; b)$ and $C_{\Psi}^{*X}(a, b)$ correspond to continuous wavelet coefficient and complex conjugate of the time series *X(t)*, respectively. Similar to wavelet spectrum, the cross-wavelet spectrum between two time series *X(t)* and *Y(t)* for detecting cross-wavelet power, phase relationships, and fluctuation periodicities can be described as (Liu 1994):

$$W_{XY}(a, \; b) = \; C_{\Psi}^{X}(a, \; b)C_{\Psi}^{*Y}(a, b) \tag{3}$$

where $C_{\Psi}^{X}(a, \; b)$ denotes the continuous wavelet spectrum of the time series *X(t)*, and $C_{\Psi}^{*Y}(a, b)$ is the complex conjugate of the time series *Y(t)*. Complex cross-wavelet spectrum which is capable of decomposing cross power and phase is defined as (Ng & Chan 2012):

$$W_{XY}(a, \; b) = |W_{XY}(a, \; b)|e^{i\varphi_{is}} \tag{4}$$

where φ_{is} is the phase difference between two time series on the scale *s* and the time t_i. However, while using a non-normalized wavelet power spectrum, XWT is merely capable of revealing high common powers and it could lead to misinterpretation of the conclusions whereas WTC as a normalized wavelet coherency can catch even low common powers (Maraun & Kurths 2004). In fact, WTC is an efficacious tool to identify the dominant periodicities within time series which can be applied as a measurement of the localized correlation between two signals, and it can be used to identify a time-frequency connection between two time series. WTC is defined as (Torrence & Webster 1999):

$$WC(a, b) = \frac{|SW_{XY}(a.b)|}{\sqrt{|SW_{XX}(a.b)|.|SW_{YY}(a.b)|}} \tag{5}$$

where $SW_{XY}(a.b)$ is the smoothing estimate of cross-wavelet spectrum for *X(t)* and *Y(t)* time series, and $SW_{XX}(a.b)$ and $SW_{YY}(a.b)$ are the smoothing estimate of the time series *X(t)* and *Y(t)*, respectively, and can be mathematically illustrated as (Torrence & Webster 1999):

$$SW_{XX}(a.b) = \int_{t-\frac{\delta}{2}}^{t+\frac{\delta}{2}} W_{XX}^{*}(a.b) W_{XX}(a.b) dadb \tag{6}$$

$$SW_{YY}(a.b) = \int_{t-\frac{\delta}{2}}^{t+\frac{\delta}{2}} W_{YY}^{*}(a.b) W_{YY}(a.b) dadb \tag{7}$$

$$SW_{XY}(a.b) = \int_{t-\frac{\delta}{2}}^{t+\frac{\delta}{2}} W_{XX}^{*}(a.b) W_{YY}(a.b) dadb \tag{8}$$

where their values range between 0 and 1. A value of 1 indicates the linear correlation between two signals, while 0 is an indicator for vanishing correlation (Luterbacher *et al.* 2002).

2.7. Artificial neural network

The ANN has been broadly used as a prediction method, and it has shown an acceptable performance for modeling nonlinear time series. A basic ANN structure consists of input, hidden, and output layers. The input layer is used to import the training data into the network, and the output layer is used to produce suitable outcomes according to the input dataset. ANN embodies neurons within different layers connected together inside the network. The structure of the ANN can be formulated as follows (Haykin & Network 2004):

$$\hat{y}_k = f_0\left[\sum_{j=1}^{M_N} w_{kj} \cdot f_h\left(\sum_{i=1}^{N_N} w_{ji} x_i + w_{j0}\right) + w_{k0}\right] \tag{9}$$

where i, j, and k indicate input, hidden, and output layer neurons, respectively. N_N and M_N, respectively, correspond to the number of neurons in input and output layers. w_{ji} is considered as a weight that is assigned to the input layer, w_{j0} is bias, and f_h and f_0 are respectively the activation function of the hidden neuron and the output layer. w_{kj} is a weight in the hidden layer, w_{k0} represents bias for the kth output neuron, x_i is the ith input in the input layer and $\hat{y}_k$ is considered as estimated output. Since preceding studies have shown the effectiveness of using three-layer feed-forward neural network (FFNN) with back propagation algorithm in modeling the hydro-climatologic processes (Maier & Dandy 2000). In this study, this algorithm is used as well. Moreover, at the training stage, the Levenberg–Marquardt algorithm due to its high convergence rate was employed to improve the performance of the ANN (Haykin & Network 2004; Kisi 2004).

2.8. Evaluation criteria

2.8.1. Correlation coefficient

The CC is a statistical measurement to identify the relationship strength between two variables. It can be defined as (Draper & Smith 1998):

$$CC = \frac{\sum (R - \bar{R}) - \sum (Z - \bar{Z})}{\sqrt{\sum (R - \bar{R})^2 - \sum (Z - \bar{Z})^2}} \tag{10}$$

where R is the predictand and Z denotes the predictor. The formula returns a value between -1 and 1, where 1 indicates a strong positive relationship, and -1 indicates a strong negative relationship.

2.8.2. Root mean square error

Root mean square error (RMSE) at training and validation stages was used to assess the skill of the proposed methodology. Legates & McCabe (1999) investigated the application of these criteria in hydro-climatology studies. The RMSE is a measure to demonstrate the standard deviation between the predictor and the predictand. It ranges from 0 to ∞ where small amounts that are closer to 0 indicate the efficiency of the model. RMSE is defined as:

$$\mathrm{RMSE} = \sqrt{\frac{\sum_{i=1}^{N} (R_i - Z_i)^2}{N}} \tag{11}$$

Where N, R_i, and Z_i are respectively the number of observed data, observed data, computed value and mean of observed data.

2.8.3. Determination coefficient

The determination coefficient (DC) measures the variability of one variable that can be affected by its relationship with another variable (Draper & Smith 1998). It is represented as a value between 0 and 1. A value of 1 shows the impeccable fit, and the model is able to accurately predict the future, while a value of 0 indicates that the model has failed to model the data. DC is formulated as follows:

$$\mathrm{DC} = 1 - \frac{\sum_{i=1}^{N} (R_i - Z_i)^2}{\sum_{i=1}^{N} (R_i - \bar{R}_i)^2} \tag{12}$$

where N, R_i, Z_i, and $\bar{R}_i$ are respectively the number of observed data, observed data, computed value and mean of observed data.

3. RESULTS AND DISCUSSIONS

In the present study, in order to project the future precipitation and temperature of the Tabriz station within the period of 2021–2060, an ANN-based downscaling method equipped with a data pre-processing wavelet technique was employed. The results of the study are presented in the three steps as follows.

3.1. First step: data screening

At the first step, WTC was used to determine the dominant periodicities in predictands and predictors, which led to the selection of dominant predictors from three CMs (i.e., BNU-ESM, Can-ESM5, and INM-CM5) based on the coherence method (WTC). To this end, first, the CWT of the two predictands was obtained. Results denoted that there is a 8–16 months periodicity for precipitation throughout 1952–1960, 1936–1968, 1969–1974, 1975–1978, 1980–1982, 1992–1994, and 2001–2004 at 95% confidence level (see Supplementary Appendix Figure 2(a)). Similarly, for average monthly temperature, there is a significant periodicity at the scale of 8–16 months (see Supplementary Appendix Figure 2(b)). Likewise, CWT for the predictors was also employed, which indicated that there is significant periodicity at the periodic scale from 8 to 16 months for the air temperature at grid point 1, geopotential height at grid point 3, eastward wind and geopotential height at grid point 4 (see Supplementary Appendix Figure 3). Subsequently, WTC between the predictors and predictands with meaningful periodic fluctuation was obtained. In order to select the dominant predictors based on the WTC technique, given that the observed precipitation and temperature have significant variability at the periodic scale from 8 to 16 months, a matrix of significance at the 95% confidence level was detached from the WTC values, and SA wavelet coherency of these values for each predictor was ranked. Thereupon maximum values of SA were selected as the dominant predictors. The results of the WTC method for selecting the predictors are tabulated in Table 2.

For the purpose of selecting dominant predictors based on the classic correlation method, CC between the predictands and predictors was calculated. After listing the predictors, the statistical student's *t*-test (Pu & Ginoux 2017) was applied to identify predictors with meaningful differences. Student's *t*-test is a helpful method to draw conclusions about the population not just samples. Therefore, predictors with the maximum value of CC were evaluated and selected. According to Table 2, selected dominant predictors for precipitation downscaling, based on the CC technique, were the variables related to humidity, namely relative humidity (hur), specific humidity (hus), and water evaporation flux (evspsbl). The selection of predictors

Table 2 | Selected dominant predictors by WTC and CC

Model	Dominant predictors for precipitation[a]		Dominant predictors for temperature[a]	
	WTC	CC	WTC	CC
BNU-ESM	tas(1)	hur1000(1)	tas(2)	tas(1)
	zg2000(2)	evspsbl(2)	zg30000(2)	tas(2)
	zg2000(3)	hur15000(3)	tas(3)	tas(3)
	zg2000(4)	hur10000(4)	tas(4)	tas(4)
CAN-ESM5	ua(2)	hur7000(3)	huss(1)	tas(1)
	zg15000(3)	hur10000(3)	tas(1)	tas(2)
	zg20000(3)	hur10000(4)	psl(3)	tas(3)
	hus100(4)	hus7000(4)	zg100(3)	tas(4)
INM-CM5	hus20000(1)	hur15000(2)	tas(3)	tas(1)
	zg15000(1)	hur3000(2)	psl(3)	uas(1)
	vas(3)	hur20000(3)	zg15000(4)	tas(3)
	zg10000(3)	hur15000(4)	zg25000(4)	zg10000(3)

[a]The number of grid points according to Figure 1 (i=1,2,3,4).

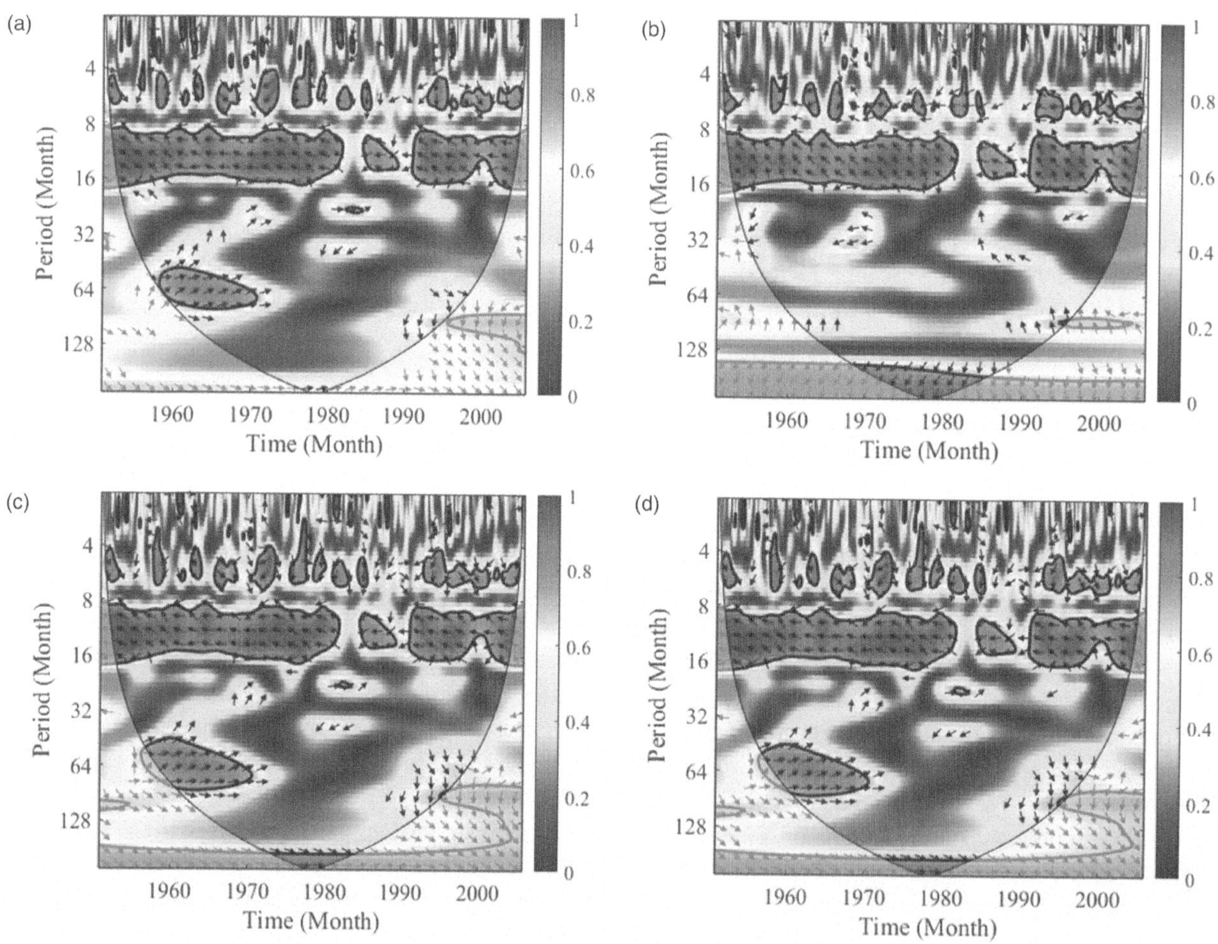

Figure 3 | WTC between precipitation for BNU-ESM and (a) tas at grid point 1, (b) zg2000 at grid point 2, (c) zg2000 at grid point 3, and (d) zg2000 at grid point 4. The thick black contour indicates the 95% confidence level using red noise as background spectrum, and the cone of influence where edge effects affect analysis is shown as a lighter shade. The relative phase relationship is shown as arrows (with in-phase pointing right, anti-phase pointing left). 1 (red) means the two signals are highly correlated and 0 (blue) means no correlation. Please refer to the online version of this paper to see this figure in colour: https://dx.doi.org/10.2166/nh.2022.094.

related to humidity variables from grid points A1, B1, A2, B2, and A3 indicates a linear relation with precipitation, which respectively originated from the substantial impact of humidity from the Caspian Sea and Urmia Lake (see Figure 1).

According to Table 2, the representative predictors based on the WTC method yielded different results compared to the CC method. Predictors related to geopotential height (zg), eastward wind (ua), northward near surface wind (vas), and specific humidity (hus) at the determined grid points in Table 2 became dominant according to nonlinear relationship. In this way, the obtained dominant variables from WTC, which have a nonlinear relationship with the precipitation, can represent the selected variables via the linear method (i.e., humidity), owing to the fact that humidity is affected by variability in sea level height, wind velocity, temperature, and air pressure. This is in line with the natural mechanism, meaning that temperature and wind velocity are driving forces to generate humidity. Figure 3 depicts the WTC between precipitation and dominant predictors of BNU-ESM. There are strong anti-phase relationships between predictors and precipitation at 8–16 months periodicity. This shows the anti-phase relationship between the condition of air layers and temperature with the occurrence of precipitation. Figure 4 shows WTC between precipitation and dominant variables of Can-ESM5, where precipitation has a statistically significant positive and negative coherence with the selected predictors according to Table 2. Figure 5 illustrates the WTC between dominant predictors and precipitation for INM-CM5 climate model. Accordingly, based on the selected

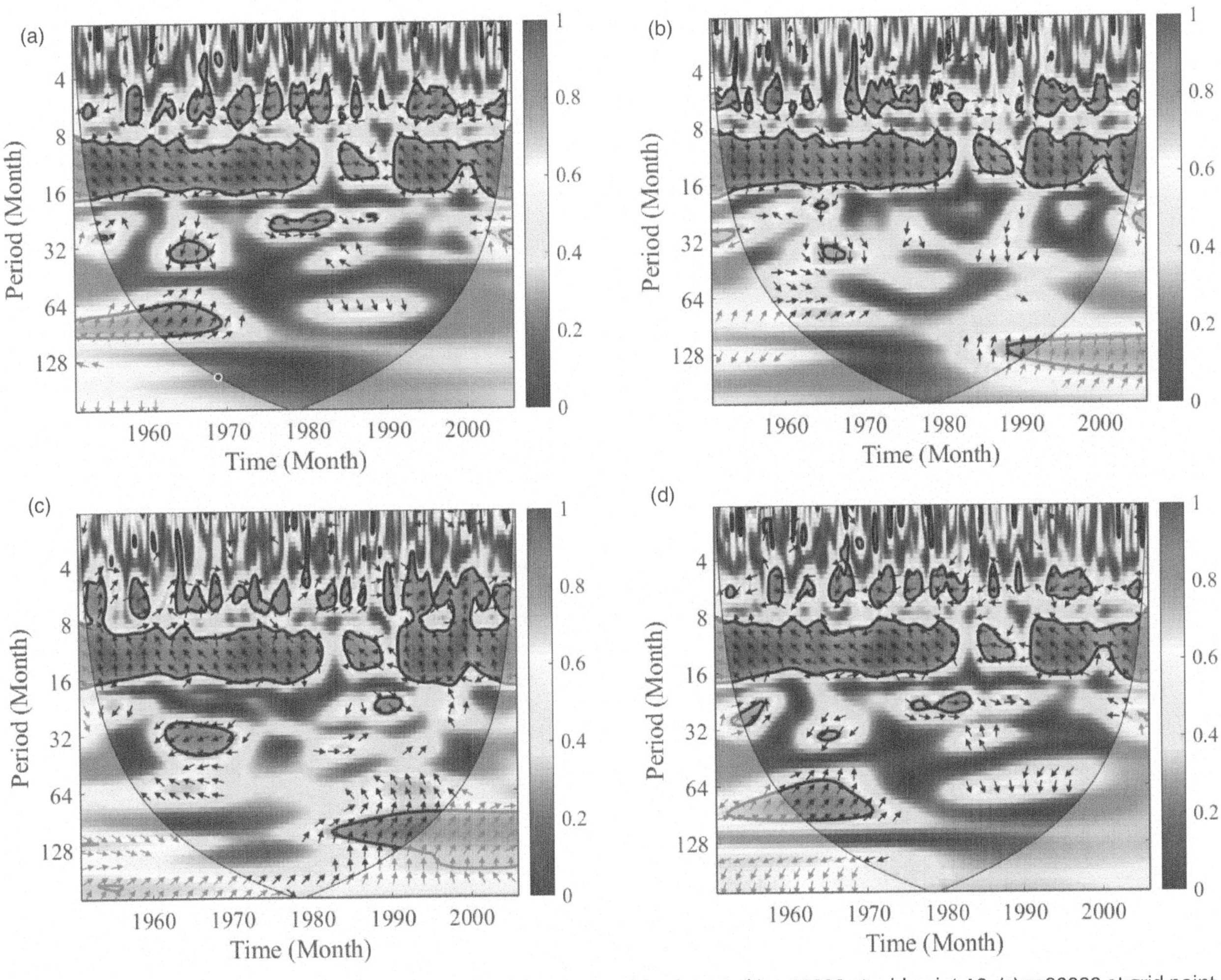

Figure 4 | WTC between precipitation for Can-ESM5 and (a) ua100 at grid point A2, (b) zg15000 at grid point A3, (c) zg20000 at grid point A3, and (d) hus100 at grid point A4. The thick black contour indicates the 95% confidence level using red noise as background spectrum, and the cone of influence where edge effects affect analysis is shown as a lighter shade. The relative phase relationship is shown as arrows (with in-phase pointing right, anti-phase pointing left). 1 (red) means the two signals are highly correlated and 0 (blue) means no correlation. Please refer to the online version of this paper to see this figure in colour: https://dx.doi.org/10.2166/nh.2022.094.

predictors, strong in-phase and anti-phase coherence are noticeable between the precipitation and the predictors at 8–16 months signal.

For downscaling the temperature, according to Table 2, based on the CC method, the dominant predictors detected for all three CMs at all four grid points were temperature-related large-scale variables (i.e., tas). This indicates a remarkable linear correlation between observed temperature and large-scale CM outputs.

Similar to the precipitation, according to Table 2, WTC for temperature demonstrates different results compared to the CC method. According to the coherence values between predictors and temperature variables, predictors related to height above the sea level (i.e., zg), air pressure (i.e., psl) and temperature (i.e., tas) and near surface humidity (i.e., huss) were identified as dominant predictors. This shows the ability of the WTC to catch nonlinear relationships between variables in addition to linear relationships. Figure 6 illustrates the WTC between temperature and BNU-ESM predictors, where in-phase coherence can be observed between the predictors and temperature. Dominant predictors for Can-ESM5 CM are illustrated in Figure 7. According to the figures, both in-phase and anti-phase relationships can be identified between the temperature and predictors. Figure 8 demonstrates the high in-phase and anti-phase correlation of INM-CM5 predictors with temperature at 8–16 months periodicity.

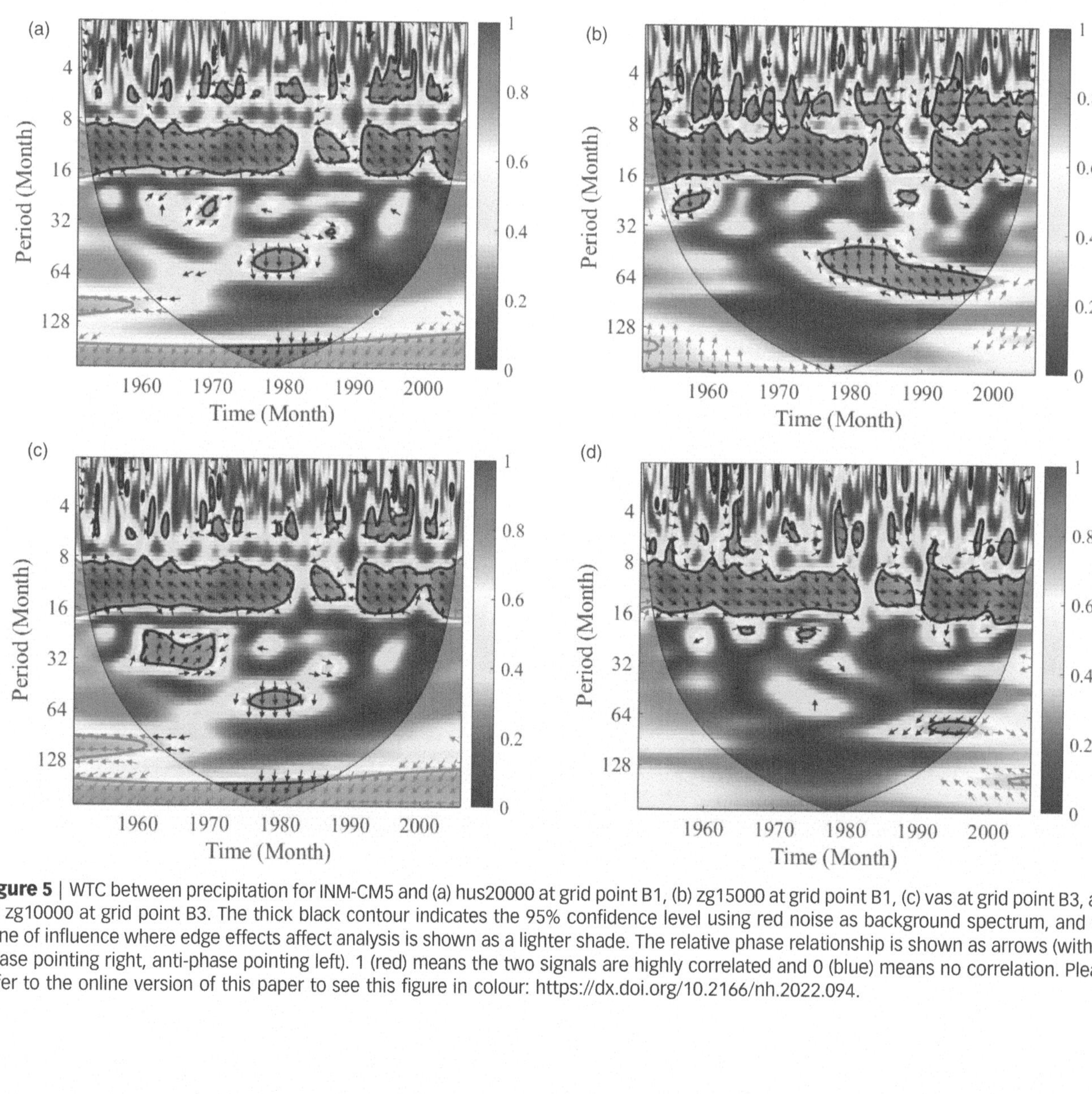

Figure 5 | WTC between precipitation for INM-CM5 and (a) hus20000 at grid point B1, (b) zg15000 at grid point B1, (c) vas at grid point B3, and (d) zg10000 at grid point B3. The thick black contour indicates the 95% confidence level using red noise as background spectrum, and the cone of influence where edge effects affect analysis is shown as a lighter shade. The relative phase relationship is shown as arrows (with in-phase pointing right, anti-phase pointing left). 1 (red) means the two signals are highly correlated and 0 (blue) means no correlation. Please refer to the online version of this paper to see this figure in colour: https://dx.doi.org/10.2166/nh.2022.094.

3.2. Second step: statistical downscaling

In the second step, the dominant inputs selected by two feature extraction methods (i.e., WTC and CC) were considered as the input of ANN as a nonlinear downscaling model. In order to put all the data in a comparable range, they were standardized during 1951–2005 in advance. The datasets were divided into two parts: the first 75% of the dominant inputs and observed data which is equal to the period 1951–1991 were used for calibration, and the 25% of them which were equivalent to the period 1992–2005 were applied to validation step. Prior studies have indicated the effectiveness of such data partitioning for the purpose of investigating the hydro-climatological processes (Baghanam *et al.* 2018; Nourani *et al.* 2019a, 2019b).

For implementing the ANN downscaling, the three-layer FFNN with back propagation training algorithm and the Levenberg–Marquardt method was used, and tangent sigmoid was considered as the activation function. The optimized number of hidden layer neurons and number of epochs were obtained through the trial and error technique. To this end, the optimum value of evaluation criteria with 360 epochs with 5 neurons in the hidden layer yielded an efficient result. RMSE and DC as the evaluation criteria were calculated for examining the effectiveness of the modeling. The results of this step are tabulated in Table 3. According to Table 3, downscaling results obtained from the WTC predictor screening method outperformed that of the CC technique. Therefore, for precipitation downscaling, WTC method performs 11–16% more effectively based on the DC

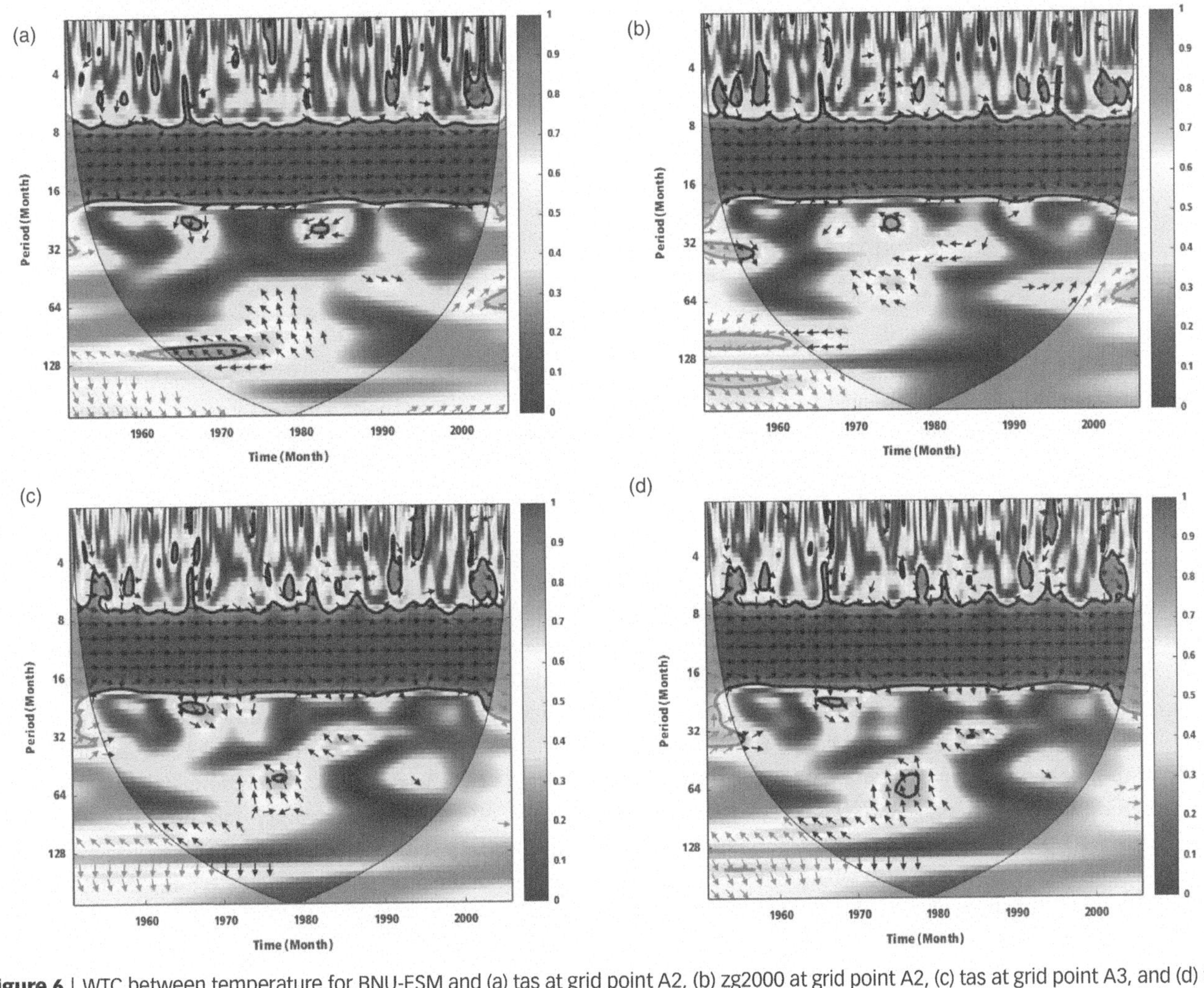

Figure 6 | WTC between temperature for BNU-ESM and (a) tas at grid point A2, (b) zg2000 at grid point A2, (c) tas at grid point A3, and (d) tas at grid point A4. The thick black contour indicates the 95% confidence level using red noise as background spectrum, and the cone of influence where edge effects affect analysis is shown as a lighter shade. The relative phase relationship is shown as arrows (with in-phase pointing right, anti-phase pointing left). 1 (red) means the two signals are highly correlated and 0 (blue) means no correlation. Please refer to the online version of this paper to see this figure in colour: https://dx.doi.org/10.2166/nh.2022.094.

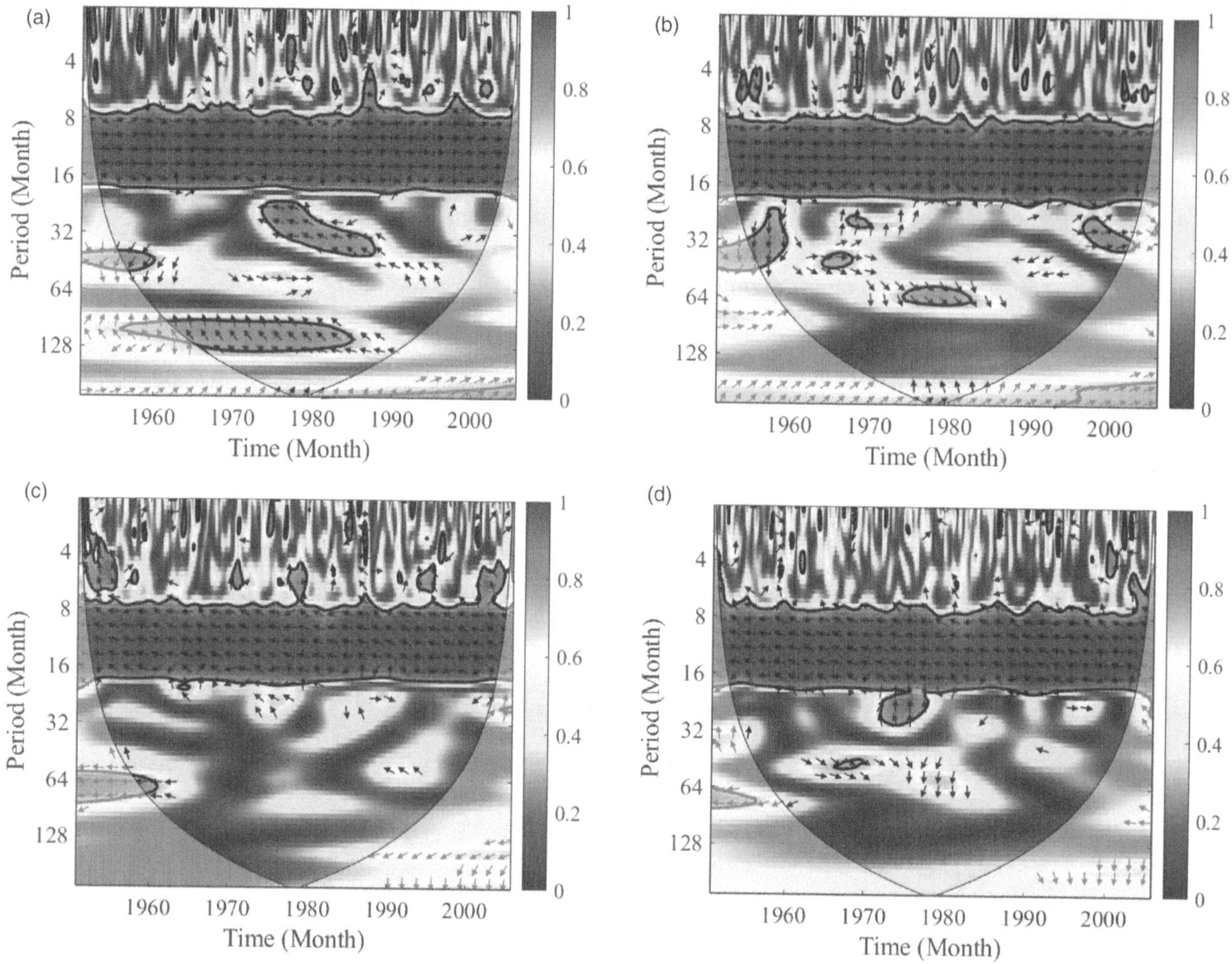

Figure 7 | WTC between temperature for Can-ESM5 and (a) huss at grid point A1, (b) tas at grid point A1, (c) psl at grid point A3, and (d) ua100 at grid point A3. The thick black contour indicates the 95% confidence level using red noise as background spectrum, and the cone of influence where edge effects affect analysis is shown as a lighter shade. The relative phase relationship is shown as arrows (with in-phase pointing right, anti-phase pointing left). 1 (red) means the two signals are highly correlated and 0 (blue) means no correlation. Please refer to the online version of this paper to see this figure in colour: https://dx.doi.org/10.2166/nh.2022.094.

validation criterion. In the same way, for temperature, downscaling based on the WTC method indicates more satisfactory results, as 35–41% efficiency for DC in validation step was obtained. This improvement appears to be rooted in the ability of WTC to opt for the linear and nonlinear features of the climate data, resulting in a more comprehensive assessment of the impacts of the driving forces on the local climate variables. Hence, the results of the ANN downscaling method based on the WTC are more promising in predicting the future climatic variables. As shown in Figures 9 and 10, it can be inferred that for both precipitation and temperature downscaling, the application of the WTC predictor screening technique outdoes the CC method. Moreover, for all three CMs (i.e., BNU-ESM, Can-ESM5, and INM-CM5), temperature downscaling denotes better performance in terms of evaluation criteria than precipitation. This is in line with the results of previous studies (Sun *et al.* 2015; Ahmadalipour *et al.* 2017) which is due to the stochastic nature of the precipitation and deterministic nature of the temperature that tends to have a steady and unvarying quality.

3.3. Third step: projection of future precipitation and temperature

Throughout the final step, the projection of the future precipitation and temperature based on the ANN optimal downscaling model was performed to predict the future oscillations (i.e., 2021–2060) under RCP4.5 and RCP8.5 scenarios for BNU-ESM and SSP2 and SSP5 scenarios for Can-ESM5 and INM-CM5 CMs.

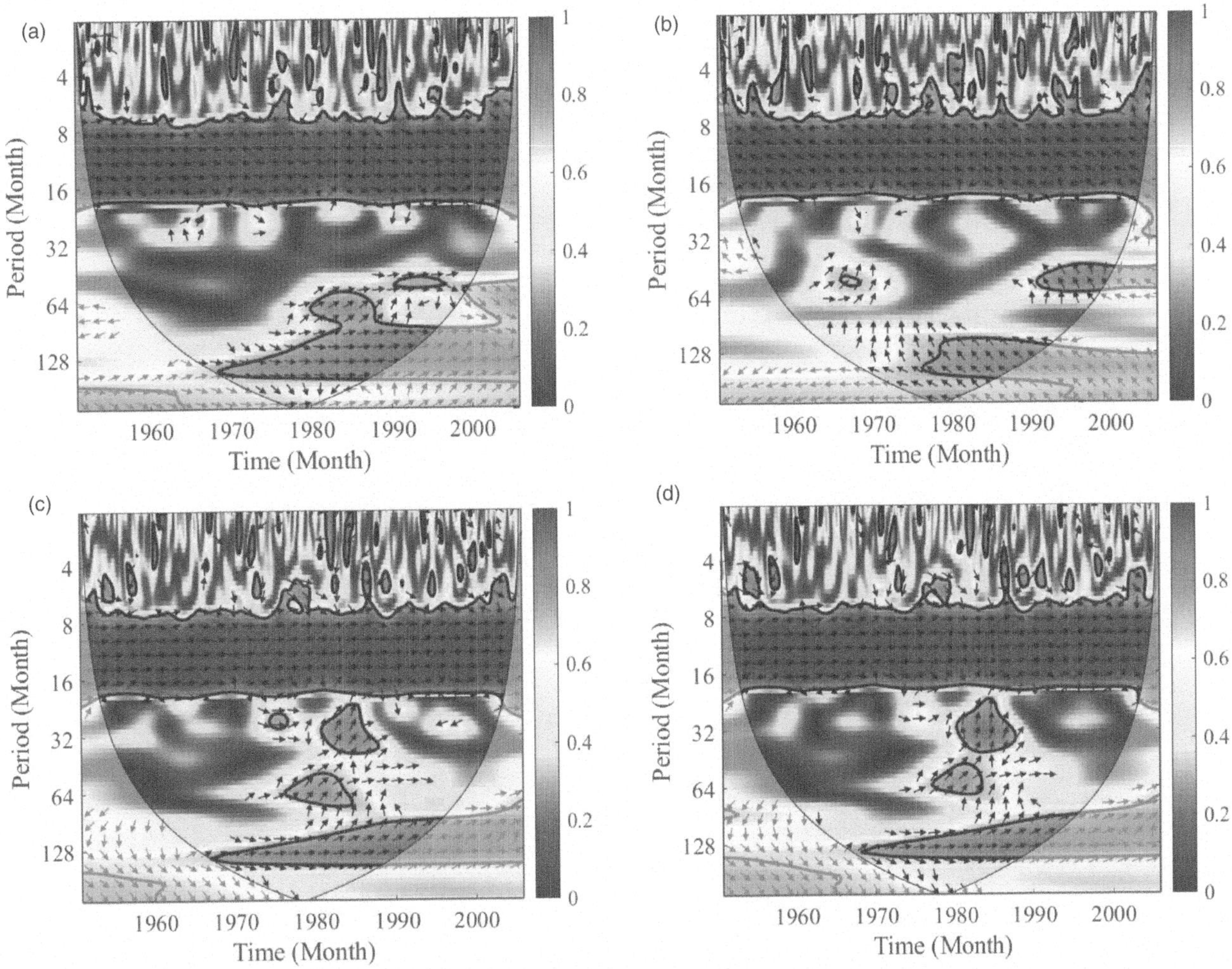

Figure 8 | WTC between temperature for INM-CM5 and (a) psl at grid point B3, (b) tas at grid point B3, (c) zg15000 at grid point B4, and (d) zg20000 at grid point B4. The thick black contour indicates the 95% confidence level using red noise as background spectrum, and the cone of influence where edge effects affect analysis is shown as a lighter shade. The relative phase relationship is shown as arrows (with in-phase pointing right, anti-phase pointing left). 1 (red) means the two signals are highly correlated and 0 (blue) means no correlation. Please refer to the online version of this paper to see this figure in colour: https://dx.doi.org/10.2166/nh.2022.094.

Table 3 | Results of downscaling dominant predictors by ANN

Feature extraction method	Variable	Model	RMSE[a] train	RMSE[a] verification	DC train	DC verification
WTC	Precipitation	BNU-ESM	0.23	0.25	0.74	0.72
		Can-ESM5	0.14	0.16	0.84	0.81
		INM-CM5	0.13	0.16	0.81	0.76
	Temperature	BNU-ESM	0.11	0.15	0.9	0.88
		Can-ESM5	0.05	0.08	0.94	0.90
		INM-CM5	0.05	0.11	0.94	0.92
CC	Precipitation	BNU-ESM	0.45	0.41	0.56	0.53
		Can-ESM5	0.35	0.39	0.62	0.65
		INM-CM5	0.22	0.26	0.71	0.65
	Temperature	BNU-ESM	0.47	0.43	0.56	0.51
		Can-ESM5	0.34	0.30	0.58	0.55
		INM-CM5	0.36	0.32	0.56	0.51

[a]Normalized RMSE.

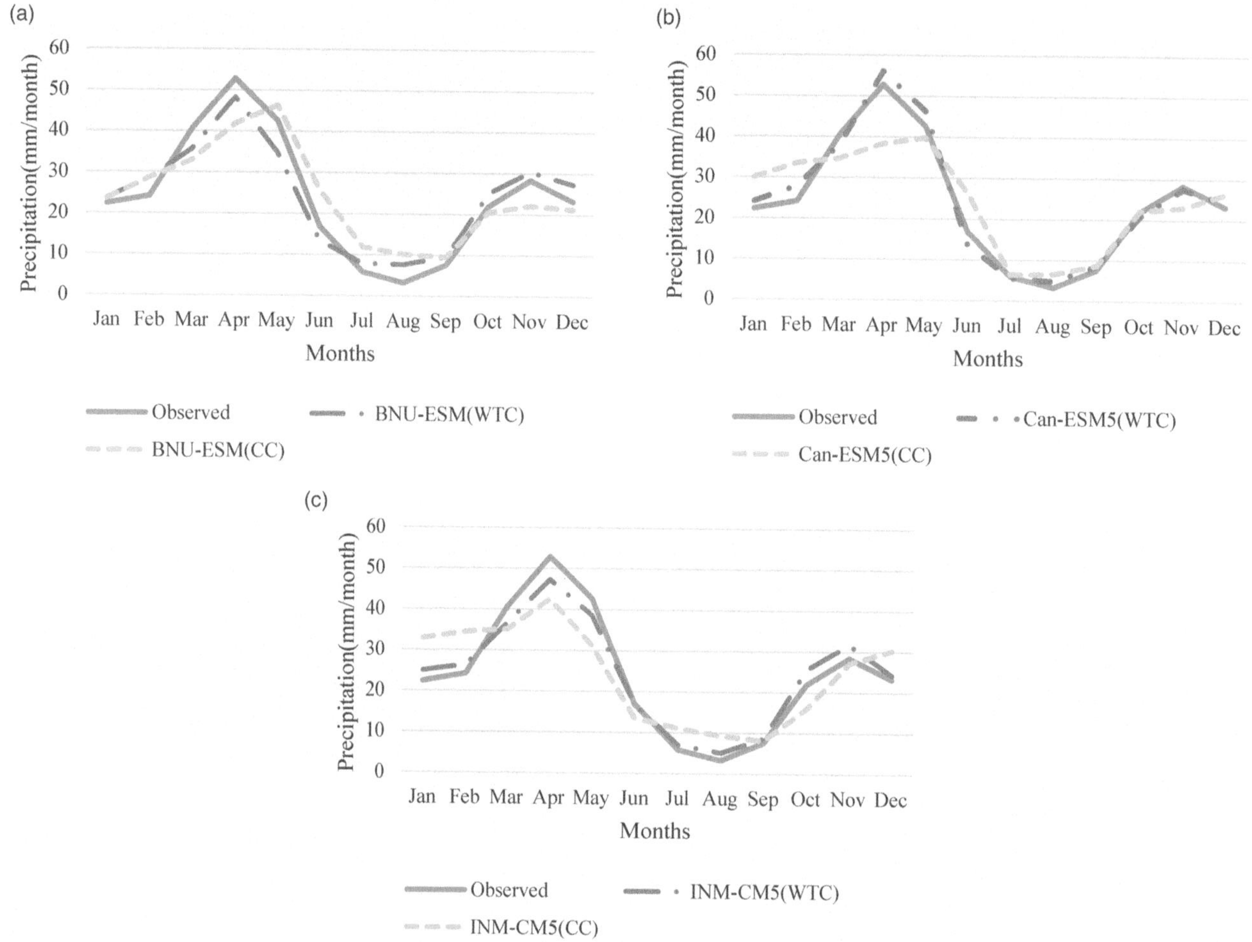

Figure 9 | Performance of ANN simulation for precipitation values during the base period (1951–2005) (a) BNU-ESM, (b) Can-ESM5, and (c) INM-CM5.

Given the results of the second step, as the downscaling based on the WTC predictor screening method indicated better results than the CC method, the projection of future climate variables (i.e., precipitation and temperature) was performed in accordance with the dominant variables selected by the WTC technique. To this end, large-scale future climate variables for each CM were used to implement the ANN simulation model. For the purpose of evaluating the ANN projection, the variation for projected monthly precipitation and average temperature and observation data at the historical period (i.e., 1951–2005) is illustrated in Figures 11 and 12 for each season of the year. According to Figure 11, there will be both rise and fall in the seasonal future precipitation of the study area. While a decrease of precipitation will be a trend in winter, summer, and fall for all three CMs under both scenarios, spring will experience a climb in precipitation. The underlying cause of this phenomenon can stem from the climate change effects due to the global warming. To elucidate, during the cold months (e.g., January, February, and December), air pollution can be an obstacle for condensation. Pollutant particles can block the incoming sunlight to the earth, thus, the differential temperature on the earth surface and the atmosphere would not be adequate for water vapor to grow and form clouds. On the other hand, in spring, it is observed that due to the decrease of contaminants, which is mostly because of the wind coming from Tabriz city northeastern (Gholampour *et al.* 2014), the pollution can be scattered. This mechanism will be accompanied by an increase in the air temperature, giving impetus for the formation and growth of clouds. The most decreasing trend in the future precipitation can be seen in Can-ESM5 under the low emission SSP2 scenario, which is 28.92%. Furthermore, BNU-ESM under the high emission RCP8.5 scenario revealed a 7.1% decrease, which is the minimum value.

As is shown in Figure 12, a growing trend for temperature is noticeable for three CMs, while RCP4.5 and SSP2 show mild increase, the growth of temperature under the RCP8.5 and SSP5 is remarkable. Furthermore, according to the difference

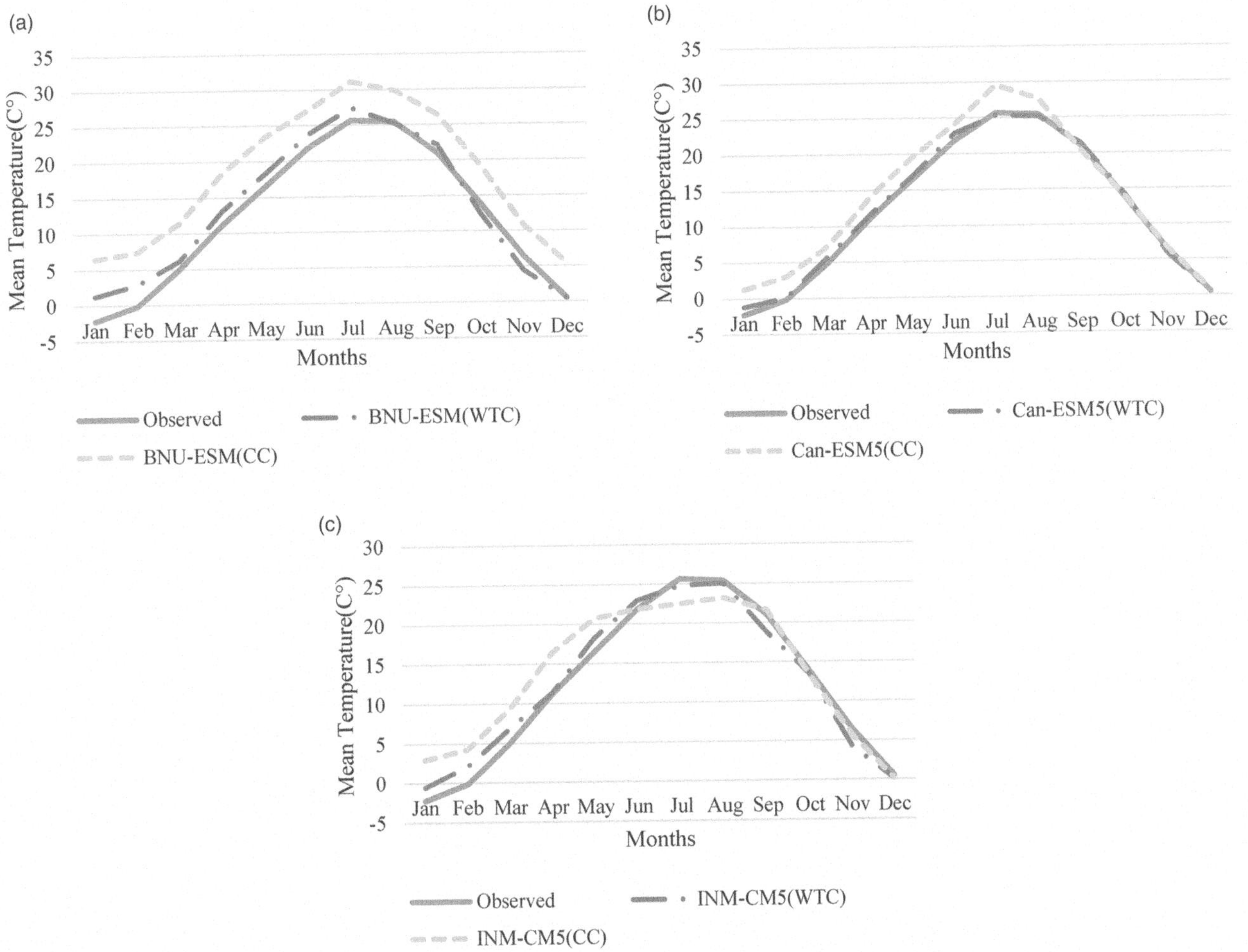

Figure 10 | Performance of ANN simulation for temperature values during the base period (1951–2005) (a) BNU-ESM, (b) Can-ESM5, and (c) INM-CM5.

between the predicted and historical values of temperature, it can be concluded that INM-CM5 under the high emission SSP5 scenario shows 4.21 °C growth in temperature, which is the highest increase for the region in the future. The minimum rate of increase is attributed to BNU-ESM under the low emission RCP4.5 scenario, numbering 2.18 °C. It should be taken into consideration that, despite the decreasing and increasing trends respectively at precipitation and temperature values, the climate of the study region will experience more extreme weather events. This is closely aligned with the Intergovernmental Panel on Climate Change 2014 report, which remarked that 'the rate of heavy precipitation and the recurrence of the heatwave is expected to go up in most areas, and Asia in particular'.

4. CONCLUDING REMARKS

This paper studied the impacts of climate change on the variation of hydrological variables (i.e., precipitation and temperature) of the Tabriz synoptic station for the period of 2021–2060. In this regard, monthly precipitation and the average temperature of the study area during 1951–2005 were used as the observation period. Three CMs, including BNU-ESM from CMIP5 models, and Can-ESM5 and INM-CM5 from the CMIP6 models were employed to implement ANN-based statistical downscaling. There are several large scale variables that can influence the climate of the area; however, the extent of this impact is not equal for all the variables, while some predictors show a strong relationship with the predictands, others do not show a significant connection. Therefore, in order to increase the accuracy of the ANN downscaling model, it is crucial to select the dominant inputs to avoid data redundancy. In this regard, WTC and CC as data-dimensionality reduction methods

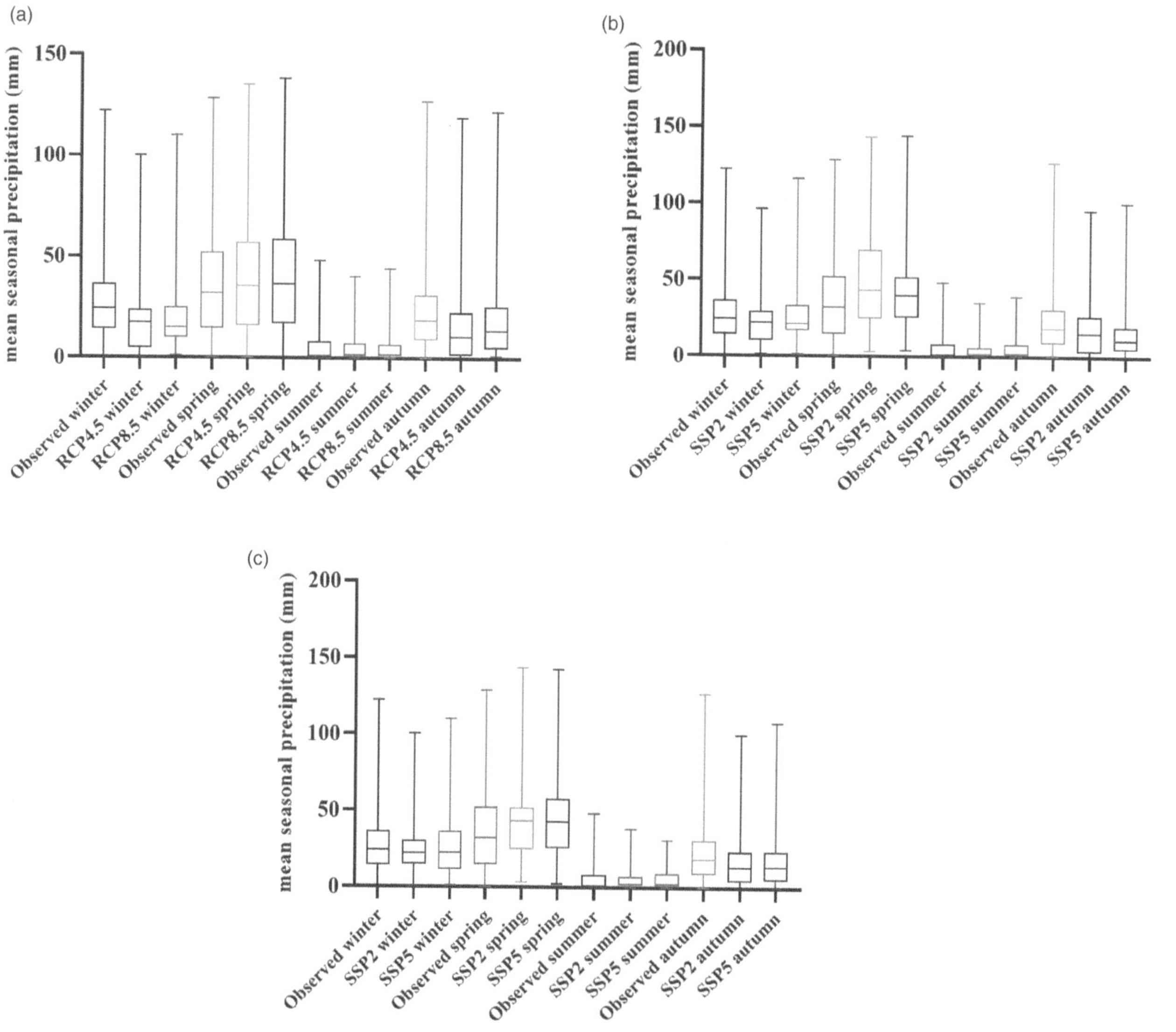

Figure 11 | Seasonal variation of precipitation based on the projected and historical period (a) BNU-ESM, (b) Can-ESM5, and (c) INM-CM5.

were used for selecting the potential variables, which led to predictor screening. The result of the ANN downscaling models indicated that the WTC predictor screening approach outperformed the CC method, due to the capability of WTC to distinguish the nonlinear relationship between predictors and predictands. WTC indicates 11–19% improvement for modeling the precipitation and 41–45% for temperature in the base period (1951–2005) in comparison to the CC approach. Finally, the future projection of precipitation and temperature was implemented and results denoted a decreasing trend in the precipitation, while the temperature will experience an increasing rate. This condition is testament to the fact that the galloping rate of GHGs emissions will lead to more global warming, and it consequently has adverse effects on the precipitation rate. All in all, it is expected that the precipitation will undergo a 7.1–28.92% decrease, while the temperature will experience a 2.18–4.21 °C increase.

In this study, a wavelet-based data screening method equipped with an ANN model was used to implement statistical downscaling of CM outputs. It is recommended to use other predictor screening methods such as supervised approaches (e.g., PCA) and unsupervised approaches (e.g., clustering) to examine the performance of different methods. Furthermore, it is suggested to incorporate more AI-based models (e.g., SVM, ANFIS, and GEP) or other statistical downscaling approaches (e.g., MLR, SDSM). This can contribute to more recognition of the capabilities and weak points of different models in downscaling the climate variables. Moreover, using different bias correction methods such as Delta Change and Quantile Mapping

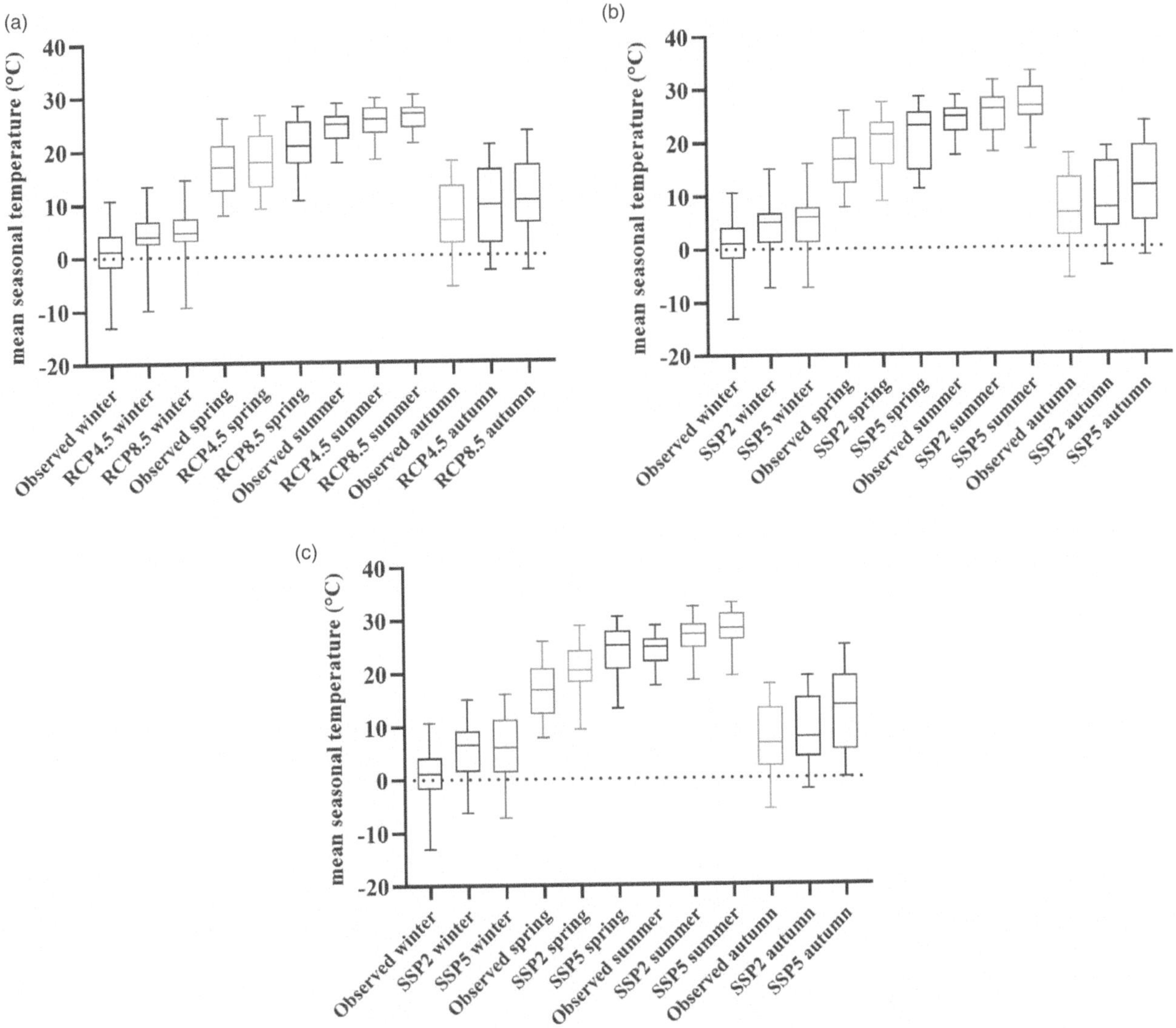

Figure 12 | Seasonal variation of temperature based on the projected and historical period (a) BNU-ESM, (b) Can-ESM5, and (c) INM-CM5.

methods could be investigated in further research works. It is proposed to utilize the ensemble of different AI models to test and compare their accuracy with the single models. Using several CMs and more scenarios could also give a more comprehensive insight into the future climate variation in a region.

DATA AVAILABILITY STATEMENT

All relevant data are available from an online repository or repositories https://esgf-node.llnl.gov/projects/esgf-llnl/.

REFERENCES

Abdellatif, M., Atherton, W. & Alkhaddar, R. 2013 A hybrid generalised linear and Levenberg–Marquardt artificial neural network approach for downscaling future rainfall in North Western England. *Hydrology Research* **44** (6), 1084–1101. https://doi.org/10.2166/nh.2013.045.

Adamowski, J. F. 2008 River flow forecasting using wavelet and cross-wavelet transform models. *Hydrological Processes* **22** (25), 4877–4891. https://doi.org/10.1002/hyp.7107.

Ahmadalipour, A., Rana, A., Moradkhani, H. & Sharma, A. 2017 Multi-criteria evaluation of CMIP5 GCMs for climate change impact analysis. *Theoretical and Applied Climatology* **128** (1), 71–87. https://doi.org/10.1007/s00704-015-1695-4.

Ahmadi, A. & Han, D. 2012 Identification of dominant sources of sea level pressure for precipitation forecasting over Wales. *Journal of Hydroinformatics* **15** (3), 1002–1021. https://doi.org/10.2166/hydro.2012.110.

Akinsanola, A. A., Kooperman, G. J., Pendergrass, A. G., Hannah, W. M. & Reed, K. A. 2020 Seasonal representation of extreme precipitation indices over the United States in CMIP6 present-day simulations. *Environmental Research Letters* **15** (9), 94003. https://doi.org/10.1088/1748-9326/ab92c1.

Anandhi, A., Srinivas, V. V., Kumar, D. N. & Nanjundiah, R. S. 2009 Role of predictors in downscaling surface temperature to river basin in India for IPCC SRES scenarios using support vector machine. *International Journal of Climatology* **29** (4), 583–603. https://doi.org/10.1002/joc.1719.

Baghanam, A. H., Nourani, V., Keynejad, M.-A., Taghipour, H. & Alami, M.-T. 2018 Conjunction of wavelet-entropy and SOM clustering for multi-GCM statistical downscaling. *Hydrology Research* **50** (1), 1–23. https://doi.org/10.2166/nh.2018.169.

Baghanam, A. H., Nourani, V., Sheikhbabaei, A. & Seifi, A. J. 2020 Statistical downscaling and projection of future temperature change for Tabriz city, Iran. **491**, 12009. https://doi.org/10.1088/1755-1315/491/1/012009.

Beecham, S., Rashid, M. & Chowdhury, R. K. 2014 Statistical downscaling of multi-site daily rainfall in a South Australian catchment using a Generalized Linear Model. *International Journal of Climatology* **34** (14), 3654–3670. https://doi.org/https://doi.org/10.1002/joc.3933.

Belayneh, A., Adamowski, J., Khalil, B. & Ozga-Zielinski, B. 2014 Long-term SPI drought forecasting in the Awash River Basin in Ethiopia using wavelet neural network and wavelet support vector regression models. *Journal of Hydrology* **508**, 418–429. https://doi.org/10.1016/j.jhydrol.2013.10.052.

Bowden, G. J., Dandy, G. C. & Maier, H. R. 2005 Input determination for neural network models in water resources applications. Part 1–background and methodology. *Journal of Hydrology* **301** (1), 75–92. https://doi.org/10.1016/j.jhydrol.2004.06.021.

Chadwick, R., Coppola, E. & Giorgi, F. 2011 An artificial neural network technique for downscaling GCM outputs to RCM spatial scale. *Nonlinear Processes in Geophysics* **18** (6), 1013–1028. https://doi.org/10.5194/npg-18-1013-2011.

Devak, M. & Dhanya, C. T. 2016 Downscaling of precipitation in Mahanadi Basin, India using support vector machine, K-nearest neighbour and hybrid of support vector machine with K-nearest neighbour. In: *Geostatistical and Geospatial Approaches for the Characterization of Natural Resources in the Environment* (Raju, N. J., ed.). Springer International Publishing, Cham, Switzerland, pp. 657–663.

Dibike, Y. B. & Coulibaly, P. 2006 Temporal neural networks for downscaling climate variability and extremes. *Neural Networks* **19** (2), 135–144. https://doi.org/10.1016/j.neunet.2006.01.003.

Draper, N. R. & Smith, H. 1998 *Applied Regression Analysis*, Vol. 326. John Wiley & Sons, Hoboken, NJ.

Eyring, V., Bony, S., Meehl, G. A., Senior, C. A., Stevens, B., Stouffer, R. J. & Taylor, K. E. 2016 Overview of the Coupled Model Intercomparison Project Phase 6 (CMIP6) experimental design and organization. *Geoscientific Model Development* **9** (5), 1937–1958. https://doi.org/10.5194/gmd-9-1937-2016.

Frost, A. J., Charles, S. P., Timbal, B., Chiew, F. H. S., Mehrotra, R., Nguyen, K. C., Chandler, R. E., McGregor, J. L., Fu, G., Kirono, D. G. C., Fernandez, E. & Kent, D. M. 2011 A comparison of multi-site daily rainfall downscaling techniques under Australian conditions. *Journal of Hydrology* **408** (1), 1–18. https://doi.org/https://doi.org/10.1016/j.jhydrol.2011.06.021.

Gangopadhyay, S., Clark, M. & Rajagopalan, B. 2005 Statistical downscaling using K-nearest neighbors. *Water Resources Research* **41** (2). https://doi.org/10.1029/2004WR003444.

Gholampour, A., Nabizadeh, R., Naseri, S., Yunesian, M., Taghipour, H., Rastkari, N., Nazmara, S., Faridi, S. & Mahvi, A. H. 2014 Exposure and health impacts of outdoor particulate matter in two urban and industrialized area of Tabriz, Iran. *Journal of Environmental Health Science and Engineering* **12** (1), 27. https://doi.org/10.1186/2052-336X-12-27.

Goyal, M. K., Burn, D. H. & Ojha, C. S. P. 2012 Evaluation of machine learning tools as a statistical downscaling tool: temperatures projections for multi-stations for Thames River Basin, Canada. *Theoretical and Applied Climatology* **108** (3), 519–534. https://doi.org/10.1007/s00704-011-0546-1.

Grinsted, A., Moore, J. & Jevrejeva, S. 2004 Application of cross wavelet transform and wavelet coherence to geophysical time series. *Nonlinear Processes in Geophysics* **11**. https://doi.org/10.5194/npg-11-561-2004.

Guo, J., Chen, H., Xu, C.-Y., Guo, S. & Guo, J. 2012 Prediction of variability of precipitation in the Yangtze River Basin under the climate change conditions based on automated statistical downscaling. *Stochastic Environmental Research and Risk Assessment* **26** (2), 157–176. https://doi.org/10.1007/s00477-011-0464-x.

Harpham, C. & Wilby, R. L. 2005 Multi-site downscaling of heavy daily precipitation occurrence and amounts. *Journal of Hydrology* **312** (1), 235–255. https://doi.org/10.1016/j.jhydrol.2005.02.020.

Haykin, S. & Network, N. 2004 A comprehensive foundation. *Neural Networks* **2** (2004), 41.

Holman, I. P., Rivas-Casado, M., Bloomfield, J. P. & Gurdak, J. J. 2011 Identifying non-stationary groundwater level response to North Atlantic ocean-atmosphere teleconnection patterns using wavelet coherence. *Hydrogeology Journal* **19** (6), 1269. https://doi.org/10.1007/s10040-011-0755-9.

Intergovernmental Panel on Climate Change 2014 *The Physical Science Basis: Working Group I Contribution to the Fifth Assessment Report of the Intergovernmental Panel on Climate Change*. Cambridge University Press, Cambridge. https://doi.org/10.1017/CBO9781107415324.

Jevrejeva, S., Moore, J. C. & Grinsted, A. 2003 Influence of the Arctic Oscillation and El Niño-Southern Oscillation (ENSO) on ice conditions in the Baltic Sea: the wavelet approach. *Journal of Geophysical Research: Atmospheres* **108** (D21). https://doi.org/10.1029/2003JD003417

Jiang, J., Zhou, T., Chen, X. & Zhang, L. 2020 Future changes in precipitation over Central Asia based on CMIP6 projections. *Environmental Research Letters* **15** (5), 54009. https://doi.org/10.1088/1748-9326/ab7d03.

Kaiser, G. 2011 Windowed Fourier transforms. In: *A Friendly Guide to Wavelets*. Birkhäuser Boston, pp. 44–59. https://doi.org/10.1007/978-0-8176-8111-1_2

Khan, M. S., Coulibaly, P. & Dibike, Y. 2006 Uncertainty analysis of statistical downscaling methods. *Journal of Hydrology* **319** (1), 357–382. https://doi.org/10.1016/j.jhydrol.2005.06.035.

Kisi, Ö. 2004 Multi-layer perceptrons with Levenberg-Marquardt training algorithm for suspended sediment concentration prediction and estimation/Prévision et estimation de la concentration en matières en suspension avec des perceptrons multi-couches et l'algorithme d. *Hydrological Sciences Journal* **49** (6), 1025–1040. https://doi.org/10.1623/hysj.49.6.1025.55720.

Kumar, S., Merwade, V., Kinter, J. L. & Niyogi, D. 2013 Evaluation of temperature and precipitation trends and long-term persistence in CMIP5 twentieth-century climate simulations. *Journal of Climate* **26** (12), 4168–4185. https://doi.org/10.1175/JCLI-D-12-00259.1.

Labat, D. 2010 Cross wavelet analyses of annual continental freshwater discharge and selected climate indices. *Journal of Hydrology* **385** (1), 269–278. https://doi.org/10.1016/j.jhydrol.2010.02.029.

Legates, D. R. & McCabe Jr., G. J. 1999 Evaluating the use of 'goodness-of-fit' measures in hydrologic and hydroclimatic model validation. *Water Resources Research* **35** (1), 233–241. https://doi.org/10.1029/1998WR900018.

Li, S. & Zheng, D. 2003 Applications of artificial neural networks to geosciences: review and prospect. *Advances in Earth Science* **18** (1), 68–76.

Liu, P. C. 1994 Wavelet spectrum analysis and ocean wind waves. In: *Wavelets in Geophysics*, Vol. 4 (Foufoula-Georgiou, E. & Kumar, P., eds). Academic Press, pp. 151–166. https://doi.org/10.1016/B978-0-08-052087-2.50012-8

Luterbacher, J., Xoplaki, E., Dietrich, D., Jones, P., Davies, T., Portis, D., González Rouco, J. F., Von Storch, H., Gyalistras, D., Casty, C. & Wanner, H. 2002 Extending North Atlantic oscillation reconstructions back to 1500. *Atmospheric Science Letters* **2**, 114–124. https://doi.org/10.1006/asle.2001.0044.

Maier, H. R. & Dandy, G. C. 2000 Neural networks for the prediction and forecasting of water resources variables: a review of modelling issues and applications. *Environmental Modelling & Software* **15** (1), 101–124. https://doi.org/10.1016/S1364-8152(99)00007-9.

Mallat, S. 1999 *A Wavelet Tour of Signal Processing*. Academic Press, London, UK.

Maraun, D. & Kurths, J. 2004 Cross wavelet analysis: significance testing and pitfalls. *Nonlinear Processes in Geophysics* **11**. https://doi.org/10.5194/npg-11-505-2004.

Moghim, S. & Bras, R. L. 2017 Bias correction of climate modeled temperature and precipitation using artificial neural networks. *Journal of Hydrometeorology* **18** (7), 1867–1884. https://doi.org/10.1175/JHM-D-16-0247.1.

Ng, E. K. W. & Chan, J. C. L. 2012 Geophysical applications of partial wavelet coherence and multiple wavelet coherence. *Journal of Atmospheric and Oceanic Technology* **29** (12), 1845–1853. https://doi.org/10.1175/JTECH-D-12-00056.1.

Ngoma, H., Wen, W., Ayugi, B., Babaousmail, H., Karim, R. & Ongoma, V. 2021 Evaluation of precipitation simulations in CMIP6 models over Uganda. *International Journal of Climatology* **41** (9), 4743–4768. https://doi.org/10.1002/joc.7098.

Nourani, V., Andalib, G. & Dąbrowska, D. 2017 Conjunction of wavelet transform and SOM-mutual information data pre-processing approach for AI-based multi-station nitrate modeling of watersheds. *Journal of Hydrology* **548**, 170–183. https://doi.org/10.1016/j.jhydrol.2017.03.002.

Nourani, V., Baghanam, A. H. & Gokcekus, H. 2018a Data-driven ensemble model to statistically downscale rainfall using nonlinear predictor screening approach. *Journal of Hydrology* **565**, 538–551. https://doi.org/10.1016/j.jhydrol.2018.08.049.

Nourani, V., Ghasemzade, M., Danandeh Mehr, A. & Sharghi, E. 2018b Investigating the effect of hydroclimatological variables on Urmia Lake water level using wavelet coherence measure. *Journal of Water and Climate Change* **10**, jwc2018261. https://doi.org/10.2166/wcc.2018.261.

Nourani, V., Jabbarian Paknezhad, N., Sharghi, E. & Khosravi, A. 2019a Estimation of prediction interval in ANN-based multi-GCMs downscaling of hydro-climatologic parameters. *Journal of Hydrology* **579**, 124226. https://doi.org/10.1016/j.jhydrol.2019.124226.

Nourani, V., Razzaghzadeh, Z., Baghanam, A. H. & Molajou, A. 2019b ANN-based statistical downscaling of climatic parameters using decision tree predictor screening method. *Theoretical and Applied Climatology* **137** (3), 1729–1746. https://doi.org/10.1007/s00704-018-2686-z.

Okkan, U. 2015 Assessing the effects of climate change on monthly precipitation: proposing of a downscaling strategy through a case study in Turkey. *KSCE Journal of Civil Engineering* **19** (4), 1150–1156. https://doi.org/10.1007/s12205-014-0052-y.

Partal, T. & Kişi, Ö. 2007 Wavelet and neuro-fuzzy conjunction model for precipitation forecasting. *Journal of Hydrology* **342** (1), 199–212. https://doi.org/10.1016/j.jhydrol.2007.05.026.

Pendergrass, A. G. & Hartmann, D. L. 2014 Two modes of change of the distribution of rain. *Journal of Climate* **27** (22), 8357–8371. https://doi.org/10.1175/JCLI-D-14-00182.1.

Pu, B. & Ginoux, P. 2017 Projection of American dustiness in the late 21st century due to climate change. *Scientific Reports* **7** (1), 5553. https://doi.org/10.1038/s41598-017-05431-9.

Rana, A. & Moradkhani, H. 2016 Spatial, temporal and frequency based climate change assessment in Columbia River Basin using multi downscaled-scenarios. *Climate Dynamics* **47** (1), 579–600. https://doi.org/10.1007/s00382-015-2857-x.

Rashid, M. M., Beecham, S. & Chowdhury, R. K. 2016 Statistical downscaling of rainfall: a non-stationary and multi-resolution approach. *Theoretical and Applied Climatology* **124** (3), 919–933. https://doi.org/10.1007/s00704-015-1465-3.

Riahi, K., van Vuuren, D. P., Kriegler, E., Edmonds, J., O'Neill, B. C., Fujimori, S., Bauer, N., Calvin, K., Dellink, R., Fricko, O., Lutz, W., Popp, A., Cuaresma, J. C., KC, S., Leimbach, M., Jiang, L., Kram, T., Rao, S., Emmerling, J., Ebi, K., Hasegawa, T., Havlik, P., Humpenöder, F., Da Silva, L. A., Smith, S., Stehfest, E., Bosetti, V., Eom, J., Gernaat, D., Masui, T., Rogelj, J., Strefler, J., Drouet, L., Krey, V., Luderer, G., Harmsen, M., Takahashi, K., Baumstark, L., Doelman, J. C., Kainuma, M., Klimont, Z., Marangoni, G., Lotze-Campen, H., Obersteiner, M., Tabeau, A. & Tavoni, M. 2017 The shared socioeconomic pathways and their energy, land use, and greenhouse gas emissions implications: an overview. *Global Environmental Change* **42**, 153–168. https://doi.org/10.1016/j.gloenvcha.2016.05.009.

Rivera, J. A. & Arnould, G. 2020 Evaluation of the ability of CMIP6 models to simulate precipitation over Southwestern South America: climatic features and long-term trends (1901–2014). *Atmospheric Research* **241**, 104953. https://doi.org/10.1016/j.atmosres.2020.104953.

Sachindra, D. A., Huang, F., Barton, A. & Perera, B. J. C. 2013 Least square support vector and multi-linear regression for statistically downscaling general circulation model outputs to catchment streamflows. *International Journal of Climatology* **33** (5), 1087–1106. https://doi.org/10.1002/joc.3493.

Sadeqi, A. & Dinpashoh, Y. 2020 Projection of precipitation and its variability under the climate change conditions in the future periods (Case study: Tabriz). *Environment and Water Engineering* **5** (4), 339–350. https://doi.org/10.22034/jewe.2020.210941.1339.

Soon, W., Velasco Herrera, V. M., Selvaraj, K., Traversi, R., Usoskin, I., Chen, C.-T. A., Lou, J.-Y., Kao, S.-J., Carter, R. M., Pipin, V., Severi, M. & Becagli, S. 2014 A review of Holocene solar-linked climatic variation on centennial to millennial timescales: physical processes, interpretative frameworks and a new multiple cross-wavelet transform algorithm. *Earth-Science Reviews* **134**, 1–15. https://doi.org/10.1016/j.earscirev.2014.03.003.

Stouffer, R. J., Eyring, V., Meehl, G. A., Bony, S., Senior, C., Stevens, B. & Taylor, K. E. 2017 CMIP5 scientific gaps and recommendations for CMIP6. *Bulletin of the American Meteorological Society* **98** (1), 95–105. https://doi.org/10.1175/BAMS-D-15-00013.1.

Su, B., Zeng, X., Zhai, J., Wang, Y. & Li, X. 2015 Projected precipitation and streamflow under SRES and RCP emission scenarios in the Songhuajiang River basin, China. *Quaternary International* **380–381**, 95–105. https://doi.org/10.1016/j.quaint.2014.03.049.

Sun, Q., Miao, C. & Duan, Q. 2015 Comparative analysis of CMIP3 and CMIP5 global climate models for simulating the daily mean, maximum, and minimum temperatures and daily precipitation over China. *Journal of Geophysical Research: Atmospheres* **120** (10), 4806–4824. https://doi.org/10.1002/2014JD022994.

Tamaddun, K. A., Kalra, A., Bernardez, M. & Ahmad, S. 2017 Multi-scale correlation between the western U.S. snow water equivalent and ENSO/PDO using wavelet analyses. *Water Resources Management* **31** (9), 2745–2759. https://doi.org/10.1007/s11269-017-1659-9.

Tan, X., Gan, T. Y. & Shao, D. 2016 Wavelet analysis of precipitation extremes over Canadian ecoregions and teleconnections to large-scale climate anomalies. *Journal of Geophysical Research: Atmospheres* **121** (24), 14414–469486. https://doi.org/10.1002/2016JD025533.

Taylor, K. E., Stouffer, R. J. & Meehl, G. A. 2012 An overview of CMIP5 and the experiment design. *Bulletin of the American Meteorological Society* **93** (4), 485–498. https://doi.org/10.1175/BAMS-D-11-00094.1.

Timbal, B., Dufour, A. & McAvaney, B. 2003 An estimate of future climate change for western France using a statistical downscaling technique. *Climate Dynamics* **20** (7), 807–823. https://doi.org/10.1007/s00382-002-0298-9.

Tisseuil, C., Vrac, M., Lek, S. & Wade, A. J. 2010 Statistical downscaling of river flows. *Journal of Hydrology* **385** (1), 279–291. https://doi.org/10.1016/j.jhydrol.2010.02.030.

Torrence, C. & Compo, G. P. 1998 A practical guide to wavelet analysis. *Bulletin of the American Meteorological Society* **79** (1), 61–78. https://doi.org/10.1175/1520-0477(1998)079 < 0061:APGTWA > 2.0.CO;2.

Torrence, C. & Webster, P. J. 1999 Interdecadal changes in the ENSO? Monsoon System. *Journal of Climate* **12** (8), 2679–2690. https://doi.org/10.1175/1520-0442(1999)012 < 2679:ICITEM > 2.0.CO;2.

Tripathi, S., Srinivas, V. V. & Nanjundiah, R. S. 2006 Downscaling of precipitation for climate change scenarios: a support vector machine approach. *Journal of Hydrology* **330** (3), 621–640. https://doi.org/10.1016/j.jhydrol.2006.04.030.

Wang, B., Jin, C. & Liu, J. 2020 Understanding future change of global monsoons projected by CMIP6 models. *Journal of Climate* **33** (15), 6471–6489. https://doi.org/10.1175/JCLI-D-19-0993.1.

Wilby, R. L. & Wigley, T. M. L. 1997 Downscaling general circulation model output: a review of methods and limitations. *Progress in Physical Geography: Earth and Environment* **21** (4), 530–548. https://doi.org/10.1177/030913339702100403.

Wilby, R. L., Dawson, C. W. & Barrow, E. M. 2002 sdsm – a decision support tool for the assessment of regional climate change impacts. *Environmental Modelling & Software* **17** (2), 145–157. https://doi.org/10.1016/S1364-8152(01)00060-3.

Xu, L., Chen, N., Xiang, Z., Chen, Z., Hu, C. & Wang, C. 2019 Improving the North American multi-model ensemble (NMME) precipitation forecasts at local areas using wavelet and machine learning. *Climate Dynamics* **53**. https://doi.org/10.1007/s00382-018-04605-z

Yu, H.-L. & Lin, Y.-C. 2015 Analysis of space–time non-stationary patterns of rainfall–groundwater interactions by integrating empirical orthogonal function and cross wavelet transform methods. *Journal of Hydrology* **525**, 585–597. https://doi.org/10.1016/j.jhydrol.2015.03.057.

Zarghami, M., Abdi, A., Babaeian, I., Hassanzadeh, Y. & Kanani, R. 2011 Impacts of climate change on runoffs in East Azerbaijan, Iran. *Global and Planetary Change* **78** (3), 137–146. https://doi.org/10.1016/j.gloplacha.2011.06.003.

First received 4 September 2021; accepted in revised form 2 February 2022. Available online 17 February 2022

© 2022 The Authors

doi: 10.2166/nh.2022.044

Evaluating the long short-term memory (LSTM) network for discharge prediction under changing climate conditions

Carolina Natel de Moura[a,*], Jan Seibert[b] and Daniel Henrique Marco Detzel[a]

[a] Department of Hydraulics and Sanitation, Federal University of Parana, Avenida Coronel Francisco Heráclito dos Santos, 100, 81531-980 Curitiba, Paraná, Brazil
[b] Department of Geography, University of Zurich, Winterthurerstrasse 190, CH-8057 Zurich, Switzerland
*Corresponding author. E-mail: carolina.natel@ufpr.br

CN, 0000-0003-3103-6789

ABSTRACT

Better understanding the predictive capabilities of hydrological models under contrasting climate conditions will enable more robust decision-making. Here, we tested the ability of the long short-term memory (LSTM) for daily discharge prediction under changing conditions using six snow-influenced catchments in Switzerland. We benchmarked the LSTM using the Hydrologiska Byråns Vattenbalansavdelning (HBV) bucket-type model with two parameterizations. We compared the model performance under changing conditions against constant conditions and tested the impact of the time-series size used in calibration on the model performance. When calibrated, the LSTM resulted in a much better fit than the HBV. However, in validation, the performance of the LSTM dropped considerably, and the fit was as good or poorer than the HBV performance in validation. Using longer time series in calibration improved the robustness of the LSTM, whereas HBV needed fewer data to ensure a robust parameterization. When using the maximum number of years in calibration, the LSTM was considered robust to simulate discharges in a drier period than the one used in calibration. Overall, the HBV was found to be less sensitive for applications under contrasted climates than the data-driven model. However, other LSTM modeling setups might be able to improve the transferability between different conditions.

Key words: climate transposability, data-driven model, differential split-sample test, model calibration, model robustness

HIGHLIGHTS

- The long short-term memory (LSTM) had good predictive accuracy in both the calibration and validation periods; however, it is always less robust than the HBV model.
- When using the maximum number of years in calibration, the LSTM was robust enough in its application under changing conditions when applied in a condition drier than the one used in calibration.

INTRODUCTION

The use of hydrological models in conditions that differ from those during model calibration is a challenging problem in hydrology and critical for application in impact studies (Blöschl *et al.* 2019). Models calibrated in certain conditions have been shown to be not always suitable for different conditions or transferable in time (Bastola *et al.* 2011; Coron *et al.* 2012; Thirel *et al.* 2015; Broderick *et al.* 2016; Dakhlaoui *et al.* 2017; Grusson *et al.* 2017; Her *et al.* 2019; Ouermi *et al.* 2019; Pan *et al.* 2019). The lack of a robust analysis of model performance under changing conditions may lead to poor water resource management.

In the context of catchment hydrology, a changing condition refers to any significant modification in land cover, climate, or water management infrastructure, potentially affecting the transformation of rainfall into runoff (Thirel *et al.* 2015). A general approach for developing hydrological models suitable for use in transient conditions is to use the differential split-sample test (DSST). The model should be calibrated and validated over contrasting periods in such a method, for instance, calibrated over a wet period and validated during a dry period (Klemeš 1986; Coron *et al.* 2012). The modeler should seek a good transferability of the calibrated parameters to a different dataset in validation, rather than only a good fit during calibration, which is often translated as model robustness. Robustness is a model's degree of insensitivity to climatic and environmental conditions (Seiller *et al.* 2012).

Model generalization for contrasting climates has been extensively explored in the literature using the DSST (Seibert 2003; Wilby 2005; Vaze *et al.* 2010; Merz *et al.* 2011; Coron *et al.* 2012; Li *et al.* 2012; Seiller *et al.* 2012; Brigode *et al.* 2013; Kling *et al.* 2015; Li *et al.* 2015; Seiller *et al.* 2015; Thirel *et al.* 2015; Broderick *et al.* 2016; Fowler *et al.* 2016; Vormoor *et al.* 2018). The results have shown that model parameters are sensitive to the climatic conditions of the calibration period (Pan *et al.* 2019), that the transfer of model parameters in time may introduce a significant level of simulation errors (Zhu *et al.* 2016), and that calibration over a wetter (drier) climate than the validation climate leads to an overestimation (underestimation) of the mean simulated runoff (Coron *et al.* 2012). Changes in mean rainfall were more likely than those in mean potential evapotranspiration or air temperature to impact performance during validation (Coron *et al.* 2012). Furthermore, Broderick *et al.* (2016) pointed out that the model transferability in contrasted climates may vary depending on the testing scenario, catchment, and evaluation criteria. Here, we argue that testing new models and new calibration protocols can help with our understanding of the modeling capabilities under changing conditions.

Although data-driven techniques have proved to outperform many traditional approaches based on conceptual or physical models for constant conditions (Dawson & Wilby 1998; Dibike & Solomatine 2001; Hu *et al.* 2018; Lee *et al.* 2018; Kratzert *et al.* 2019a; Rafaeli Neto *et al.* 2019; Xu *et al.* 2020), and the models are reliable in out-of-sample generalization (Shen 2018), little work has been carried out to test the capabilities of data-driven methods to make reasonable predictions under changing conditions. A significant limitation of data-driven models may be that they do not benefit from our understanding of physical phenomena and instead rely on the data provided during optimization. Shortridge *et al.* (2016) argued that data-driven models could only generate reliable predictions for conditions comparable to those experienced historically. Otherwise, the models are likely to introduce considerable uncertainty into their projections.

The long short-term memory (LSTM), a particular type of recurrent neural network (RNN), has been shown to be promising in capturing the hydrological behavior from the learning process (Xu *et al.* 2020). Lees *et al.* (2021) showed that the LSTM simulates discharge with a consistent high model performance in a large range of catchments in Great Britain, including catchments typically considered difficult to model with four lumped conceptual models. Kratzert *et al.* (2019b) applied the LSTM model over 531 basins over the USA and found a high correlation between the values of the internal cells of an LSTM network and natural processes.

Recently, O & Orth (2020) evaluated state-of-the-art models to changing conditions, calibrating an LSTM network and two process-based models in 161 catchments distributed across Europe. In their modeling setup, the LSTM model and the process-based models had different calibration approaches. The LSTM was calibrated over all catchments at once using two approaches: calibrating on an extreme reference period (365 days) and calibrating with one randomly selected year from each catchment rather than the respective extreme reference year. In contrast, the process-based models were calibrated in individual catchments and only using the extreme reference period. The models were then used to simulate in the remaining years characterized by a transient condition. The models showed overall performance loss, which generally increased the more conditions deviated from the reference climate, and overall, relatively high robustness was demonstrated by the physically-based model.

In light of the discussion above, in this paper we tested new calibration protocols and extended the scope of the model evaluation, with focus on the LSTM model. This is done by (a) benchmarking the LSTM using the same modeling setup for both data-driven and process-based models (which includes calibrating one model to each catchment instead of calibrating the LSTM over all catchments), (b) testing if increasing the number of years in model calibration would lead to better model performance and robustness, in contrast to only 1 year used in the previous study), and finally (c) calibrating the models in constant conditions as comparison. We then evaluated the robustness of the LSTM for contrasted conditions compared to both its application in constant conditions and the robustness obtained by the process-based model.

STUDY AREA AND DATA

For our study, we used six snow-influenced catchments located in Switzerland, ranging from ~60 to 400 km^2, with a mean altitude between ~500 and 1,200 m.a.s.l. The location and description of the catchments (location, area, altitude, daily mean temperature, annual precipitation, mean daily discharge, and snow fraction) are presented in Figure 1 and Table 1. Our catchment choice aimed to select catchments mainly located in the Swiss plateau, within a climate homogeneous area, and considered nearly natural (i.e., there is negligible impact on runoff from human activity) (Orth *et al.* 2015).

The data needed to model the daily discharge were air temperature (°C), precipitation (mm d^{-1}), and the estimates of long-term monthly potential evapotranspiration (mm month^{-1}). Precipitation and air temperature data were obtained from the

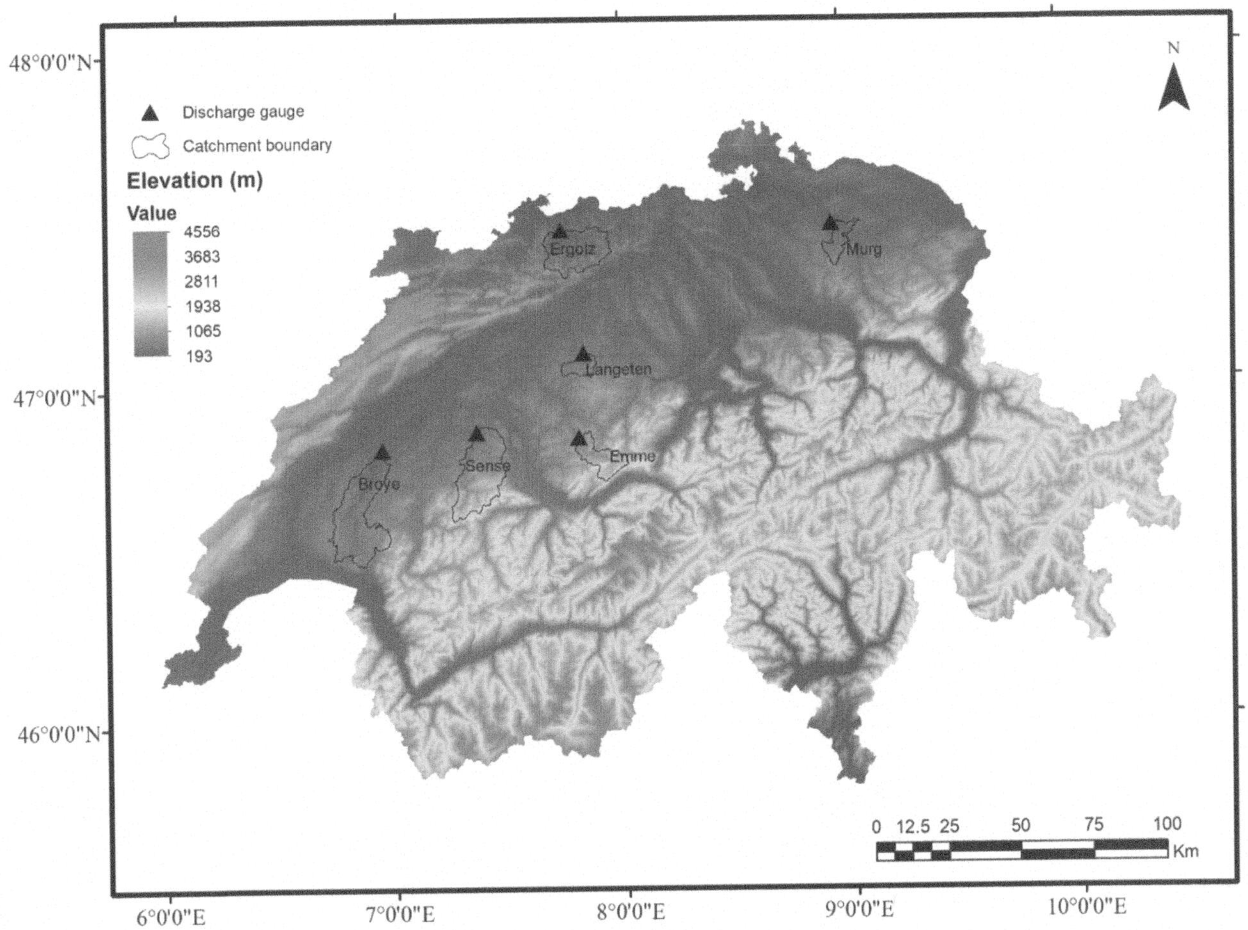

Figure 1 | Location of the study catchments.

Table 1 | Properties of the study catchments

Catchment	Mean altitude (m)	Area (km^2)	Daily mean temperature (°C)	Total precipitation (mm $year^{-1}$)	Mean discharge (mm d^{-1})	Snow fraction[a] (%)
Broye	710	392	8.6	1,190	1.6	5
Emme	1,189	124	5.6	1,692	3.0	19
Ergolz	590	261	8.6	1,091	1.2	6
Langeten	766	60	7.5	1,305	1.8	10
Murg	650	79	8.0	1,313	2.0	7
Sense	1,068	352	6.3	1,445	2.1	13

[a]Snow fraction (%): fraction of precipitation falling with temperature below 0 °C.

gridded meteorological forcing data at the spatial resolution of 2 km × 2 km from the Swiss Federal Office of Meteorology and Climatology (MeteoSwiss). We obtained daily discharge measurements from the Swiss Federal Office for the Environment (FOEN).

METHODS

Long short-term memory (LSTM)

The LSTM is a particular type of RNN used to process long time-sequences of data (Hochreiter & Schmidhuber 1997), in which the output of each time-step is fed as input to the next time-step. The control of the information flow is managed in units called gates and memory cells. The cell remembers values over arbitrary time intervals, and three gates regulate the flow of information into and out of the cell: the forget gate, the input gate, and the output gate. At every time-step t, each of the three gates is presented with the input $x[t]$ (i.e., explanatory variables) as well as the output $h[t-1]$ of the memory cells at the previous time-step $[t-1]$.

The first gate is the forget gate, which controls what information is removed from the cell state vector (Equation (1)). The hidden state h is initialized in the first time-step by a vector of zeros. In the next step, a potential update vector for the cell state is computed from the current input $x[t]$ and the last hidden state $h[t-1]$ by Equation (2). Additionally, the second gate is computed, the input gate (Equation (3)), defining which (and to what degree) information of $\bar{c}[t]$ is used to update the cell state in the current time-step. With the results of the forget gate and input gate, the cell state $c[t]$ is updated by Equation (4). Like the hidden state vector, the cell state is initialized by a vector of zeros in the first time-step. The last gate is the output gate, which controls the information of the cell state $c[t]$ that flows into the new hidden state $h[t]$. The output gate is calculated by Equation (5). Finally, the hidden state $h[t]$ is calculated by using the current cell state and the output gate value (Equation (6)). The model output is a linear combination of hidden states at the last time-step (Kratzert *et al.* 2018).

$$f[t] = \sigma(W_f x[t] + U_f h[t-1] + b_f) \tag{1}$$

$$\bar{c}[t] = \tan h(W_g x[t] + U_g h[t-1] + b_g) \tag{2}$$

$$i[t] = \sigma(W_i x[t] + U_i h[t-1] + b_i) \tag{3}$$

$$c[t] = f[t] \otimes c[t-1] + i[t] \otimes \bar{c}[t] \tag{4}$$

$$o[t] = \sigma(W_o x[t] + U_o h[t-1] + b_o) \tag{5}$$

$$h[t] = o[t] \otimes \tan h\ (c[t]) \tag{6}$$

where $f[t]$, $i[t]$, and $o[t]$ are the forget, input, and output gates represented by vectors with values in the range (0, 1), $\bar{c}[t]$ is a vector with values in the range $(-1, 1)$, $x[t]$ is the input vector (forcings and static attributes), $\tan h\ (\cdot)$ is the hyperbolic tangent, $\sigma\ (\cdot)$ represents the logistic sigmoid function, $\otimes$ denotes element-wise multiplication, and Ws, Us, and bs are sets of learnable parameters (i.e., two adjustable weight matrices and a bias vector).

In this work, we used a network consisting of a single LSTM layer with one hidden unit and a dense layer that connects the output of the LSTM at the last time-step to a single output neuron with linear activation. The LSTM model was implemented using the Keras package in Python, the Adam activation function, and the mean-squared error as loss function. To predict the discharge of a single time-step (day), we provided as input the last t consecutive time-steps of independent meteorological variables (daily precipitation [mm d^{-1}] and air temperature [°C]). We obtained the best hyperparameters of the LSTM model through a trial-and-error tuning approach. We varied the values of the following hyperparameters: length of the input sequence (time-steps), number of neurons in the hidden layer, and number of epochs. Our analysis resulted in the selection of 50 neurons, 50 epochs, and 365 days as time-steps.

HBV model

We benchmarked the performance of the LSTM model against the bucket-type HBV-Light version model (Seibert & Vis 2012). The HBV model consists of four routines including the snow routine, the soil routine, the groundwater routine, and the routing routine. This model usually simulates daily discharge based on daily precipitation, daily air temperature, and estimates of long-term monthly potential evapotranspiration rates. The HBV was used as both a lower and upper benchmark with two different parameterization methods (Seibert *et al.* 2018). As a lower benchmark, we used the ensemble mean of simulations with 1,000 randomly selected parameter sets, referred to hereafter as 'uncalibrated HBV'. For the upper benchmark, we calibrated the HBV model using an automatic genetic algorithm and the Nash–Sutcliff efficiency (NSE) as objective function, referred to hereafter as 'calibrated HBV'. In both cases, we specified feasible parameter ranges based on previous model applications.

Calibration procedure

The LSTM and HBV were calibrated individually for each one of the catchments resulting in six LSTM models and six HBV models. We calibrated and validated the models according to the DSST proposed by Klemeš (1986) for changing conditions.

According to Klemeš (1986), if the model is intended to simulate streamflow in a wet climate scenario, then it should be calibrated on a dry period of the historical record and validated on a wet period and *vice versa*. Additionally, we calibrated and validated a model under constant conditions.

Selection of the calibration and validation periods

The period between 1961 and 2018 was used to select the constant and changing condition periods. We mimicked the changing conditions by selecting two continuous periods in the time series with different hydrological conditions in the historical record. The dry and wet periods were chosen as the annual discharge below and above the long-term average discharge, respectively. The discharge changes between the periods were on average 50%. This is similar to the future hydrological changes expected for Switzerland of an increase in mean and maximum floods of 5–24% in the near future and of 25–49% in the far future, with exception to the Southern alpine catchments, where the mean annual floods may decrease in the far future (Köplin *et al.* 2014). For the constant conditions, we selected continuous periods containing both dry and wet years.

We also selected calibration periods with different time-series sizes, ranging from 2 to 6 years (2, 3, 4, and 6 years) for each catchment and condition (constant and changing), to test the influence of the amount of data used in the calibration on the model performance. We limited this analysis to 6 years due to data availability. We needed continuous periods with only low or high discharge, which were limited, on average, to 6 years across all the catchments.

Evaluation metrics and robustness

We evaluated the model performance using the NSE (Nash & Sutcliffe 1970), Kling–Kupta efficiency (KGE) (Gupta *et al.* 2009), non-parametric efficiency (NPE) (Pool *et al.* 2018), and mean absolute relative error (MARE) (Staudinger *et al.* 2011). The metrics range from $-\infty$ to 1, where 1 indicates perfect agreement between simulations and observations, and values lower than zero indicate very poor performance. These metrics were chosen to evaluate different hydrograph phases, the NSE focus on peaks and discharge dynamics, the KGE focus on the mean, variability, and dynamic, the NPE is the non-parametric version of the KGE, and finally, the MARE focus on low-to-medium flows. The robustness was calculated as the difference between the efficiency in calibration and validation (Hallouin *et al.* 2020). The independent two-sample *t*-test was used to evaluate whether the LSTM mean robustness was equal to the robustness obtained with the HBV model, and to compare the mean robustness of the LSTM under changing and constant conditions, at the significance level (α) of 0.05.

RESULTS

Model performance

In the calibration mode, the LSTM performed better than the HBV model for all criteria as expected, since it is more flexible (it has more degrees of freedom) than the conceptual model. However, the performance of the LSTM decreased more than the calibrated HBV when switching to the validation periods (Figure 2). The uncalibrated HBV model performed less well, but the performance was still better than what one might expect from a model run with random parameters and/or no local information. Therefore, we considered that a model performance of about 0.5 for NSE basically indicates that a model has no skill. By definition, its performance did not systematically differ between the calibration and validation periods for the uncalibrated model. For KGE, the patterns were roughly similar, whereas for NPE and MARE, which are more different from the NSE used for calibration, the calibrated models (LSTM and HBV) were less superior compared to the uncalibrated HBV model, especially when using fewer years during calibration.

The effect of the time-series size used in calibration on the performance of the models is represented in the *x*-axis of Figure 2. There was a positive correlation between the time-series length and model performances, which was more pronounced for the LSTM model. When evaluating the model's performance against metrics not used for the optimization of the model (i.e., KGE, NPE, and MARE), the increase in the time-series length used in calibration is essential to obtain LSTM performances comparable to the HBV model during the validation for contrasted conditions. Simulations for changing conditions performed less well than those for constant conditions in validation. However, the differences were less pronounced using the maximum number of years in calibration (i.e., 6 years).

The hydrographs and scatter plots of observed and estimated discharge using the best configuration, that is, using 6 years in calibration, for one of the study catchments are presented in Figures 3 and 4, respectively. The hydrograph shows the

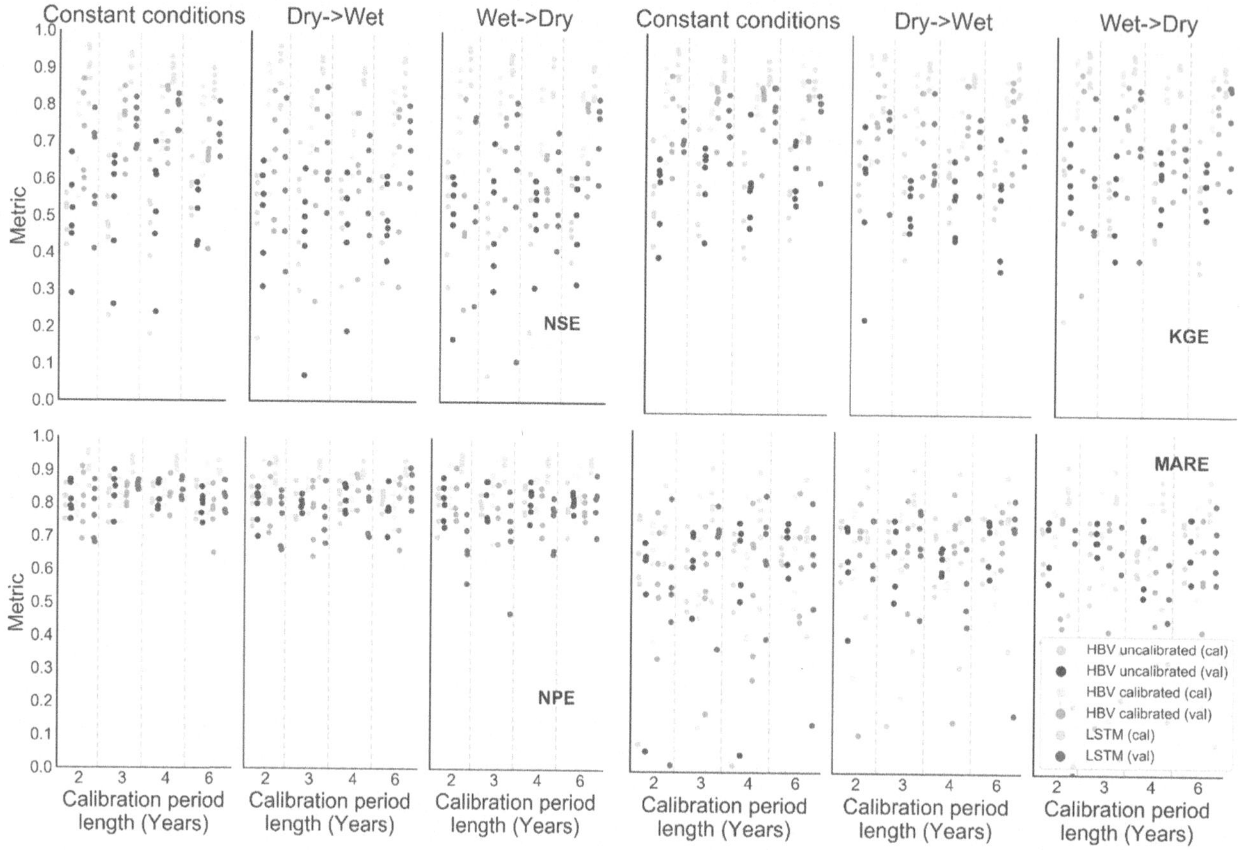

Figure 2 | Dot plots of model performance in calibration (cal) and validation (val) periods for the six catchments under the study. Each grid is one metric, and within each grid, each subgrid is one condition, from left to right: constant conditions, Dry → Wet (calibration in a dry period and validation in a wet one), and Wet → Dry (calibration in a wet period and validation in a dry one). The *y*-axis was limited to the interval between 0 and 1.

underestimation of the peaks, especially those in spring (when the snow accumulated during the winter starts to melt) by all models. However, most low- and mid-flows were predicted well. This is clearly shown in the scatter plots of the observed and simulated flows in Figure 4. The scatter plots also indicate that the predictions deviate more from the observed values in the uncalibrated HBV model. There is an underestimation of the peaks when applying the model in conditions wetter than those it was calibrated in, and the LSTM model simulations are slightly less spread than those of the calibrated HBV model.

Model robustness

The model robustness was evaluated as the difference in performance between calibration and validation periods (Table 2). The LSTM was considered robust enough for generalization in changing conditions when the LSTM mean robustness did not significantly differ from both the mean robustness of the bucket-type model and of the constant period for most of the metrics.

The calibrated HBV was always more robust than the LSTM model for both constant and non-constant conditions. The LSTM was robust enough for changing conditions only when the model was applied in a drier period than that used in calibration and using the maximum number of years during calibration (6 years). While a good indication of robustness was already observed with a shorter time series used in the calibration for the HBV, a longer dataset length was needed for the LSTM.

DISCUSSION

The LSTM had poorer performance under changing conditions. Others have found similar results when applying process-based models under changing conditions (Refsgaard & Knudsen 1996; Xu 1999; Seibert 2003; Wilby 2005; Chiew *et al.* 2009; Vaze *et al.* 2010; Bastola *et al.* 2011).

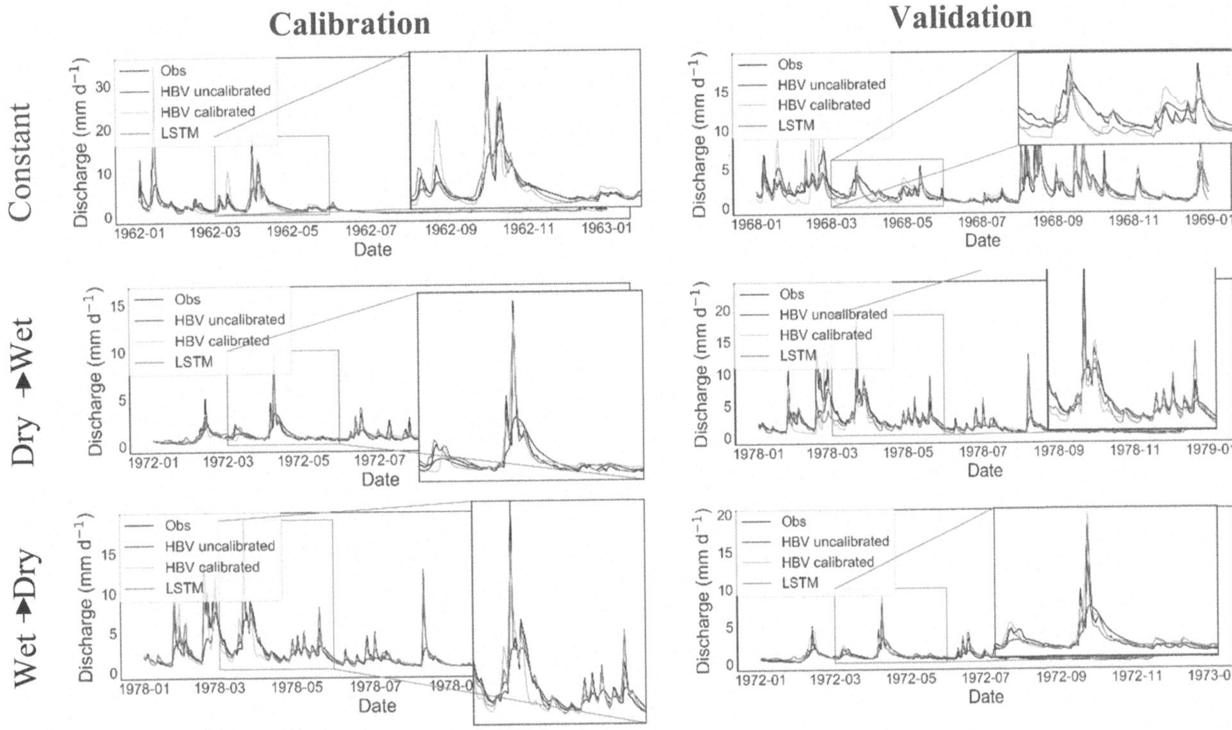

Figure 3 | Observed and simulated hydrographs.

Overall, when calibrated, the LSTM resulted in a much better fit than the HBV. However, the performance drop when going into a validation mode is also much larger for the LSTM (less robust). For the validation period, the LSTM was at best as good as the HBV (especially for other criteria than used in calibration and for changing conditions).

The LSTM was shown to be more dependent on dataset length to perform as well as the bucket-type model. The improvement in model performance/robustness with the increase of the time-series size used in calibration was also observed by Ayzel & Heistermann (2021) and Gauch *et al.* (2019) while testing the performance of the LSTM networks for streamflow prediction in constant conditions. Here, this positive correlation was more pronounced for the constant conditions than that for changing conditions. The lesser contribution of the time-series size in model performance under changing conditions may be explained by the limitation of the data provided for the calibration (only dry or wet periods used in calibration), that is, less information about the hydrological processes was provided to the model. The physical constraints of the HBV model made the need for longer data series in calibration less important, indicating the suitability of this model for predictions when data are limited. The same was observed by Ayzel & Heistermann (2021) while comparing an LSTM network to the GR4H conceptual model.

The robustness analysis showed that the LSTM is robust enough for climate transposability to a drier period. The generalization from a dry period to a wetter period is less satisfactory, mainly because in this case, the model needs to extrapolate to a discharge range not used in calibration, as also reported by Pan *et al.* (2019) and Wilby (2005) employing traditional hydrological models.

It is important to highlight that in this model setup, the LSTM was trained on individual catchments and its calibration over a large number of catchments and with a larger data series can yield better results and should be explored further. However, comparing models with different structures is not an easy task, especially when trying to keep a fair comparison between the models. More sophisticated hyperparameter tuning techniques may also improve the LSTM model's simulations, as well as coupling the model with process-based models.

CONCLUSION

In this work, we tested the predictive ability of the LSTM for daily discharge prediction in snow-influenced catchments under changing conditions. When calibrated, the LSTM resulted in a much better fit than the HBV, however, in a validation mode,

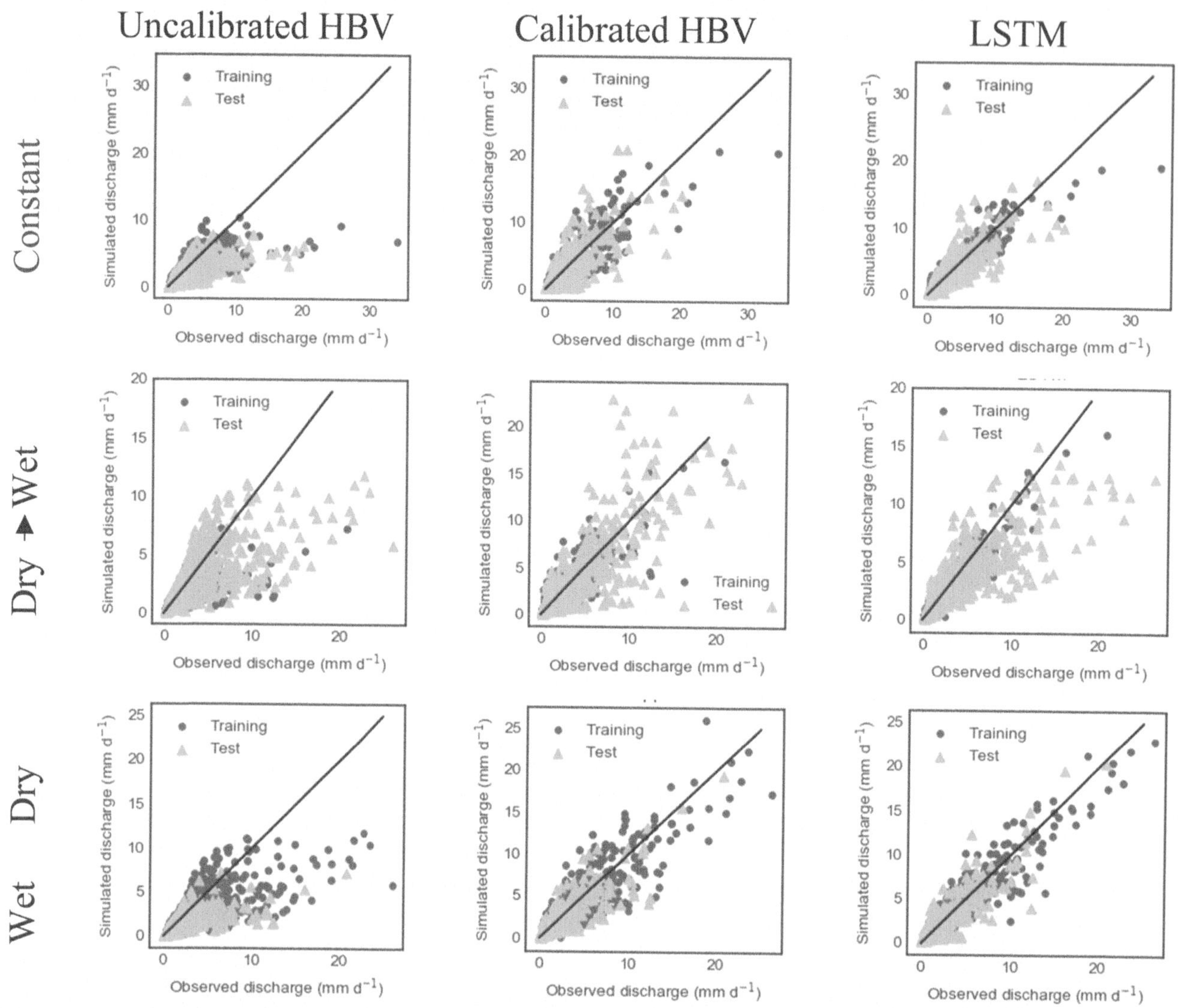

Figure 4 | Scatter plot of the observed and the simulated discharge.

Table 2 | Average robustness of the LSTM (left number) and the calibrated HBV model (right number) defined by the subtraction between the efficiency in calibration and validation

		Constant			
		NSE	**KGE**	**NPE**	**MARE**
Calibration period length (years)	2	0.28|0.04	**0.16|0.03**	0.14|0.01	0.27|0.06
	3	0.16|0.00	0.17|0.00	0.07| − 0.01	0.1| − 0.02
	4	0.12|0.02	**0.09|0.04**	**0.07|0.03**	**0.08|0.04**
	6	**0.14|0.10**	**0.10|0.04**	**0.08|0.07**	**0.12|0.07**
	Dry → Wet				
	2	**0.28|0.13**	**0.17|0.04**	0.14|0.04	**0.18|0.00**
	3	**0.24|0.13**	0.21|0.08	**0.13|0.05**	0.06| − 0.06
	4	0.26|0.09	0.18|0.02	0.11| − 0.01	0.06| − 0.18
	6	**0.19|0.15**	0.14|0.04	**0.05|0.00**	0.07| − 0.04
	Wet → Dry				
	2	**0.31|0.24**	**0.28|0.20**	0.23|0.03	0.54|0.14
	3	**0.31|0.09**	**0.21|0.07**	0.21|0.05	**0.34|0.18**
	4	**0.32|0.18**	**0.22|0.14**	0.17|0.07	0.41|0.20
	6	0.13|0.04	**0.12|0.07**	**0.11|0.06**	**0.16|0.10**

Bold values are showed when the LSTM did not differ significantly from the calibrated HBV. Underlined values are showed when the LSTM under a changing condition did not differ significantly from the constant condition ($\alpha = 0.05$).

the LSTM often performed worse than the HBV (especially for other criteria than used in calibration and for changing conditions). The performance drop when going into a validation mode was larger for the LSTM, indicating less robustness, and the data-driven model was shown to be more dependent on dataset length used in calibration to deliver robustness comparable to a bucket-type model.

Despite this, the results indicate that using longer data series in calibration can benefit the use of the LSTM in contrasting conditions. We recommend that other LSTM modeling setups should be studied further to improve the model performance in such conditions.

ACKNOWLEDGEMENTS

The authors thank the research funding agencies CAPES-Brazil and ESKAS – Swiss Government Excellence Scholarship for the scholarships granted to the first author at different periods of time during this research.

DATA AVAILABILITY STATEMENT

Data cannot be made publicly available; readers should contact the corresponding author for details.

REFERENCES

Ayzel, G. & Heistermann, M. 2021 The effect of calibration data length on the performance of a conceptual hydrological model versus LSTM and GRU: a case study for six basins from the CAMELS dataset. *Computers & Geosciences* **149**, 104708. https://doi.org/10.1016/j.cageo.2021.104708.

Bastola, S., Murphy, C. & Sweeney, J. 2011 The role of hydrological modelling uncertainties in climate change impact assessments of Irish river catchments. *Advances in Water Resources* **34**(5), 562–576. https://doi.org/10.1016/j.advwatres.2011.01.008.

Blöschl, G., Bierkens, M. F., Chambel, A., Cudennec, C., Destouni, G., Fiori, A. *et al.* 2019 Twenty-three unsolved problems in hydrology (UPH) – a community perspective. *Hydrological Sciences Journal* **64**(10), 1141–1158. https://doi.org/10.1080/02626667.2019.1620507.

Brigode, P., Oudin, L. & Perrin, C. 2013 Hydrological model parameter instability: a source of additional uncertainty in estimating the hydrological impacts of climate change? *Journal of Hydrology* **476**, 410–425. https://doi.org/10.1016/j.jhydrol.2012.11.012.

Broderick, C., Matthews, T., Wilby, R. L., Bastola, S. & Murphy, C. 2016 Transferability of hydrological models and ensemble averaging methods between contrasting climatic periods. *Water Resources Research* **52**(10), 8343–8373. https://doi.org/10.1002/2016WR018850.

Chiew, F. H. S., Teng, J., Vaze, J., Post, D. A., Perraud, J. M., Kirono, D. G. C. & Viney, N. R. 2009 Estimating climate change impact on runoff across southeast Australia: method, results, and implications of the modeling method. *Water Resources Research* **45**(10). https://doi.org/10.1029/2008WR007338.

Coron, L., Andreassian, V., Perrin, C., Lerat, J., Vaze, J., Bourqui, M. & Hendrickx, F. 2012 Crash testing hydrological models in contrasted climate conditions: an experiment on 216 Australian catchments. *Water Resources Research* **48**(5). doi:10.1029/2011wr011721.

Dakhlaoui, H., Ruelland, D., Tramblay, Y. & Bargaoui, Z. 2017 Evaluating the robustness of conceptual rainfall-runoff models under climate variability in northern Tunisia. *Journal of Hydrology* **550**, 201–217. https://doi.org/10.1016/j.jhydrol.2017.04.032.

Dawson, C. W. & Wilby, R. 1998 An artificial neural network approach to rainfall-runoff modelling. *Hydrological Sciences Journal* **43**(1), 47–66. https://doi.org/10.1080/02626669809492102.

Dibike, Y. B. & Solomatine, D. P. 2001 River flow forecasting using artificial neural networks. *Physics and Chemistry of the Earth, Part B: Hydrology, Oceans and Atmosphere* **26**(1), 1–7. https://doi.org/10.1016/S1464-1909(01)85005-X.

Fowler, K. J. A., Peel, M. C., Western, A. W., Zhang, L. & Peterson, T. J. 2016 Simulating runoff under changing climatic conditions: revisiting an apparent deficiency of conceptual rainfall-runoff models. *Water Resources Research* **52**, 1820–1846. https://doi.org/10.1002/2015WR018068.

Gauch, M., Mai, J. & Lin, J. 2021 The proper care and feeding of CAMELS: How limited training data affects streamflow prediction. *Environmental Modelling & Software* **135**, 104926. https://doi.org/10.1016/j.envsoft.2020.104926.

Grusson, Y., Anctil, F., Sauvage, S. & Sánchez Pérez, J. M. 2017 Assessing the climatic and temporal transposability of the SWAT model across a large contrasted watershed. *Journal of Hydrologic Engineering* **22**(6), 04017004. doi:10.1061/(ASCE)HE.1943-5584.0001491.

Gupta, H. V., Kling, H., Yilmaz, K. K. & Martinez, G. F. 2009 Decomposition of the mean squared error and NSE performance criteria: implications for improving hydrological modelling. *Journal of Hydrology* **377**(1–2), 80–91. https://doi.org/10.1016/j.jhydrol.2009.08.003.

Hallouin, T., Bruen, M. & O'Loughlin, F. E. 2020 Calibration of hydrological models for ecologically relevant streamflow predictions: a trade-off between fitting well to data and estimating consistent parameter sets? *Hydrology and Earth System Sciences* **24**(3), 1031–1054. https://doi.org/10.5194/hess-24-1031-2020.

Her, Y., Yoo, S. H., Cho, J., Hwang, S., Jeong, J. & Seong, C. 2019 Uncertainty in hydrological analysis of climate change: multi-parameter vs. multi-GCM ensemble predictions. *Scientific Reports* **9**(1), 1–22. https://doi.org/10.1038/s41598-019-41334-7.

Hochreiter, S. & Schmidhuber, J. 1997 LSTM can solve hard long time lag problems. In: *Advances in Neural Information Processing Systems*, pp. 473–479. https://doi.org/10.3390/w10111543.

Hu, C., Wu, Q., Li, H., Jian, S., Li, N. & Lou, Z. 2018 Deep learning with a long short-term memory networks approach for rainfall-runoff simulation. *Water* **10**(11), 1543. https://doi.org/10.3390/w10111543.

Klemeš, V. 1986 Operational testing of hydrological simulation models. *Hydrological Sciences Journal* **31**(1), 13–24. https://doi.org/10.1080/02626668609491024.

Kling, H., Stanzel, P., Fuchs, M. & Nachtnebel, H.-P. 2015 Performance of the COSERO precipitation–runoff model under non-stationary conditions in basins with different climates. *Hydrological Sciences Journal* **60**, 1374–1393. https://doi.org/10.1080/02626667.2014.959956.

Köplin, N., Schädler, B., Viviroli, D. & Weingartner, R. 2014 Seasonality and magnitude of floods in Switzerland under future climate change. *Hydrological Processes* **28**(4), 2567–2578. https://doi.org/10.1002/hyp.9757.

Kratzert, F., Klotz, D., Brenner, C., Schulz, K. & Herrnegger, M. 2018 Rainfall–runoff modelling using long short-term memory (LSTM) networks. *Hydrology and Earth System Sciences* **22**(11), 6005–6022. https://doi.org/10.5194/hess-22-6005-2018.

Kratzert, F., Klotz, D., Shalev, G., Klambauer, G., Hochreiter, S. & Nearing, G. 2019a Towards learning universal, regional, and local hydrological behaviors via machine learning applied to large-sample datasets. *Hydrology and Earth System Sciences* **23**(12). https://doi.org/10.5194/hess-23-5089-2019.

Kratzert, F., Herrnegger, M., Klotz, D., Hochreiter, S. & Klambauer, G. 2019b Neural hydrology – interpreting LSTMs in hydrology. In: *Explainable AI: Interpreting, Explaining and Visualizing Deep Learning. Lecture Notes in Computer Science*, Vol. 11700 (Samek, W., Montavon, G., Vedaldi, A., Hansen, L. & Müller, K. R., eds.). Springer, Cham. https://doi.org/10.1007/978-3-030-28954-6_19.

Lee, G., Jung, S. & Lee, D. 2018 Comparison of physics-based and data-driven models for streamflow simulation of the Mekong river. *Journal of Korea Water Resources Association* **51**(6), 503–514.

Lees, T., Buechel, M., Anderson, B., Slater, L., Reece, S., Coxon, G. & Dadson, S. J. 2021 Benchmarking data-driven rainfall-runoff models in Great Britain: a comparison of LSTM-based models with four lumped conceptual models. *Hydrology and Earth System Science Discussions* [preprint]. https://doi.org/10.5194/hess-2021-127, in review.

Li, C. Z., Zhang, L., Wang, H., Zhang, Y. Q., Yu, F. L. & Yan, D. H. 2012 The transferability of hydrological models under nonstationary climatic conditions. *Hydrology and Earth System Science* **16**, 1239–1254. https://doi.org/10.5194/hess-16-1239-2012.

Li, H., Beldring, S. & Xu, C.-Y. 2015 Stability of model performance and parameter values on two catchments facing changes in climatic conditions. *Hydrological Sciences Journal* **60**, 1317–1330. https://doi.org/10.1080/02626667.2014.978333.

Merz, R., Parajka, J. & Blöschl, G. 2011 Time stability of catchment model parameters: implications for climate impact analyses. *Water Resources Research* **47**, W02531. https://doi.org/10.1029/2010WR009505.

Nash, J. E. & Sutcliffe, J. V. 1970 River flow forecasting through conceptual models part I – a discussion of principles. *Journal of Hydrology* **10**(3), 282–290. https://doi.org/10.1016/0022-1694(70)90255-6, 1970.

O., S., Dutra, E. & Orth, R. 2020 Robustness of process-based versus data-driven modeling in changing climatic conditions. *Journal of Hydrometeorology* **21**(9), 1929–1944. https://doi.org/10.1175/JHM-D-20-0072.1.

Orth, R., Staudinger, M., Seneviratne, S. I., Seibert, J. & Zappa, M. 2015 Does model performance improve with complexity? A case study with three hydrological models. *Journal of Hydrology* **523**, 147–159. https://doi.org/10.1016/j.jhydrol.2015.01.044.

Ouermi, K. S., Paturel, J. E., Adounpke, J., Lawin, A. E., Goula, B. T. A. & Amoussou, E. 2019 Comparison of hydrological models for use in climate change studies: a test on 241 catchments in West and Central Africa. *Comptes Rendus Geoscience* **351**(7), 477–486. https://doi.org/10.1016/j.crte.2019.08.001.

Pan, Z., Liu, P., Gao, S., Xia, J., Chen, J. & Cheng, L. 2019 Improving hydrological projection performance under contrasting climatic conditions using spatial coherence through a hierarchical Bayesian regression framework. *Hydrology and Earth System Sciences* **23**(8), 3405–3421. https://doi.org/10.5194/hess-23-3405-2019.

Pool, S., Vis, M. & Seibert, J. 2018 Evaluating model performance: towards a non-parametric variant of the Kling-Gupta efficiency. *Hydrological Sciences Journal* **63**(13–14), 1941–1953. https://doi.org/10.1080/02626667.2018.1552002.

Rafaeli Neto, S. L., Sá, E. A. S., Debastiani, A. B., Padilha, V. L. & Antunes, T. A. 2019 Efficacy of rainfall-runoff models in loose coupling spacial decision support systems modelbase. *Water Resources Management* **33**(3), 889–904. https://doi.org/10.1007/s11269-018-2086-2.

Refsgaard, J. C. & Knudsen, J. 1996 Operational validation and intercomparison of different types of hydrological models. *Water Resources Research* **32**(7), 2189–2202. https://doi.org/10.1029/96WR00896.

Seibert, J. 2003 Reliability of model predictions outside calibration conditions: paper presented at the Nordic Hydrological Conference (Røros, Norway 4–7 August 2002). *Hydrology Research* **34**(5), 477–492. https://doi.org/10.2166/nh.2003.0019.

Seibert, J. & Vis, M. J. 2012 Teaching hydrological modeling with a user-friendly catchment-runoff-model software package. *Hydrology and Earth System Sciences* **16**(9), 3315–3325. https://doi.org/10.5194/hess-16-3315-2012.

Seibert, J., Vis, M. J., Lewis, E. & Meerveld, H. V. 2018 Upper and lower benchmarks in hydrological modelling. *Hydrological Processes* **32**(8), 1120–1125. https://doi.org/10.1002/hyp.11476.

Seiller, G., Anctil, F. & Perrin, C. 2012 Multimodel evaluation of twenty lumped hydrological models under contrasted climate conditions. *Hydrology and Earth System Sciences* **16**(4), 1171–1189. https://doi.org/10.5194/hess-16-1171-2012.

Seiller, G., Hajji, I. & Anctil, F. 2015 Improving the temporal transposability of lumped hydrological models on twenty diversified U.S. watersheds. *Journal of Hydrology: Regional Studies* **3**, 379–399. https://doi.org/10.1016/j.ejrh.2015.02.012.

Shen, C. 2018 A transdisciplinary review of deep learning research and its relevance for water resources scientists. *Water Resources Research* **54**(11), 8558–8593. https://doi.org/10.1029/2018WR022643.

Shortridge, J. E., Guikema, S. D. & Zaitchik, B. F. 2016 Machine learning methods for empirical streamflow simulation: a comparison of model accuracy, interpretability, and uncertainty in seasonal watersheds. *Hydrology and Earth System Sciences* **20**(7). doi: 10.5194/hess-20-2611-2016.

Staudinger, M., Stahl, K., Seibert, J., Clark, M. P. & Tallaksen, L. M. 2011 Comparison of hydrological model structures based on recession and low flow simulations. *Hydrology and Earth System Sciences* **15**(11), 3447–3459. https://doi.org/10.5194/hess-15-3447-2011.

Thirel, G., Andréassian, V. & Perrin, C. 2015 On the need to test hydrological models under changing conditions. *Hydrological Sciences Journal* **60**(7–8), 1165–1173. doi:10.1080/02626667.2015.1050027.

Vaze, J., Post, D. A., Chiew, F. H. S., Perraud, J. M., Viney, N. R. & Teng, J. 2010 Climate non-stationarity–validity of calibrated rainfall–runoff models for use in climate change studies. *Journal of Hydrology* **394**(3–4), 447–457. https://doi.org/10.1016/j.jhydrol.2010.09.018.

Vormoor, K., Heistermann, M., Bronstert, A. & Lawrence, D. 2018 Hydrological model parameter (in)stability – 'Crash testing' the HBV model under contrasting flood seasonality conditions. *Hydrological Sciences Journal* **63**(991), 1007. https://doi.org/10.1080/02626667.2018.1466056.

Wilby, R. L. 2005 Uncertainty in water resource model parameters used for climate change impact assessment. *Hydrological Processes* **19**(16), 3201–3219. https://doi.org/10.1002/hyp.5819.

Xu, C. Y. 1999 Operational testing of a water balance model for predicting climate change impacts. *Agricultural and Forest Meteorology* **98**, 295–304. doi:10.1016/S0168-1923(99)00106-9.

Xu, W., Jiang, Y., Zhang, X., Li, Y., Zhang, R. & Fu, G. 2020 Using long short-term memory networks for river flow prediction. *Hydrology Research* **51**(6), 1358–1376. https://doi.org/10.2166/nh.2020.026.

Zhu, Q., Zhang, X., Ma, C., Gao, C. & Xu, Y. P. 2016 Investigating the uncertainty and transferability of parameters in SWAT model under climate change. *Hydrological Sciences Journal* **61**(5), 914–930. https://doi.org/10.1080/02626667.2014.1000915.

First received 4 May 2021; accepted in revised form 10 March 2022. Available online 8 April 2022

doi: 10.2166/nh.2022.093

The need for training and benchmark datasets for convolutional neural networks in flood applications

Abdou Khouakhi *, Joanna Zawadzka and Ian Truckell

School of Water, Energy and Environment, Centre for Environmental and Agricultural Informatics, Cranfield University, Cranfield, UK
*Corresponding author. E-mail: a.khouakhi@cranfield.ac.uk

AK, 0000-0002-1224-3208; JZ, 0000-0002-6172-901X

ABSTRACT

Flood-related image datasets from social media, smartphones, CCTV cameras, and unmanned aerial vehicles (UAVs) present valuable data for the management of flood risk, and particularly for the application of modern convolutional neural networks (CNNs) to specific flood-related problems such as flood extent detection and flood depth estimation. This review discusses the increasing role of CNNs in flood research with a growing number of published datasets, particularly since 2018. We note the lack of open and labelled flood image datasets and the growing need for an open, benchmark data library for image classification, object detection, and segmentation relevant to flood management. Such a library would provide benchmark datasets to advance CNN flood applications in general and serve as a resource, providing data scientists and the flood research community with the necessary data for model training and validation.

Key words: convolutional neural networks, flood images, floods

HIGHLIGHTS

- There is a growing need for an open, benchmark image library for CNNs relevant to flood applications.
- The majority of papers reviewed used relatively small sets of data often manually annotated for training which leads to weak models, reproducibility, and replicability issues.
- Building an open dataset is a community effort and requires close collaboration with researchers and organizations that generate the data.

1. INTRODUCTION

Recent developments in cloud computing, artificial intelligence (AI) algorithms, and image processing have impacted many research fields, such as flood detection and prediction. Technologies such as high-resolution smartphone cameras, connectivity, social media, unmanned aerial vehicle (UAV), and satellite imagery have led to valuable and new types of visual data sources relevant to flood applications (Fohringer *et al.* 2015; Tkachenko *et al.* 2017; Hashemi-Beni & Gebrehiwot 2021). The use of these data to develop predictive models, however, remains an ongoing challenge, due to their diverse and disparate nature.

Visual data (i.e., images), image processing, and computer vision algorithms in flood application have been used for decades. For example, image processing and edge detection techniques have been used to estimate information from images, including water level (Park *et al.* 2009; Yang *et al.* 2014), flow velocity (Le Coz *et al.* 2010), and flood inundation extent (Horritt *et al.* 2001). Recently, convolutional neural networks (CNNs) have demonstrated unprecedented performance for image detection and segmentation. CNNs are a type of deep neural network which is in turn a specific type of machine learning belonging to the broader field of AI. Like classic machine learning algorithms, CNNs learn features from labelled training data to predict an accurate output from an unknown input. CNNs are formed of a network of stacked layers, which are used to extract richer features from the input data. The model becomes deeper as more layers are stacked leading to the increasing complexity of the network allowing for features to be learned (Krizhevsky *et al.* 2012). Therefore, the collection and labelling of training data are crucial components of the CNN training process. For an in-depth review on recent advances in deep learning including CNNs and their architectures, we refer the reader to the literature (Hoeser & Kuenzer 2020; Alzubaidi *et al.* 2021).

Historically, satellite and aerial images are two of the most extensively used visual data types for flood-related applications. Landsat MultiSpectral Scanner was one of the first sources of imagery used for flood delineation, for instance, to delineate flood-affected areas in regions of the United States (Deutsch & Ruggles 1974; Rango & Anderson 1974). Many other satellite missions followed, such as Landsat thematic mapper, SPOT (Satellite Pour l'Observation de la Terre) imagery, and Very High-Resolution Radiometer (AVHRR), were used extensively for flood management (Brown *et al.* 1987; Profeti & Macintosh 1997; Jain *et al.* 2006; Qi *et al.* 2009; Inman & Lyons 2020). Recent missions, such as Sentinel-1 and -2, provide open and even higher spatial resolution imagery opening many opportunities to apply machine learning and CNNs for improved flood-related research. More recently, the UAV technology is being used to generate on-demand and high-resolution imagery at lower costs for rapid flood assessment. UAVs can acquire high-resolution data (compared to satellite imagery) for the fast and accurate detection of inundated areas under complex landscapes and inaccessible flooded areas.

Despite efforts in recent years to exploit the power of CNNs, machine learning, and visual data for flood risk management, there are still many challenges, such as the limited availability of labelled images to train and validate CNN models (Iqbal *et al.* 2021). Labelling images is a tedious, manual process in most cases without collaboration and sharing of annotated images, and the creation of benchmark datasets within the flood application community, a significant amount of time will be spent by individual research groups in the labelling process. For example, approximately 13 h was required to label just 100 UAV images as water, buildings, road, and vegetation (Gebrehiwot *et al.* 2019a).

Here, we present a review of different initiatives using CNNs for flood applications; we highlight the recent advances and provide benefits of increasing the availability of annotated visual datasets for deep CNNs applied to flood research. The main research questions that we address are: (1) what is the current usage of CNNs for flood research? (2) what are the training datasets and techniques used to address flood-related problems as related to CNNs? and (3) what is the benefit of open training data and data annotation for flood applications?

2. FLOOD APPLICATIONS AND DEEP CNNS

To review existing approaches using CNNs for flood application, we first searched the Scopus database using the search terms *'deep learning' OR 'fully convolutional' OR 'convolutional neural network' OR 'CNN' OR 'convolutional network' AND 'flood'*. We performed the search on titles, abstracts, and keywords and retained only journal articles. The results (236 journal papers) were filtered to retain only journal articles published after 2010 (222 journal papers), given that advances in CNN algorithm development have only occurred since 2010. Figure 1 shows the number of articles indexed in Scopus published from 2010

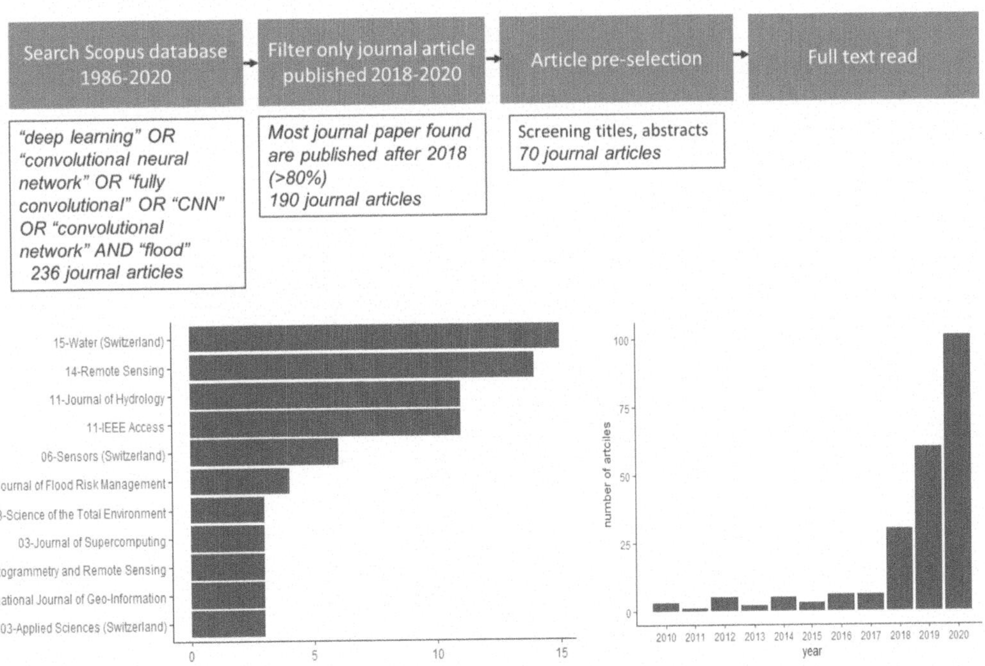

Figure 1 | Summary of the review process (top), number of published papers on deep CNNs and floods across (right panel) different journals and peer-reviewed papers and their corresponding journals (left panel). Numbers to the left of each journal indicate the number of published papers in each journal.

and reveals that most papers have been published in the 3 years only. We, therefore, focus the review only on papers published since 2018 (190 journal articles). Several papers were omitted from the list after the first examination of abstracts.

A preliminary examination of the papers confirms that the role of CNNs in flood research has been growing steadily, particularly from 2018 with the number of Scopus journal papers almost doubling every year. For the purpose of this work and given the importance of the datasets in training CNN models, we structured this review in the following four main aspects: (1) sources of visual data (images) used for model training; (2) labelling strategy; (3) CNN algorithms employed; and (4) types of flooding problems solved (e.g., detection and extent mapping).

2.1. Sources of visual data

Satellite imagery, airborne imagery, CCTV footage, ground camera photographs and images shared on social media, and the internet are the main sources of data used across the reviewed papers. We found about 43% of the 70 reviewed papers used ground camera images, whereas 37% of papers used satellite images such as Landsat and Sentinel imagery. Other authors used a combination of both satellite and camera images. Only ~7% of reviewed papers used UAV imagery (Figure 2). To provide further details, we clustered the reviewed papers into studies that used ground RGB (red, green, blue) camera images (i.e., ground camera photos including social media and internet images) and those used satellite and airborne images including multispectral imagery. We also highlighted studies that used both datasets.

2.1.1. RGB ground images

RGB ground camera images shared on social media have been widely retrieved and used to aid with flood detection and delineation. For example, Sazara *et al.* (2019) used 253 images of size 385 × 512 pixels with a flood and 238 without floods, and shuffled and split the data into training and testing subsets to train and evaluate a CNN classifier of images containing a scene with floodwater, before segmenting the flooded area. Moy De Vitry *et al.* (2019) used existing CCTV footage to provide qualitative flood level information over time. The authors first delineated flooded areas using CNNs and developed a proxy named 'static observer flooding index' indicating water level fluctuations visible from a surveillance camera's viewpoint. Photos of floods from social media typically contain geolocation, date-time, and other metadata about floods relevant for model training. Crowdsourced photos from predefined online applications and from Twitter have also been used to locate the flooded area and can be used to validate numerical models (Wang *et al.* 2018). Another example where images of drains and gullies from Google Images and YouTube were employed to train a CNN model to detect and classify drainage blockages (Mishra *et al.* 2020). The images were classified according to different blockage severity classes. The authors suggested that the trained model could then be used with an Internet of things (IoT)-enabled camera to monitor gullies and drainage. Finally, Chaudhary *et al.* (2020) gathered images containing flooded objects to estimate the flood level based on features partially submerged in water. They employed a CNN model on images with the object of known dimensions (e.g., car, bus, and bicycle) to infer the water level. These examples highlight the promise of social media images for the development of CNNs.

2.1.2. Satellite and airborne imagery

Images retrieved from satellites are widely used for flood risk management. With recent algorithmic advances in image detection and segmentation, many recent studies have been conducted taking advantage of combining CNNs and satellite imagery for flood applications.

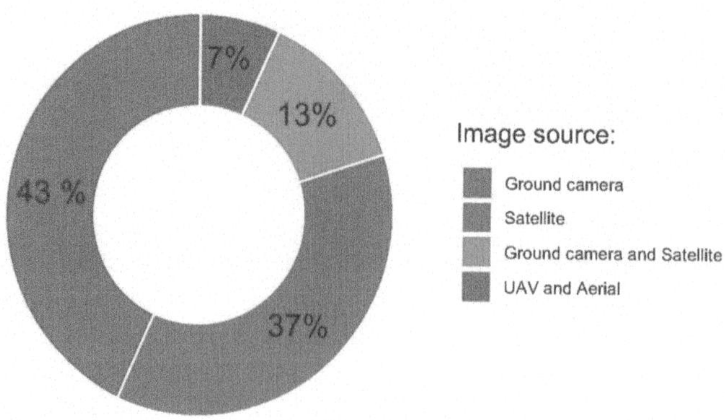

Figure 2 | Sources of visual data (Image source) used in the 70 reviewed papers.

In a recent study, Sentinel-2 images acquired from the MediaEval 2019 competition were used to detect floods (Jain *et al.* 2020). The data consisted of 335 sets with 267 images used for training and 68 for testing. Each set contained varying 1- to 24-day time series of images prior and after flood events; i.e., 2,770 images in total. The images were of 512 × 512 pixels in size from all the 12 bands of sentinel-2 data. Indices such as the Normalized Difference Water Index (NDWI), the modified NDWI, and the Automated Water Extraction Index (AWEI) were computed and added to the initial Sentinel-2 images to train a CNN (VGG16) network with a fully connected layer. Another study proposed a novel approach named H2O-Net, integrating Sentinel-2 and height resolution commercial imagery from PlanetScope to delineate floodwater (Akiva *et al.* 2020). The Sentinel training dataset was comprised of 3,676 images of 512 × 512 pixels covering RGB and SWIR (shortwave infrared) data resampled to 10 m per grid cell. PlanetScope is a high-resolution optical satellite providing RGB and near infrared (NIR) bands at 3 m resolution.

Many other studies have harnessed Landsat and Sentinel-1 and -2 images, by deriving indices used to train CNN models to assess flood characteristics and susceptibility (e.g., Tien Bui *et al.* 2020; Wang *et al.* 2020a; Zhao *et al.* 2020; Panahi *et al.* 2021). For example, the Normalized Difference Vegetation Index (NDVI) derived from Landsat 8 images was used alongside maps of flood-generating factors to train a CNN for flash flood susceptibility (Panahi *et al.* 2021). Similarly, Zhao *et al.* (2020) integrated Landsat images and other factors such as elevation, slope, the NDVI, and rainfall to assess flood susceptibility within an urban catchment.

Other studies have integrated both satellite and camera-based images to strengthen the model training process. For example, crowdsourced images and satellite images were used in conjunction to detect the passability of roads during floods (Ahmad *et al.* 2019).

Synthetic aperture radar imagery (SAR) is another valuable data source that has been used in flood delineation and detection, as it has the capability to see through clouds and darkness. For instance, co-registered optical Sentinel-2 and SAR Sentinel-1 image time series were recently employed for the detection of floods using combined CNNs and recurrent neural networks (Rambour *et al.* 2020a).

Finally, the UAV technology has recently been recognized as an efficient and rapid way to quickly acquire high-resolution imagery for flood applications (Rivas Casado *et al.* 2018; Gebrehiwot *et al.* 2019b). Deep CNN approaches have been used to extract flooded areas from high-resolution UAV images (Gebrehiwot *et al.* 2019b). The authors used a manually labelled dataset alongside transfer learning, i.e., a pre-trained model.

In general, satellite-based imagery data have been used for flood detection, flood extent mapping, or flood damage assessment due to their wide spatial coverage but also the spectral information they contain beyond the visible bands. Most research examined here relayed on freely available mid-resolution data such as Sentinel-2 and-1 and Landsat products (Peng *et al.* 2019; Wieland & Martinis 2019; Rambour *et al.* 2020b). Ground images, on the other hand, are mainly harnessed for flood depth estimation as well as for detection and mapping of floods due to the resolution and the apparent flooded features that can be used as proxies for flood depth estimation (Chaudhary *et al.* 2020; Huang *et al.* 2020).

Both satellite and ground camera imagery have been used for damage detection and assessment (Rubio *et al.* 2019; Gupta & Shah 2020; Muhadi *et al.* 2020), but they both have associated challenges and limitations. While satellite imagery provides a valuable resource of images for CNN flood applications, the images are often provided in large image tiles and require pre-processing and transformation to smaller subsets of imagery (image chips) to be used as CNN inputs. Additionally, small spatial features of flooded areas and floods in small streams are difficult to map using low-resolution satellite imagery. The quality and availability of flood images, particularly optical data, are also hampered by sky conditions such as cloud coverage. For ground imagery data, finding a sufficient number of images relating to a certain flood event can be challenging (Barz *et al.* 2019).

One of the most challenging phases in developing a CNN flood application from both ground and satellite image data is the labelling of images for training. Different strategies have been developed to increase the number of annotated images.

2.2. Labelling strategies

Generally, the training datasets required for different types of deep CNN problems, such as image classification, semantic segmentation, and object detection, are composed of pairs of data (images) and labels, then the images are fed to the CNN model alongside their corresponding labels as outputs to train the model. Labels are generated from ground reference data and/or annotated imagery. Very few pre-labelled and benchmark datasets are currently available for visual flood data. Most (~57%) of the studies reviewed here used their own manually labelled images assisted by tools such as markGT,

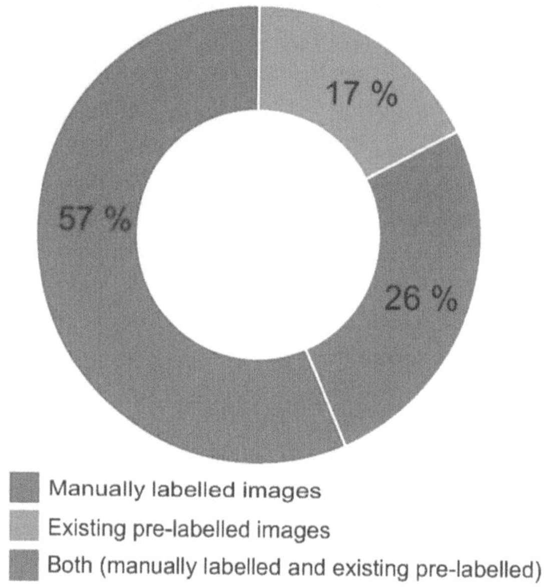

Figure 3 | Training data-labelling approaches employed in the reviewed literature (70 papers).

LabelMe, and DarkLabel (Figure 3). Some studies used a combination of manually annotated images and existing benchmark pre-labelled data (Supplementary Material, Table S1), such as MediaEval (Bischke *et al.* 2017, 2018, 2019) which is a benchmark dataset of satellite imagery and social media images and the European Flood 2013 datasets (Barz *et al.* 2019) which is an annotated dataset of images from the central European flood 2013 (Table 1).

Some studies used automated techniques to minimize the effect of a limited annotated training dataset. For example, to examine the role of SAR intensity and interferometric coherence combined with CNN in urban flood detection, Li *et al.* (2019) introduced an active self-learning CNN framework that selects informative unlabelled samples based on a temporal-ensembling CNN model. Subsequently, these samples are pseudo-labelled by a multi-scale spatial filter. Chaudhary *et al.* (2020) proposed a multi-task deep CNN approach for water depth estimation based on social media images for flood mapping. They trained the model for a small set of annotated water levels (regression task) and a larger set of weak annotated dataset (ranking task) to save annotation effort. The authors suggest that weak supervision using pairwise ranking can be a promising alternative to costly and time-consuming, fine-grained labelling.

2.3. CNN algorithms

CNNs are one of the most successful algorithms for tackling computer vision problems and have achieved high performance for image-related problems such as the detection of features on images. Although modern CNNs can be traced back to the 1990 s following the work of LeCun *et al.* (1989) and Lecun *et al.* (1998), their wide expansion came after 2012 as the data availability and computation capacities increased, but also as deeper architectures improved their representational capacity. This is marked by the work of Krizhevsky *et al.* (2012), who developed the AlexNet architecture that achieved notable performance labelling pictures in the ImageNet challenge (Russakovsky *et al.* 2015).

An example of a simple CNN architecture includes one input layer, one convolutional layer, one pooling layer, and one fully connected neural network layer. There are many variants of CNNs as their architectures vary with different aspects such as processing units, parameter and hyper-parameter optimization strategies, design patterns, and connectivity of layers. For an in-depth comparison of various CNN architectures and their evolution, see the review survey conducted by Khan *et al.* (2020). The vast majority of the reviewed work here (~80%) used existing CNN pre-trained architectures such as Alexnet (Krizhevsky *et al.* 2012), VGG (Simonyan & Zisserman 2015), ResNet (He *et al.* 2016), DenseNet (Huang *et al.* 2017), GoogleNet (Szegedy *et al.* 2015), InceptionV3 (Szegedy *et al.* 2016), and Xception (Chollet 2017) (Supplementary Material, Table S1). Some studies developed problem-specific CNN models (e.g., H2O-Network), with the aim of achieving better predictive performance (Li *et al.* 2019; Akiva *et al.* 2020). One of the most used pre-trained networks is the VGG-16 network (Simonyan & Zisserman 2015) that is trained on the ImageNet dataset which contains over a million

Table 1 | Summary of publicly-accessible benchmark datasets employed in the literature for flood applications

Data name	Types	References	
MediaEval Multimedia Satellite task 2017	Satellite Imagery and Social Media	Bischke *et al.* (2017)	Combination of satellite and social media images and segmentation masks.
MediaEval Multimedia Satellite task 2018	Satellite Imagery and Social Media	Bischke *et al.* (2018)	Combination of satellite and social media images and segmentation masks.
MediaEval Multimedia Satellite task 2019	Satellite Imagery and Social Media	Bischke *et al.* (2019)	Combination of satellite and social media images and segmentation masks.
European Flood 2013	RGB ground images	Barz *et al.* (2019)	Annotated dataset of images from the central European flood 2013.
SEN12-FLOOD	SAR and Multispectral images	Rambour *et al.* (2020b)	Optical Sentinel-2 and SAR Sentinel-1 image sequences for the detection of flooded areas.
Ms COCO	RGB ground images	Lin *et al.* (2014)	General object recognition datasets.
Cityscapes	RGB ground images	Cordts *et al.* (2016)	Semantic understanding of urban street scenes.

Ms COCO and Cityscapes are general purpose training images used to support model training.

images and 1,000 labels (Russakovsky *et al.* 2015). VGG16 is composed of 13 convolutional layers, 5 max-pooling layers, and 3 fully connected layers. Therefore, the number of layers having tunable parameters is 16 with 13 convolutional layers and 3 fully connected layers. A pre-trained network is used on new data by using layers as feature extractors where each data point can be mapped into a new representation. Lower layers can be considered as low-level feature extractors and are applicable across domains. Whereas higher layers are more domain-specific (Figure 4). Another approach is to use existing weights of the pre-trained network as initialization weights for the new dataset. This approach could prevent overfitting but it is computationally demanding.

For example, Gebrehiwot *et al.* (2019b) used a pre-trained VGG-based fully CNN model to map flood extent using UAV images. The classification results showed classification accuracy with more than 90% compared to 89% using the SVM classification approach.

Some studies used an ensemble of networks to improve performances, for example, Lopez-Fuentes *et al.* (2020) used ensemble of InceptionV3, Xception, VGG16, VGG19, InceptionResNetV2, MobileNet, DenseNet121, DenseNet201, and

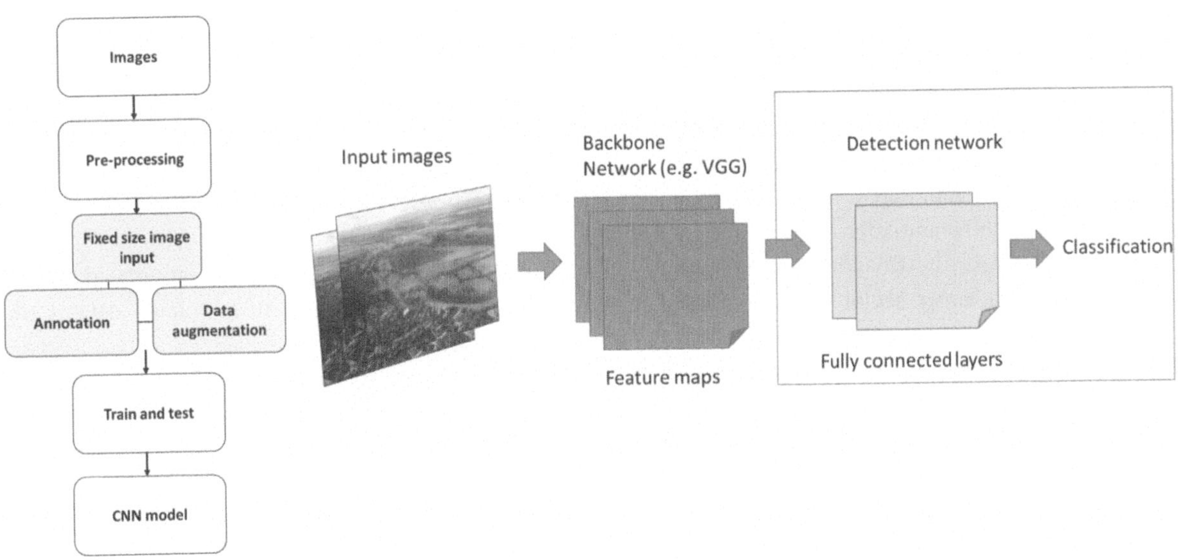

Figure 4 | Typical workflow and CNN architecture.

NaSNetLarge to improve predictive performances. U-net (Ronneberger *et al.* 2015) is also one of the popular architectures used for image segmentation to map flood extent (e.g., Moy De Vitry *et al.* 2019; Wieland & Martinis 2019).

2.4. Types of flood problems

Deep CNNs have increasingly been used in many areas of flood management research in the last years. We identified the following three main flood-related problems where CNNs have been actively employed: (1) the detection of flood presence and extent; (2) flood depth or water level estimation; (3) flood susceptibility mapping; and (4) damage assessment of flood protection structures (Supplementary Material, Table S1).

2.4.1. Detection of flood presence and extent

Distinguishing flooded and non-flooded areas and their extent in image data is one of the widely used applications of CNNs in the reviewed papers. Witherow *et al.* (2019) proposed an image processing pipeline for detecting inundated roadways using crowdsourced pairs of dry and flooded images. The pre-processing approach includes training a region-based R-CNN for vehicle detection on images before applying the water edge detection, image inpainting, and contrast correction algorithms to segment inundated areas. The authors obtained satisfactory performances but they highlighted challenges in recreating the dry images from the same image angle in addition to the drawback of several manually selected model parameters in their proposed approach. Wieland & Martinis (2019) introduced a processing workflow for automated flood monitoring from multispectral satellite data selected from various parts of the world. They used CNN U-Net architecture to discriminate water from shadow, snow, land, and cloud classes and then evaluated the performance of the model with other benchmark approaches such as random forests and a simple NDWI thresholding method. The water class is then further analyzed to separate floodwater from permanent water. Multispectral imagery is also used by Peng *et al.* (2019) to develop a CNN-based data fusion framework for mapping urban flood extent with pre- and post-flooding surface reflectance imagery. The authors proposed the patch similarity CNN with two variants to estimate the similarity between pre- and post-flooding satellite multispectral surface reflectance image patches, and then to determine whether the post-flooding patch under test is flooded. The results indicated good performances and showed that surface reflectance-based data are better than digital numbers-based data as surface reflectance is more stable under varied inconsistent illumination conditions. Similarly, a CNN classification model to map the extent of flooded areas from Landsat satellite images is proposed by Sarker *et al.* (2019). The authors utilized the spatial information from the neighbouring area of target pixel in classification. The proposed classification model was compared to a conventional SVM classification model with the CNN model outperforming the Support Vector Machines (SVM) classification model. Extracting flooded areas from high-resolution UAV imagery is gaining popularity in recent years, Gebrehiwot *et al.* (2019a) fine-tuned a pre-trained CNN model (FCN-16 s) using a small UAV dataset to distinguish floodwater from other classes. They compared CNN performance with more classic machine learning models, such as SVM, and found CNN-based classifier more suitable in flood imagery segmentation with an overall accuracy of 95%.

2.4.2. Flood depth or water level estimation

Several approaches have been developed to estimate food depth from visual images with CNNs. Chaudhary *et al.* (2019) proposed a method to estimate the flood-water level from images gathered from social media by assessing features in the images that are submerged in water. Their workflow includes a first Mask R-CNN to define a set of regions of interest within the image then, for the objects belonging to specific categories (objects with known dimensions) estimate how much objects are submerged in water. From that, a dataset was built with annotating images with flood-water level information that was used to train another CNN model for the instance segmentation and flood level in the image.

This work was extended by Chaudhary *et al.* (2020) by training the model for a small set of annotated water levels (regression task) and a larger set of weak annotated dataset (ranking task) generating an annotated dataset of 8,145 images. The water level was estimated using a multi-task approach with the VGG16 CNN.

CCTV systems and CNNs have been also used to provide qualitative flood level trend information. For example, Moy De Vitry *et al.* (2019) proposed an approach to detect floodwater in CCTV footage and devised a qualitative flood index as a proxy for water level fluctuations visible from a surveillance camera's view point. They first trained a CNN model on 1,218 flooding images collected from the Internet to detect flooded extent area and applied the model to six CCTV videos representing different flooding and lighting conditions. The authors then inferred the water level fluctuations using an index named *static observer flooding index* which is a dimensionless proxy that can be extracted from segmented images of stationary surveillance cameras. The index was computed as the ratio of flooded pixels to the total pixels in the segmented stationary scene. The index value can vary between 0% with no flooding visible and 100% with only flooding visible. When this index is evaluated at multiple consecutive moments in time, the variation of its value provides information about fluctuations of the actual water level.

2.4.3. Flood susceptibility mapping

The CNN has also been employed for flood susceptibility mapping, for instance, Wang *et al.* (2020b) used a spatial database of 13 flood influencing factors such as land use, rainfall, lithology, slop, and 108 historical flood events to develop CNN models to classify and map flood susceptibility in Shangyou, China. The influencing factors layers were converted into a stack of raster of 30 × 30 m grid size then to small 2D and 3D matrices. Similar susceptibility mapping approach is used by Zhao *et al.* (2020) to assess flood susceptibility for an urban catchment and by Tien Bui *et al.* (2020) for flash floods susceptibility.

2.4.4. Damage assessment of flood protection structures

Damage assessment of flooding structures is one the growing areas in CNN flood applications. Rubio *et al.* (2019) used CNNs (VGG16) to automatically assess damage on concrete bridges. The authors used a database of 734 images bridge damage to detect 'delamination' and 'rebar exposure'. Their model showed mean accuracy of 89.7% for delamination and 78.4% for rebar exposure. The results suggest also that a VGG-based network pre-trained on ImageNet is well suited for texture recognition in the scope of civil infrastructure surface damage semantic segmentation.

More recently, Mishra *et al.* (2020) trained a CNN model (VGG16) to detect and classify drainage blockages based on images of drains and gullies from Google Images and YouTube. The images were classified according to different blockage severity classes. The authors suggested that the trained model could then be used with an IoT-enabled camera to monitor gullies and drainage.

3. BENEFITS OF OPEN ANNOTATED DATA FOR CNNS FLOOD APPLICATIONS

This mini-review demonstrates the growing importance of deep CNNs applied to flood problems. A steady increase of deep CNNs use cases in the literature covering different problems ranging from surface water detection, flood extent delineation, depth or water level estimation, flood prediction, and damage assessment of flood protection structures. The review reveals also various challenges of existing approaches and highlights a need for community collaborations to provide dataset benchmarks that can be used to train and validate models, and improve the model performance and generalization.

3.1. Lack of visual open datasets

Availability of benchmark image datasets is found lacking across the literature. The majority of papers reviewed employ relatively small sets (<1,000 images) of data often manually annotated for training. Though deficiency in benchmark images is a wide limitation in machine learning and CNN applications across fields, it is particularly obvious in flooding applications, and there few libraries of training images for community collaboration. This lack of open datasets leads to reproducibility and replicability issues: currently it is difficult to reproduce CNN-flood research given the inaccessibility of much of the data. Additionally, it is difficult to benchmark the performances of developed models.

3.2. Benefits of an open library

CNNs and machine learning research in general depends on both the availability of training datasets and of objectively interpretable, comparable, and reproducible algorithm benchmarks. Availability of training and benchmark datasets are critical for generating long-term advances in deep learning methods, and their application to flood risk management challenges. The creation of an open flood library for model training is likely to facilitate the creation of datasets, increase collaboration, and improve CNN applications to flooding challenges. Such a library could be used to standardize the setup of data and reporting of benchmarks. Datasets could be uniformly formatted in standardized formats and could be easily downloaded programmatically through Application Programming Interface (APIs) and client libraries. The library could provide standardized train-test to ensure that results can be objectively compared. Testing different models on the same datasets is crucial, users could have possibility to upload their own datasets to the database. Such an approach would reduce the time and efforts that researchers spend implementing existing baselines to conduct comparative experiments.

In terms of the economic benefits of open data, many countries are assessing the benefits of open data in general and recent research showing that higher economic benefits are expected from open data. For example, the European data market was estimated at 54,351 and 59,539 million Euros in 2015 and 2016, respectively (Berends *et al.* 2020). The same study estimated a Gross Domestic Product (GDP) growth of 36.9% in 2016–2020, which saves 2,549 man-hours and reduces energy use by 16%. Furthermore, open data jobs in Europe are estimated to increase from 75,000 to 100,000 jobs between 2016 and 2020 (Berends *et al.* 2020). Open data do not just drive the economic growth but also enables for transparency, accountability, innovation, and knowledge. In the UK, the generated economic output of the data economy was £73.3 billion which is set to grow to £94.6 billion in 2025 (Waterman *et al.* 2021). Public data such as open flood training data can have a high potential for re-use in new services and applications, addressing societal challenges, increasing transparency and accountability, and achieving efficiency gains by sharing data.

There are many flood management applications that would benefit from the availability of open and benchmarked datasets for instance, the detection of blockage of drainage in urban areas. A rich and easily accessed training and benchmark dataset would improve the model performance and enable the automatic and real-time detection of blockages. Another example of a use case where a collaborative and an open library might be beneficial is in the monitoring of the performance of flood prevention structures such as dams. Enhanced availability of such datasets could also improve the performance of the use of UAV technologies for flood detection; for the identification of humans during the search and rescue; or the detection of damage for insurance purposes.

4. CONCLUSIONS

In this review, we provide an overview of the use cases of the CNN for flood management with a focus on the evaluation of the availability visual (images) annotated data. We reviewed the types of visual datasets used and common model architectures applied. We highlighted the need of an open library of training datasets for flood applications and discussed the benefits of such open library. The main highlights from this review are as follows:

- Most CNN flood applications that are widely used across the literature are the detection of flood presence and extent, flood depth or water level estimation, and damage assessment of flood protection structures.
- As training datasets, about 43% of the 70 reviewed papers used ground camera images, while ~37% of papers used satellite images such as Landsat and Sentinel imagery. Other papers used a combination of both satellite and camera images. Only few (7%) of reviewed papers used UAV imagery.
- The majority of papers reviewed employ relatively small sets of data often manually annotated for training. This lack of open datasets leads to reproducibility and replicability issues: currently it is difficult to reproduce CNN-flood research given the inaccessibility of much of the data. Additionally, it is difficult to benchmark the performances of developed models.
- Very few pre-labelled and benchmark datasets are currently available for flood analyses. Most of the studies reviewed used their own manual labelling. Some studies used a combination of existing benchmark pre-labelled data such as MediaEval and the European Flood 2013 datasets. Few studies employed techniques to help automate image annotation.
- Most reviewed studies used existing pre-trained CNN models to extract main image features with few studies employed problem-specific CNN architectures.

Finally, this review highlights the need for more training datasets to accelerate the progress of CNNs applications for flood management. This can be done through the creation of collaborative open training datasets for flood management applications. Building an open dataset is a community effort and requires close collaboration with researchers and organizations generating the data. Furthermore, such an open library would improve access, sharing of datasets, and contribute to overcoming some of the challenges in a collaborative manner, thereby accelerating future advances of CNNs for flood application.

DATA AVAILABILITY STATEMENT

All relevant data are included in the paper or its Supplementary Information.

REFERENCES

Ahmad, K., Pogorelov, K., Riegler, M., Ostroukhova, O., Halvorsen, P., Conci, N. & Dahyot, R. 2019 Automatic detection of passable roads after floods in remote sensed and social media data. *Signal Processing: Image Communication* **74**, 110–118. https://doi.org/10.1016/j.image.2019.02.002.

Akiva, P., Purri, M., Dana, K., Tellman, B. & Anderson, T. 2020 H2O-Net: Self-supervised flood segmentation via adversarial domain adaptation and label refinement. In *Proceedings of the IEEE/CVF Winter Conference on Applications of Computer Vision*. pp. 111–122.

Alzubaidi, L., Zhang, J., Humaidi, A. J., Al-Dujaili, A., Duan, Y., Al-Shamma, O., Santamaría, J., Fadhel, M. A., Al-Amidie, M. & Farhan, L. 2021 Review of deep learning: concepts, CNN architectures, challenges, applications, future directions. *Journal of Big Data* **8** (1). https://doi.org/10.1186/s40537-021-00444-8.

Barz, B., Schröter, K., Münch, M., Yang, B., Unger, A., Dransch, D. & Denzler, J. 2019 Enhancing flood impact analysis using interactive retrieval of social media images. *ArXiv*. https://doi.org/10.5445/KSP/1000087327/06.

Berends, J., Wendy, C. & Cosmina, R. 2020 *Analytical Report 9: The Economic Benefits of Open Data*. https://doi.org/10.2830/081754.

Bischke, B., Helber, P., Schulze, C., Srinivasan, V., Dengel, A. & Borth, D. 2017 The multimedia satellite task at mediaeval 2017: emergency response for flooding events. *CEUR Workshop Proceedings* **1984**, 13–15.

Bischke, B., Helber, P., Zhao, Z., De Bruijn, J. & Borth, D. 2018 The multimedia satellite task at MediaEval 2018 emergency response for flooding events. *CEUR Workshop Proceedings* **2283**, 4–6.

Bischke, B., Helber, P., Brugman, S., Basar, E., Zhao, Z., Larson, M. & Pogorelov, K. 2019 The multimedia satellite task at mediaeval 2019 estimation of flood severity. *CEUR Workshop Proceedings* **2670**, 4–6.

Brown, A. G., Gregory, K. J. & Milton, E. J. 1987 The use of Landsat multispectral scanner data for the analysis and management of flooding on the river Severn, England. *Environmental Management* **11** (5). https://doi.org/10.1007/BF01880168.

Chaudhary, P., D'Aronco, S., Moy de Vitry, M., Leitão, J. P. & Wegner, J. D. 2019 Flood-water level estimation from social media images. *ISPRS Annals of the Photogrammetry, Remote Sensing and Spatial Information Sciences* **IV-2/W5** (2/W5), 5–12. https://doi.org/10.5194/isprs-annals-IV-2-W5-5-2019.

Chaudhary, P., D'Aronco, S., Leitão, J. P., Schindler, K. & Wegner, J. D. 2020 Water level prediction from social media images with a multi-task ranking approach. *ISPRS Journal of Photogrammetry and Remote Sensing* **167**, 252–262. https://doi.org/10.1016/j.isprsjprs.2020.07.003.

Chollet, F. 2017 Xception: Deep learning with depthwise separable convolutions. In *Proceedings of the IEEE Conference on Computer Vision and Pattern Recognition*. pp. 1251–1258.

Cordts, M., Omran, M., Ramos, S., Rehfeld, T., Enzweiler, M., Benenson, R., Franke, U., Roth, S. & Schiele, B. 2016 The cityscapes dataset for semantic urban scene understanding. In *Proceedings of the IEEE Computer Society Conference on Computer Vision and Pattern Recognition, 2016-Decem*, pp. 3213–3223. https://doi.org/10.1109/CVPR.2016.350.

Deutsch, M. & Ruggles, F. 1974 Optical data processing and projected applications of the erts-1 imagery covering the 1973 Mississippi river valley floods. *JAWRA Journal of the American Water Resources Association* **10** (5), 1023–1039. https://doi.org/https://doi.org/10.1111/j.1752-1688.1974.tb00622.x.

Fohringer, J., Dransch, D., Kreibich, H. & Schröter, K. 2015 Social media as an information source for rapid flood inundation mapping. *Natural Hazards and Earth System Sciences* **15** (12), 2725–2738. https://doi.org/10.5194/nhess-15-2725-2015.

Gebrehiwot, A., Hashemi-Beni, L., Thompson, G., Kordjamshidi, P. & Langan, T. E. 2019a Deep convolutional neural network for flood extent mapping using unmanned aerial vehicles data. Sensors *(*Switzerland*)* **19** (7). https://doi.org/10.3390/s19071486.

Gebrehiwot, A., Hashemi-Beni, L., Thompson, G., Kordjamshidi, P. & Langan, T. E. 2019b Deep convolutional neural network for flood extent mapping using unmanned aerial vehicles data. Sensors *(*Switzerland*)*. https://doi.org/10.3390/s19071486.

Gupta, R. & Shah, M. 2020 Rescuenet: Joint building segmentation and damage assessment from satellite imagery. In *2020 25th International Conference on Pattern Recognition (ICPR)*, IEEE, pp. 4405–4411.

Hashemi-Beni, L. & Gebrehiwot, A. 2021 Flood extent mapping: an integrated method using deep learning and region growing using UAV optical data. *IEEE Journal of Selected Topics in Applied Earth Observations and Remote Sensing* **14**, 2127–2135. https://doi.org/10.1109/JSTARS.2021.3051873.

He, K., Zhang, X., Ren, S. & Sun, J. 2016 Deep residual learning for image recognition. In *Proceedings of the IEEE Conference on Computer Vision and Pattern Recognition*. pp. 770–778.

Hoeser, T. & Kuenzer, C. 2020 Object detection and image segmentation with deep learning on earth observation data: a review-part I: evolution and recent trends. *Remote Sensing* **12** (10), 1667. https://doi.org/10.3390/rs12101667.

Horritt, M. S., Mason, D. C. & Luckman, A. J. 2001 Flood boundary delineation from synthetic aperture radar imagery using a statistical active contour model. *International Journal of Remote Sensing* **22** (13). https://doi.org/10.1080/01431160116902.

Huang, G., Liu, Z., Van Der Maaten, L. & Weinberger, K. Q. 2017 Densely connected convolutional networks. In *Proceedings of the IEEE Conference on Computer Vision and Pattern Recognition*. pp. 4700–4708.

Huang, J., Kang, J., Wang, H., Wang, Z. & Qiu, T. 2020 A novel approach to measuring urban waterlogging depth from images based on mask region-based convolutional neural network. *Sustainability* **12** (5), 2149. https://doi.org/10.3390/su12052149.

Inman, V. L. & Lyons, M. B. 2020 Automated inundation mapping over large areas using landsat data and google earth engine. *Remote Sensing* **12** (8). https://doi.org/10.3390/RS12081348.

Iqbal, U., Perez, P., Li, W. & Barthelemy, J. 2021 How computer vision can facilitate flood management: a systematic review. *International Journal of Disaster Risk Reduction* **53**, 102030. https://doi.org/10.1016/j.ijdrr.2020.102030.

Jain, S. K., Saraf, A. K., Goswami, A. & Ahmad, T. 2006 Flood inundation mapping using NOAA AVHRR data. *Water Resources Management* **20** (6). https://doi.org/10.1007/s11269-006-9016-4.

Jain, P., Schoen-Phelan, B. & Ross, R. 2020 Automatic flood detection in Sentinel-2 images using deep convolutional neural networks. In: *Proceedings of the 35th Annual ACM Symposium on Applied Computing*, pp. 617–623. https://doi.org/10.1145/3341105.3374023.

Khan, A., Sohail, A., Zahoora, U. & Qureshi, A. S. 2020 A survey of the recent architectures of deep convolutional neural networks. *Artificial Intelligence Review* **53** (8), 5455–5516. https://doi.org/10.1007/s10462-020-09825-6.

Krizhevsky, A., Sutskever, I. & Hinton, G. E. 2012 Imagenet classification with deep convolutional neural networks. *Advances in Neural Information Processing Systems* **25**, 1097–1105.

Le Coz, J., Hauet, A., Pierrefeu, G., Dramais, G. & Camenen, B. 2010 Performance of image-based velocimetry (LSPIV) applied to flash-flood discharge measurements in Mediterranean rivers. *Journal of Hydrology* **394** (1–2). https://doi.org/10.1016/j.jhydrol.2010.05.049.

LeCun, Y., Boser, B., Denker, J., Henderson, D., Howard, R., Hubbard, W. & Jackel, L. 1989 Handwritten digit recognition with a back-propagation network. *Advances in Neural Information Processing Systems* **2**.

Lecun, Y., Bottou, L., Bengio, Y. & Haffner, P. 1998 Gradient-based learning applied to document recognition. *Proceedings of the IEEE* **86** (11), 2278–2324. https://doi.org/10.1109/5.726791.

Li, Y., Martinis, S. & Wieland, M. 2019 Urban flood mapping with an active self-learning convolutional neural network based on TerraSAR-X intensity and interferometric coherence. *ISPRS Journal of Photogrammetry and Remote Sensing* **152**, 178–191. https://doi.org/10.1016/j.isprsjprs.2019.04.014.

Lin, T. Y., Maire, M., Belongie, S., Hays, J., Perona, P., Ramanan, D., Dollár, P. & Zitnick, C. L. 2014 Microsoft COCO: common objects in context. Lecture Notes in Computer Science *(*Including Subseries Lecture Notes in Artificial Intelligence and Lecture Notes in Bioinformatics*)* **8693** (PART 5), 740–755. https://doi.org/10.1007/978-3-319-10602-1_48.

Lopez-Fuentes, L., Farasin, A., Zaffaroni, M., Skinnemoen, H. & Garza, P. 2020 Deep learning models for road passability detection during flood events using social media data. *Applied Sciences* **10** (24), 8783. https://doi.org/10.3390/app10248783.

Mishra, B. K., Thakker, D., Mazumdar, S., Neagu, D., Gheorghe, M. & Simpson, S. 2020 A novel application of deep learning with image cropping: a smart city use case for flood monitoring. *Journal of Reliable Intelligent Environments*. https://doi.org/10.1007/s40860-020-00099-x.

Moy De Vitry, M., Kramer, S., Dirk Wegner, J. & Leitao, J. P. 2019 Scalable flood level trend monitoring with surveillance cameras using a deep convolutional neural network. *Hydrology and Earth System Sciences* **23** (11), 4621–4634. https://doi.org/10.5194/hess-23-4621-2019.

Muhadi, N. A., Abdullah, A. F., Bejo, S. K., Mahadi, M. R. & Mijic, A. 2020 Image segmentation methods for flood monitoring system. *Water (Switzerland)* **12** (6), 1–10. https://doi.org/10.3390/w12061825.

Panahi, M., Jaafari, A., Shirzadi, A., Shahabi, H., Rahmati, O., Omidvar, E., Lee, S. & Bui, D. T. 2021 Deep learning neural networks for spatially explicit prediction of flash flood probability. *Geoscience Frontiers* **12** (3), 1076. https://doi.org/10.1016/j.gsf.2020.09.007.

Park, S., Lee, N., Han, Y. & Hahn, H. 2009 The water level detection algorithm using the accumulated histogram with band pass filter. *World Academy of Science, Engineering and Technology* **56**. https://doi.org/10.5281/zenodo.1072871.

Peng, B., Meng, Z., Huang, Q. & Wang, C. 2019 Patch similarity convolutional neural network for urban flood extent mapping using bi-temporal satellite multispectral imagery. *Remote Sensing* **11** (21). https://doi.org/10.3390/rs11212492.

Profeti, G. & Macintosh, H. 1997 Flood management through landsat TM and ERS SAR data: a case study. *Hydrological Processes* **11** (10). https://doi.org/10.1002/(sici)1099-1085(199708)11:10<1397::aid-hyp530>3.0.co;2-y.

Qi, S., Brown, D. G., Tian, Q., Jiang, L., Zhao, T. & Bergen, K. M. 2009 Inundation extent and flood frequency mapping using LANDSAT imagery and digital elevation models. *GIScience and Remote Sensing* **46** (1). https://doi.org/10.2747/1548-1603.46.1.101.

Rambour, C., Audebert, N., Koeniguer, E., Le Saux, B., Crucianu, M. & Datcu, M. 2020a Flood detection in time series of optical and sar images. *International Archives of the Photogrammetry, Remote Sensing and Spatial Information Sciences – ISPRS Archives* **43** (B2), 1343–1346. https://doi.org/10.5194/isprs-archives-XLIII-B2-2020-1343-2020.

Rambour, C., Audebert, N., Koeniguer, E., Le Saux, B., Crucianu, M. & Datcu, M. 2020b *SEN12-FLOOD: A SAR and Multispectral Dataset for Flood Detection*. IEEE Dataport. https://doi.org/https://dx.doi.org/10.21227/w6xz-s898.

Rango, A. & Anderson, A. T. 1974 Flood hazard studies in the Mississippi river basin using remote sensing. *JAWRA Journal of the American Water Resources Association* **10** (5). https://doi.org/10.1111/j.1752-1688.1974.tb00625.x.

Rivas Casado, M., Irvine, T., Johnson, S., Palma, M. & Leinster, P. 2018 The use of unmanned aerial vehicles to estimate direct tangible losses to residential properties from flood events: a case study of Cockermouth following the Desmond storm. *Remote Sensing* **10** (10), 1548. https://doi.org/10.3390/rs10101548.

Ronneberger, O., Fischer, P. & Brox, T. 2015 U-net: Convolutional networks for biomedical image segmentation. In *International Conference on Medical Image Computing and Computer-Assisted Intervention*. Springer, Cham, pp. 234–241.

Rubio, J. J., Kashiwa, T., Laiteerapong, T., Deng, W., Nagai, K., Escalera, S., Nakayama, K., Matsuo, Y. & Prendinger, H. 2019 Multi-class structural damage segmentation using fully convolutional networks. *Computers in Industry* **112**. https://doi.org/10.1016/j.compind.2019.08.002.

Russakovsky, O., Deng, J., Su, H., Krause, J., Satheesh, S., Ma, S., Huang, Z., Karpathy, A., Khosla, A., Bernstein, M., Berg, A. C. & Fei-Fei, L. 2015 Imagenet large scale visual recognition challenge. *International Journal of Computer Vision* **115** (3). https://doi.org/10.1007/s11263-015-0816-y.

Sarker, C., Mejias, L., Maire, F. & Woodley, A. 2019 Flood mapping with convolutional neural networks using spatio-contextual pixel information. *Remote Sensing* **11** (19), 1–25. https://doi.org/10.3390/rs11192331.

Sazara, C., Cetin, M. & Iftekharuddin, K. M. 2019 Detecting floodwater on roadways from image data with handcrafted features and deep transfer learning. In: *2019 IEEE Intelligent Transportation Systems Conference, ITSC 2019*, pp. 804–809. https://doi.org/10.1109/ITSC.2019.8917368.

Simonyan, K. & Zisserman, A. 2015 Very deep convolutional networks for large-scale image recognition. In *3rd International Conference on Learning Representations, ICLR 2015 – Conference Track Proceedings*.

Szegedy, C., Liu, W., Jia, Y., Sermanet, P., Reed, S., Anguelov, D., Erhan, D., Vanhoucke, V. & Rabinovich, A. 2015 Going deeper with convolutions. *Proceedings of the IEEE Conference on Computer Vision and Pattern Recognition* 1–9.

Szegedy, C., Vanhoucke, V., Ioffe, S., Shlens, J. & Wojna, Z. 2016 Rethinking the inception architecture for computer vision. *Proceedings of the IEEE Conference on Computer Vision and Pattern Recognition* 2818–2826.

Tien Bui, D., Hoang, N. D., Martínez-Álvarez, F., Ngo, P. T. T., Hoa, P. V., Pham, T. D., Samui, P. & Costache, R. 2020 A novel deep learning neural network approach for predicting flash flood susceptibility: a case study at a high frequency tropical storm area. *Science of the Total Environment* **701**, 134413. https://doi.org/10.1016/j.scitotenv.2019.134413.

Tkachenko, N., Jarvis, S. & Procter, R. 2017 Predicting floods with Flickr tags. *PLOS ONE* **12** (2), e0172870. https://doi.org/10.1371/journal.pone.0172870.

Wang, R.-Q., Mao, H., Wang, Y., Rae, C. & Shaw, W. 2018 Hyper-resolution monitoring of urban flooding with social media and crowdsourcing data. *Computers & Geosciences* **111**, 139–147. https://doi.org/10.1016/j.cageo.2017.11.008.

Wang, Y., Fang, Z., Hong, H. & Peng, L. 2020a Flood susceptibility mapping using convolutional neural network frameworks. *Journal of Hydrology* **582**, 124482. https://doi.org/10.1016/j.jhydrol.2019.124482.

Wang, Y., Fang, Z., Hong, H. & Peng, L. 2020b Flood susceptibility mapping using convolutional neural network frameworks. *Journal of Hydrology* **582**, 124482. https://doi.org/10.1016/j.jhydrol.2019.124482.

Waterman, L., Casado, R., Bergin, E. & Mcinally, G. 2021 *A Mixed-Methods Investigation into Barriers for Sharing Geospatial and Resilience Flood Data in the UK*, pp. 1–16.

Wieland, M. & Martinis, S. 2019 A modular processing chain for automated flood monitoring from multi-spectral satellite data. *Remote Sensing* **11** (19). https://doi.org/10.3390/rs11192330.

Witherow, M. A., Sazara, C., Winter-Arboleda, I. M., Elbakary, M. I., Cetin, M. & Iftekharuddin, K. M. 2019 Floodwater detection on roadways from crowdsourced images. *Computer Methods in Biomechanics and Biomedical Engineering: Imaging and Visualization* **7** (5–6), 529–540. https://doi.org/10.1080/21681163.2018.1488223.

Yang, H. C., Wang, C. Y. & Yang, J. X. 2014 Applying image recording and identification for measuring water stages to prevent flood hazards. *Natural Hazards* **74** (2). https://doi.org/10.1007/s11069-014-1208-2.

Zhao, G., Pang, B., Xu, Z., Peng, D. & Zuo, D. 2020 Urban flood susceptibility assessment based on convolutional neural networks. *Journal of Hydrology* **590**, 125235. https://doi.org/10.1016/j.jhydrol.2020.125235.

First received 1 September 2021; accepted in revised form 26 April 2022. Available online 17 May 2022

doi: 10.2166/nh.2021.050

Application of red edge band in remote sensing extraction of surface water body: a case study based on GF-6 WFV data in arid area

Zhao Lu[a], Daqing Wang [a,*], Zhengdong Deng[a], Yue Shi[a], Zhibin Ding[a], Hao Ning[a], Hongfei Zhao[b], Jiazheng Zhao[c], Haoli Xu[a] and Xiaoning Zhao[a]

[a] Defense Engineering College, Army Engineering University, Nanjing 210007, China
[b] Postdoctoral Workstation of Jinling Hospital, Nanjing 210016, China
[c] Wuhu Environmental Supervision Detachment, Wuhu 241000, China
*Corresponding author. E-mail: wangdaqing2020@126.com

DW, 0000-0003-2470-4371

ABSTRACT

This paper mainly researches the application method of red edge band in water body remote sensing extraction. Gaofen-6 (GF-6) WFV data were chosen for the high spatial resolution, more bands, and wide width. Two new methods were proposed: the single-band threshold method based on the red edge 2 band and the decision tree model method based on the combined operation of the green band, red band, near infrared band, red edge 1 band, and red edge 2 band. Four traditional methods were used for comparing the extraction accuracy. Two study areas with different characteristics were chosen to analyze the reliability of the proposed method, one mountainous and one urban region, both located in Minqin, Gansu, China, which is a typical arid area. The results showed that the two red edge bands of the GF-6 WFV data can be utilized to extract water body information. Kappa coefficients extracted from the single-band threshold method based on the red edge 2 band in water bodies in mountainous and urban areas reached 96.18% and 93.21%, respectively. The decision tree method has the best extraction effect. Kappa coefficients of this method in mountain and urban water bodies were 97.73% and 94.41%, respectively.

Key words: decision tree, GF-6 WFV data, red edge band, surface water body extraction

HIGHLIGHTS

- The red edge band of Chinese Gaofen-6 WFV date was found to be applicable to the extraction of surface water.
- A decision tree model based on the joint operation of five bands was established, which can effectively extract surface water.

ABBREVIATIONS

AWEI	Automatic Water Extraction Index
CIWI	Combined Index of NVDI and NIR for Water Identification
EWI	Enhanced Water Index
FAI	Floating Algae Index
GF-6	GaoFen-6 satellite
ISODATA	Iterative Selforganizing Data Analysis Techniques algorithm
KNN	K-nearest neighbor
K-means	K-means clustering algorithm
MAWEI	Modified Automatic Water Extraction Index
MOWI	Modified Optimization Water Index
NDWI	Normalized Difference Water Index
NWI	New Water Body Index
PMS	Panchromatic/Multi-spectral cameras
PCA	Principal component analysis
SWI	Shadow Water Index
SVM	Support vector machine
WFV	Wide field view cameras

1. INTRODUCTION

As a very precious natural resource, surface water occupies a very important position in the ecological environment, directly affecting human production and life (Gong *et al.* 2018), especially the development of agriculture (Pierpaolo 2018). In arid and semi-arid areas where agriculture competes keenly with ecosystem for water, integrated management of both surface water and groundwater resources at basin scale is crucial (Wu *et al.* 2016). Therefore, accurately obtaining surface water information is very meaningful research. Remote sensing technology has the advantages of fast imaging, wide range, and short observation period. It can provide significant technical support for long-term, rapid and objective acquisition of surface water information. Extracting surface water information through remote sensing technology is currently one of the principal contents of remote sensing application research (Zhao *et al.* 2017).

Studies have shown that non-water features (such as terrain shadows, vegetation shadows, etc.) on remote sensing images affect the accuracy of water body extraction (Kasvi *et al.* 2019; Yue *et al.* 2020). In order to effectively extract real water bodies, researchers have proceeded to do a great deal of research and explore some methods (Bijeesh & Narasimhamurthy 2020). According to the characteristics of the extraction methods, these methods can fall into three types.

The first type of remote sensing extraction method is built on the spectral characteristics of water. This type of method mainly uses a single band or multiple bands in remote sensing images to construct models, which extracts water bodies by suppressing remote sensing information of non-water body features such as vegetation, buildings, and soil. Such methods include single-band threshold method (Liu *et al.* 2019), spectral relationship method (Zhang 2008), and so on. Water body indexes based on spectral characteristics have also been proposed and applied to water body extraction research, such as: Floating Algae Index (FAI) (Xu *et al.* 2015), Normalized Difference Water Index (NDWI) (Yang *et al.* 2017; Acharya *et al.* 2018), Modified Normalized Difference Water Index (MNDWI) (Sarp & Ozcelik 2016), Enhanced Water Index (EWI) (Jason & Du 2017), New Water Body Index (NWI) (Ding 2009), Shadow Water Index (SWI) (Wang *et al.* 2017), Automatic Water Extraction Index (AWEI) (Feyisa *et al.* 2014), Modified Automatic Water Extraction Index (MAWEI) (Nie *et al.* 2019), and Combined Index of NDVI and NIR for Water Body Identification (CIWI) (Li *et al.* 2015). Although these methods have shown good extraction capabilities in numerous studies, some researchers have pointed out the problem of fragmentation and discontinuity in the results extracted by these methods (Zhou *et al.* 2012).

The second type of method is the pixel-based classification method. This type of method mainly includes unsupervised classification and supervised classification (Kaplan & Avdan 2017). Unsupervised classification is a means of image classification without prior classification criteria. The main algorithms are Iterative Self Organizing Data Analysis Techniques Algorithm (ISODATA) and K-means clustering algorithm (K-Means). Supervised classification algorithms are tantamount to selecting a certain number of samples as training samples based on visual judgment and on-site investigation, and then use the decision function to classify the data in the image. The core algorithms include minimum distance method, parallelepiped method, characteristic window curve method, maximum likelihood method, etc. (Liang *et al.* 2018).

The third type of method is the object-oriented classification method. This type of method is to segment the image into objects composed of several homogeneous pixels, and these objects will be used as processing units. These processing units include not only the spectral information of the feature, but also comprehensively contain the spatial structure information, texture information, and so on (Tong *et al.* 2017). Although the object-oriented classification methods were proposed earlier and there are many related research papers (Kettig & Landgrebe 1976; Câmara *et al.* 1996), they have not served in water body extraction until recent years (He *et al.* 2016; Huang *et al.* 2016). K-nearest neighbor (KNN) pixel method (Wang *et al.* 2019), support vector machine (SVM) method (Zhang *et al.* 2019), and principal component analysis (PCA) (Duan *et al.* 2015) are the primary commonly used algorithms of this type of method. Regardless of the fact that the object-oriented classification method has been a hot topic in water research in recent years, Blaschke (2010) pointed out that the spatial resolution of satellite images was an important limiting factor for object-oriented classification techniques. This type of method had limited effect on water body extraction from remote sensing images of medium and low spatial resolution.

In addition to the above types of methods, researchers have also attempted to introduce machine learning technology to extract water bodies. Based on the OSTU algorithm, Zhao (2016) used a new water body index to complete the extraction of four main types of water body information about oceans, rivers, lakes, and reservoirs. Yu *et al.* (2017) used convolutional neural networks and logistic regression to extract water bodies. Chen *et al.* (2019) used convolutional neural network and Deeplabv3 semantic segmentation neural network to extract water bodies, respectively. The results of these studies showed that the deep learning method could effectively remove the influence of shadows and buildings.

With the optimization of algorithms, water body extraction methods are developing continuously. In order to achieve higher accuracy, improving the spatial resolution of remote sensing data and enriching image information are important ways to extract water bodies effectively. Data from satellites such as Landsat8 OLI (Hao & Wan 2019), Landsat-7 ETM+ (Xi *et al.* 2009), Sentinel (Huang *et al.* 2018), GF-1 (Deng *et al.* 2017), GF-2 (Wang *et al.* 2018), have been widely used by researchers in the extraction of water bodies. Among these satellite data, some data have low spectral spatial resolution and time resolution, some data have high spatial resolution but low time resolution, and some data have low spectral resolution. As a newly launched satellite in recent years, WFV data collected and produced by Gaofen-6 satellite (GF-6) have several advantages. Compared with the satellite data mentioned above, WFV data can realize continuous dynamic observation, with high spatial resolution (up to 16 m) and high repeated coverage (every 2 days). GF-6 WFV data contain eight bands, which are rich in band information, which provides the possibility to explore new water body methods. In addition, the width of WFV data reaches 800 km, which can realize information extraction in a larger scale. Of course, like many multispectral remote sensing data, these data are easily disturbed by the natural environment. At present, there is little research literature about data modification by the application. A few studies are mainly used in the extraction of vegetation coverage (Deng *et al.* 2021), coastal sounding mapping (Sun *et al.* 2021), rice mapping, and growth monitoring (Jiang *et al.* 2021), etc.

In this paper, first, the reflectivity of different ground objects to eight bands of GF-6WFV were extracted respectively. Through analysis, it is found that the reflectivity of the water body to the red edge band was significantly different from that of other ground objects to red edge band. Secondly, by calculating the reflectivity of different bands, it is found that different bands' operational results can produce different extraction results, and a decision tree model is constructed on the basis of constant debugging. Finally, the extraction results of our method were compared with those of traditional methods. In order to analyze the reliability of the proposed method, rivers and reservoirs in mountainous areas and scenic lakes in urban areas were both taken as research objects.

2. RESEARCH AREA AND DATA SOURCE

2.1. Study area

The study area (38°19′47″–38°44′50″N, 102°48′15″–103°19′1″E) is situated in Minqin County, Wuwei City, Gansu Province, China, with an area of about 2,100 km^2 (Figure 1). It is adjacent to the Tengger Desert in the east and the Badain Jaran Desert in the west, which belongs to a typical arid desert climate. The geographic types are mountains, plains and sand dunes, with an altitude of 1,400–2,100 m. Before the comprehensive management of the Shiyang River Basin, the total surface water in the area was less than 100 million m^3, which seriously affected people's lives and agricultural production. The geographical location of the study area is presented in Figure 1. There were two main surface water bodies with different characteristic in the study area. The first area was located near Hongya Mountain (Figure 1(d)). The water bodies in this area mainly include Shiyang River and Asia's largest desert reservoir – Hongya Mountain Reservoir. The typical ground features in this area were vegetation, farmland, highways, mountains, etc. River mudflats, mountain shadows, and vegetation shadows may affect the water body extraction effect. The second area was situated in Minqin County (Figure 1(c)). The main water bodies in this area included Qingquan Lake and the viewing lake in People's Park. The typical ground features were roads, buildings, farmland, etc. Buildings and building shadows may affect the water body extraction results.

In order to visually identify the water body more conveniently, the false color image of the remote sensing image can be obtained by combining B2, B3, and B4 in the GF-6 WFV image, as shown in Figure 2. On false color images, most water bodies appeared in a dark tone. According to the location and characteristics of the water body, the water bodies in the mountainous areas were marked and made up of four areas (Figure 2(a)). The water body of No. 1 area was Shiyang River, and the water body contains a great deal of sediment, which was light green on the false color image. The water body of No. 2 area was the tidal flat of Shiyang River, with trees interspersed with many shadows, and appears dusky black on the false color image. The water body of No. 3 area was Hongyashan Reservoir. The reservoir was deeper and the water quality was relatively clear. The false color image appeared dark blue. The water body in No. 4 area was a canal that flowed out of the Hongyashan Reservoir. The water depth was relatively shallow. Many trees grew on both sides of the canal. The false color image was also somber black. The water bodies in the urban area were mainly Qingquan Lake and the artificial lake in the People's Park, and their locations were marked as shown in Figure 2(b). The water body of No. 5 area was an artificial lake in the People's Park of Minqin County, and the water body of No. 6 area was the Qingquan Lake.

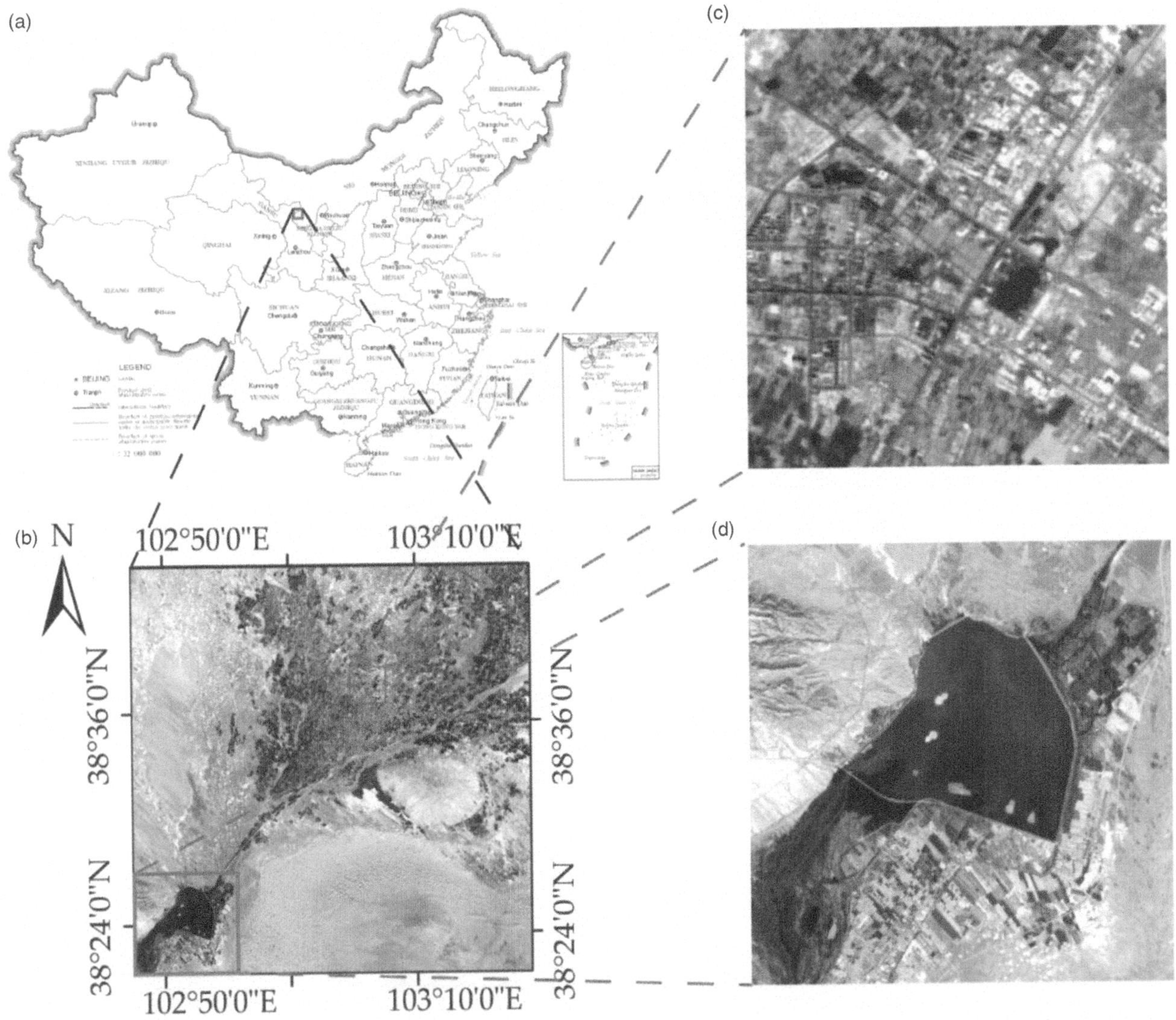

Figure 1 | The location of the study area. The red rectangle in the upper left figure (a) was the geographic location of the study area. The image on the lower left (b) was the true color image of GF-6 satellite image after preprocessing. The image on the lower right (d) was the first area. The image on the upper right figure (c) was the second area. Please refer to the online version of this paper to see this figure in color: http://dx.doi.10.2166/nh.2021.050.

2.2. GF-6 WFV data

The Chinese GF-6 satellite was officially called into use on March 21, 2019. The satellite is equipped with a 2 meter panchromatic/8 meter multi-spectral high-resolution camera (PMS) and a 16 meter multi-spectral medium-resolution WFV camera. The observation width of PMS is 90 km, and that of WFV is 800 km. The revisit period is 2 days. The GF-6 WFV sensor covers eight frequency bands: blue, green, red, near infrared, red edge 1, red edge 2, purple, and yellow. The technical specification for GF-6 WFV data is presented in Table 1.

GF-6 WFV data used throughout this study came from the Land Survey Satellite Data Service Platform of the China Resources Satellite Application Center. The satellite image was acquired on September 29, 2019, and the weather in the study area was sunny and cloudless. Affected by atmospheric conditions, topography, and other factors, remote sensing images may cause image distortion due to assembly deformation, noise interference, and other reasons. Before conducting research on water body extraction, preprocessing such as radiation calibration, atmospheric correction, geometric correction, and image cropping must be done to reduce the adverse effects caused by factors such as the atmosphere, light, and terrain. Figure 1(b) is the true color image of GF-6 satellite image after preprocessing.

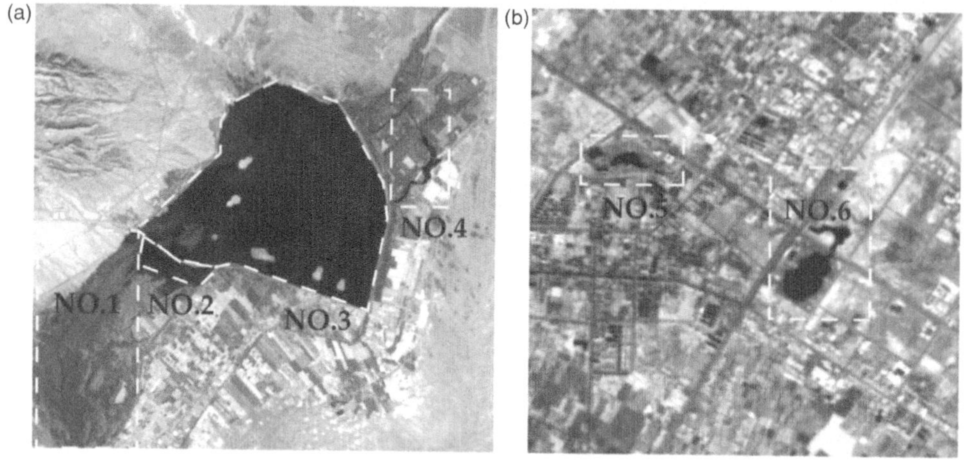

Figure 2 | False color display images of two water bodies. The image on the left (a) was false color images of the water bodies in mountainous areas. The image on the right (b) was false color images of the water bodies in urban areas.

Table 1 | Technical specification of GF-6 WFV data used in this study

Band	Channel	Central wavelength (nm)	Band width (nm)
Band 1 (B1)	Blue	485	70
Band 2 (B2)	Green	555	70
Band 3 (B3)	Red	660	60
Band 4 (B4)	Near infrared	830	120
Band 5 (B5)	Red edge 1	710	40
Band 6 (B6)	Red edge 2	750	40
Band 7 (B7)	Purple	425	50
Band 8 (B8)	Yellow	610	40

3. METHODOLOGY

GF-6 WFV data had not been applied to the study of remote sensing extraction of surface water body. Before studying the extraction method of water body, it was necessary in order to analyze the waveband emission characteristics of typical features in the study area. On the basis of field investigation and verification, the spectral features of typical feature pixels such as vegetation, lakes, rivers, farmland, roads, buildings, and shadows in the two areas of the study area were extracted. In mountainous areas, the reflectivity of pixels, mainly including vegetation, reservoir water, river water, farmland, roads, buildings, and mountain shadows were statistically analyzed. The results are presented in Figure 3. In urban areas, statistical analysis was mainly performed on the reflectance of the pixels of vegetation, lakes, farmland, buildings, and architectural shadows, as shown in Figure 4.

In Figure 3, compared with other ground features, the water body exhibited strong absorption and weak reflection characteristics for all wavebands, and the reflectivity decreased as the wavelength increases. However, the reflectivity of each waveband of the water body pixels in the Shiyang River was greater than that of the water body pixels in the Hongyashan Reservoir. The reason was that the high sediment content and turbid water quality in the Shiyang River enhanced the reflection of the wavebands. In all wavebands, the reflection differences between water bodies and non-water bodies in the two wavebands of B4 (red band) and B6 (red edge 2 band) were very significant. For the water body pixel of the reservoir, the reflectivity of the two bands B4 and B6 were significantly lower than that of other ground feature pixels. By setting reasonable thresholds in the two bands, the water body of the reservoir could be totally extracted. For the river water pixels, the reflectivity of the two bands B4 and B6 were also lower than that of other ground object pixels except the shadow pixels, but it was

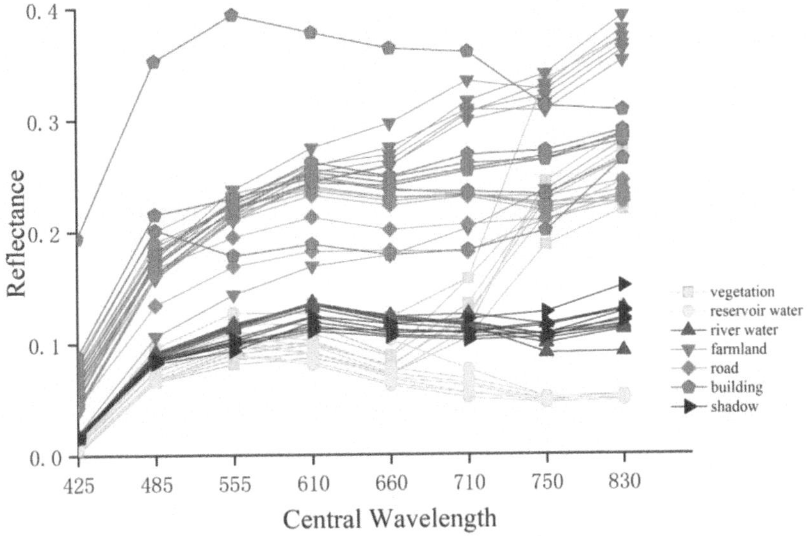

Figure 3 | Spectral curves of typical features in mountainous areas.

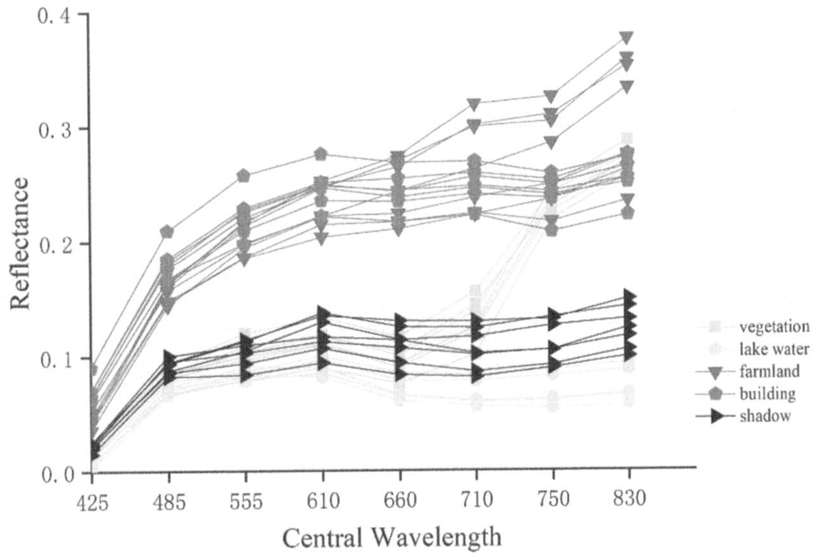

Figure 4 | Spectral curve of typical features in urban areas.

similar to the reflectivity of the shadow pixels. Therefore, when these two bands were utilized to extract pure river water pixels, the interference of shadow pixels may be greater.

In Figure 4, the reflectivity change trend of the water body pixel and non-water body pixel for the two wavebands B4 and B6 was basically similar to that in Figure 3. Different from the water pixels of the reservoir, the reflectivity of water pixels in the urban lake was very close to that of the building shadow pixels. When using the single-band threshold method, in order to extract more real water bodies pixels, the erroneous mention of shadow pixels was inevitable.

According to the above extraction results of spectral curves of separate ground objects, it can be found that the reflectivity differences of distinct bands of GF-6 WFV data to ground objects were very obvious. Abundant band information provides the possibility to explore extraction methods based on new bands, and is also beneficial to explore the application prospect of classification algorithms based on machine learning.

To validate the accuracy of the method proposed, other common methods such as single-band threshold method based on near infrared band, water index method based on NDWI, unsupervised classification method, and supervised classification method based on SVM were mainly used to calibrate innovative methods. The research technical route of this paper is shown in Figure 5. The related methods and mechanisms is discussed in the following text.

3.1. Water body extraction based on single-band threshold method

The single-band threshold method is a classification method based on the reflection characteristics of a certain band of distinctive features in remote sensing images. The B4 band has served in numerous studies of water bodies extraction (Mondejar & Tongco 2019). But from the above analysis, it can be seen that the B6 band also could be tried to be used in water body extraction. In this study, both B4 and B6 bands were selected for the single-band threshold method. The single-band threshold method was expressed as:

$$B < T. \tag{1}$$

In Formula (1), B is the gray value of B4 or B6 band in remote sensing image. T is the gray threshold to be set.

3.2. Water body extraction based on NDWI

NDWI is a water index constructed using the green band and near infrared band. This index was considered to be able to effectively suppress vegetation and soil information, and it has also been used in surface water body extraction in many studies (Ali *et al.* 2019; Yue *et al.* 2020). The formula of NDWI was expressed as (Mondejar & Tongco 2019):

$$\mathrm{NDWI} = (\mathrm{B2} - \mathrm{B4})/(\mathrm{B2} + \mathrm{B4}) \tag{2}$$

In Formula (2), B2 is the gray value of the green light band; B4 is the gray value of the near infrared band.

The threshold of NDWI was essential for extracting water bodies. In this study, referring to the research conclusion of McFeeters (1996), 0 was used as the classification threshold. That was, pixels with NDWI values greater than 0 were considered to be water bodies, and pixels with NDWI values less than 0 were classified as non-water bodies.

3.3. Water body extraction based on decision tree of multi-band combined operation

Many scholars had noted that ground features such as shadows and buildings seriously interfere with the extraction of water bodies (Bijeesh & Narasimha 2019; Li *et al.* 2019; Li *et al.* 2020). In further analysis of the spectral reflection characteristics of the ground object pixels, the gray level results between the B2, B3, B4, and B5 bands had definite correlation with the ground object pixels. By subtracting the gray values of the B2 and B4 bands, the difference of most water pixels was greater

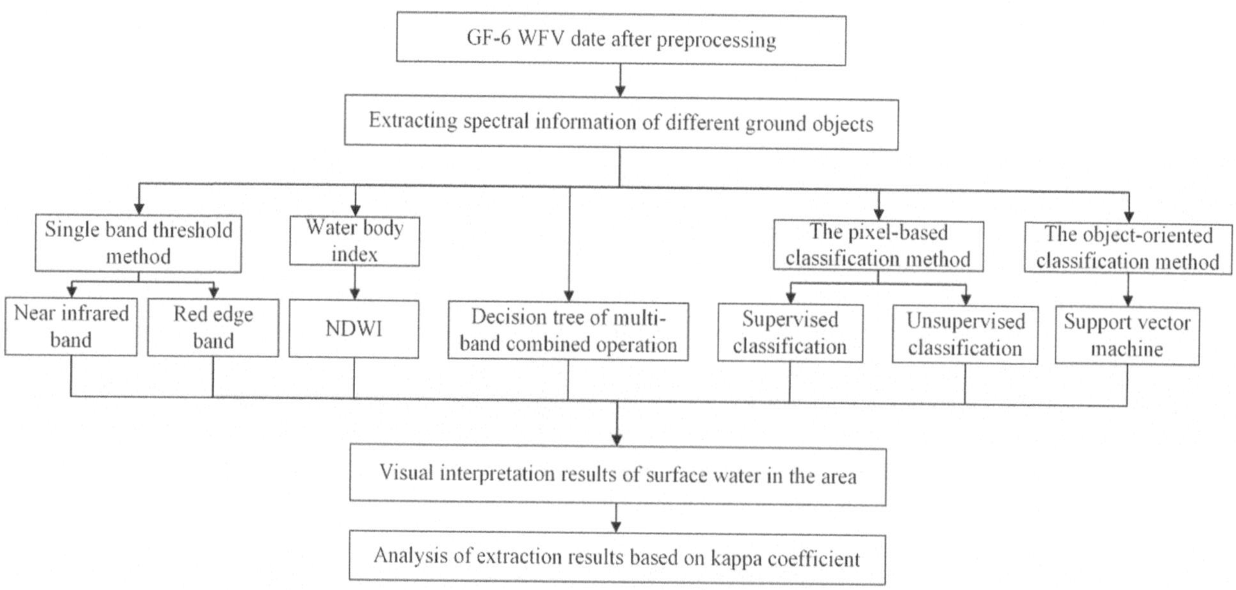

Figure 5 | Technical route of surface water body extraction based on various methods.

than 0, and the difference of other non-water feature pixels was mostly less than 0. The difference of a few road pixels was greater than 0, and the difference of most shadow pixels was less than −110. By subtracting the gray values of the B5 and B3 bands, the difference between most water pixels and shadow pixels was less than 50, and the difference between extra feature pixels was mostly greater than 0. Combined with the debugging results in the single-band threshold method, when the gray value of the B6 band was less than 1,100, the removal of non-aqueous features such as buildings and roads could be achieved. Therefore, the non-water pixels could be removed step by step by making effective use of the reflection characteristics of different bands for the combined operation. In the study, a decision tree model was constructed built on the above-mentioned discovered rules, which removed the interference of shadows and vegetation pixels, then removed the interference of road pixels and building pixels, and finally leaves the surface water pixels. The decision tree model is presented in Figure 6.

3.4. Water body based on supervised classification and unsupervised classification

In the study, both unsupervised classification and supervised classification were used for water body extraction in two areas.

The two main algorithms ISODATA and K-means were both used for unsupervised separation. Unsupervised classification was based on the pixel characteristics of satellite images. The number of classification types had a definite impact on the classification results. In the study, 15 types of features were set for unsupervised classification. After the classification results were obtained, through the visual interpretation, the pixels in the classified image were sorted into two categories: water and non-water.

When using supervised classification, it was necessary to select appropriate training samples. The types of ground features in the study area were complex. If the training sample type was not properly selected, some features with complex pixel features may be ignored, which would affect the extraction effect. The strategy included in this study was to simplify the sample types. In mountainous areas, the water bodies of reservoirs, rivers, canals, and floodplains were regarded as one type of sample type, and vegetation, buildings, farmland, roads, shadows, and other features were collectively classified as another type of sample type. In urban areas, the water bodies of Qingquan Lake and artificial lakes were considered as one type of sample type, and buildings, roads, vegetation, shadows, and other features were collectively classified as another type of sample type. When sampling, for water bodies, there must be typical pixels selected as samples in the six regions in Figure 2.

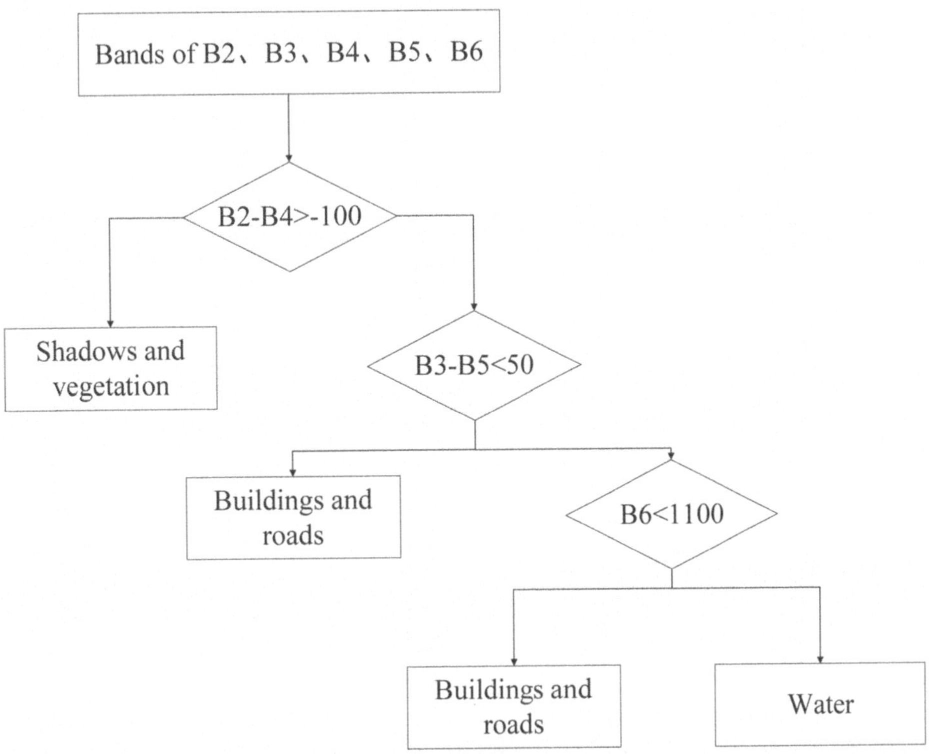

Figure 6 | Decision tree model based on multi-band combined operation.

For non-aqueous bodies, the sample sampling volume should be increased, and the various non-aqueous features on the image should be covered as much as possible. Among several algorithms for supervised classification, the support vector machine (SVM) algorithm is essentially a binary classification algorithm that fundamentally works to separate data points into two classes, and it is more suitable for solving the problems such as few samples, nonlinearity, and local minima (Zhang *et al.* 2019). Thus, the SVM algorithm was selected as the supervised classification method in this study.

3.5. Accuracy evaluation

In the study, the real water boundary in most areas was established by field survey. By plotting the measured water boundary on the remote sensing image, the actual water pixel can be interpreted intuitively on the satellite image. Because the spatial resolution of GF-6 WFV data is 16 meters, some pixels will contain many different types of ground objects such as water, land, and vegetation. For these pixels, in visual interpretation, if the area of non-water features in the pixel exceeds 1/2, the pixel is regarded as non-water. In the study, some areas were not in a position for the boundary of a water body to be measured. In these few areas, the real water pixels are determined in GF-6 WFV images by comparing with the visually interpreted water pixels in GF-2 PMS images with higher spatial resolution obtained in the same period. The total number of pixels in the mountain image was 548,340, and the total number of real water body pixels was 89,271. The total number of pixels in the urban image was 96,408, and the total number of real water pixels was 878.

At present, there is no uniform standard for the comparison of water body extraction accuracy. Some studies have pointed out that the construction of confusion matrix to calculate the overall classification accuracy, Kappa coefficient, overall accuracy, misclassification error, missing error, and so on are often used by remote sensing processing software (Wang & Qin 2018), which are also used to compare the accuracy of water extraction methods (Du *et al.* 2014; Sekertekin 2021).

Kappa coefficient was utilized to test the consistency of the classification effect. The calculation of Kappa coefficient is based on a confusion matrix. In the study, the water body extraction results obtained by different methods and the classification results of visual interpretation were utilized to construct the confusion matrix and calculate the corresponding Kappa coefficient. In order to better analyze the differences between the extraction methods, the overall accuracy, the omission error, and the misclassification error of each extraction result were also calculated. Overall accuracy is the percentage of all correctly identified pixels in remote sensing images. Omission error is the probability that a real water body pixel is incorrectly classified into other categories, and the misclassification error is the probability that a non-water body pixel is incorrectly classified as a water body. The extraction of water body in the study was a two-class problem, so only a two-order confusion matrix was required to be constructed. Kappa calculation formula is as follows:

$$kappa = \frac{p_0 - p_e}{1 - p_0} \tag{3}$$

$$p_0 = \frac{\sum_{i=1}^{n} a_{ii}}{N} \tag{4}$$

$$p_e = \frac{\left(\sum_{j=1}^{n} a_{1j} \times \sum_{i=1}^{n} a_{i1} + \sum_{j=1}^{n} a_{2j} \times \sum_{i=1}^{n} a_{21}\right)}{N \times N} \tag{5}$$

In the above formula, N is the total number of pixels in the study area. The n is the order of the confusion matrix, and the value in the study is 2; a_{ij} is the constituent element of the confusion matrix.

4. RESULTS AND DISCUSSION

4.1. Extraction of water bodies in mountainous areas

Extraction results of water bodies in mountainous areas based on different methods are illustrated in Figure 7, where the water body pixels are marked in blue. Figure 7(a) is the real water body result of visual interpretation. Figure 7(b) and 7(c) are the water body extraction results based on the single-band threshold method of B4 and B6, respectively. The water body extraction results of the two bands were relatively similar, but the extraction of water pixels based on the B6 band was greater than the extraction results based on the B4 band. The Shiyang River in No. 1 area was a continuous river, but the pixels in the extraction results were broken, and only some real water pixels were extracted. Extraction results of the

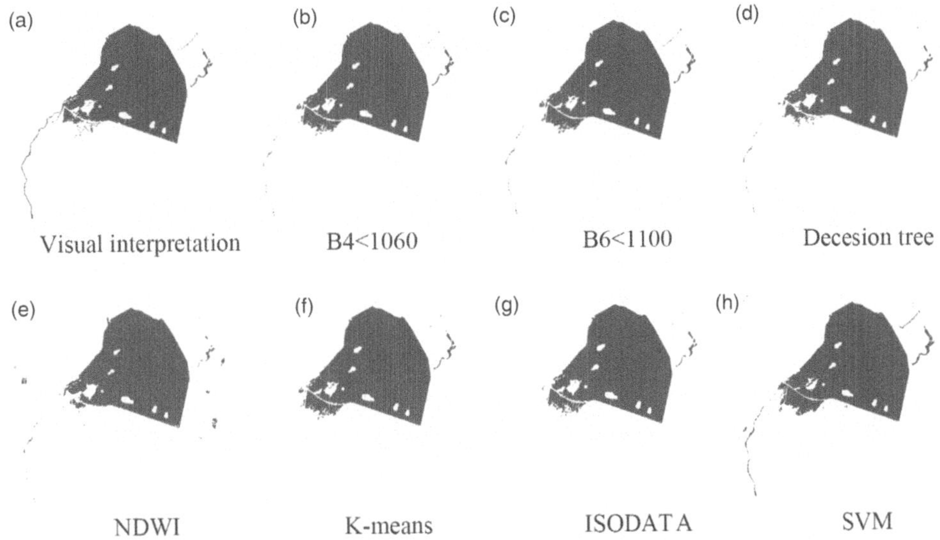

Figure 7 | Extraction results of water bodies in mountainous areas based on different methods. Please refer to the online version of this paper to see this figure in color: http://dx.doi.10.2166/nh.2021.050.

single-band threshold method based on the B6 band were slightly better than those based on the B4 band. The water bodies of No. 2 and No. 3 areas had been over-extracted, and a large number of non-water body pixels were incorrectly classified as water bodies. The feature types of the pixels extracted by mistake were mainly vegetation and shadows on the floodplain in No. 2 area and aquatic plants in No. 3 area. The spectral information of pixels in these places was complicated, especially the presence of shadows seriously interferes with the extraction results. Finally, the threshold method based on the B4 band had a limited effect on the extraction of the small water channel in the upper right corner of the water body in No. 4 area, and the threshold method based on the B6 band had a better effect. Figure 7(d) is the extraction result obtained by the decision tree method based on multi-band combination operation. Compared with the results in Figure 7(b) and 7(c), the water body pixels of Shiyang River extracted from No. 1 area in Figure 7(d) were almost the same as the results in Figure 7(c). The water pixels extracted in area No. 2 were significantly less than the results in Figure 7(b) and 7(d). By comparing with real water body pixels visually interpreted in Figure 7(a), this method was more effectively avoiding the interference of vegetation and shadows in the floodplain. However, when extracting small water bodies in No. 4 area, the extraction effect of this method was not as good as the single-band threshold method. Figure 7(e) shows the extracted results based on the NDWI. Only the water body extraction result in No. 3 area was closer to the result of visual interpretation, and many real water body pixels in the other three areas were omitted. At the same time, many pixels of buildings and roads in the extraction results were incorrectly classified as water bodies. Figure 7(f) is the extraction result of unsupervised classification based on K-means algorithm, and Figure 7(g) is the extraction result of unsupervised classification based on ISODATA algorithm. Regardless of whether the ISODATA or K-means algorithm was used, the water pixels of the Shiyang River in the No. 1 area had not been effectively extracted. Based on these two algorithms, the water body extraction results in the other three regions were very similar. The reason why the Shiyang River in No. 1 area was not effectively extracted may be that the sediment content in the river water was higher, causing the pixels to be more akin to the soil. From the extracted results, it can be concluded that the unsupervised classification method was not applicable to areas where the spectral information of pixels was significantly altered due to the difference in water quality.

Figure 7(h) is the extraction result of the supervised classification method based on the SVM algorithm. This method was quite effective for the extraction of real water pixels in all four areas. Especially, the water body extraction results in No. 1 and No. 4 areas were closer to the results of visual interpretation. However, it is worth noting that this method incorrectly classifies many non-water pixels in the four areas as water pixels. The reason for this situation may be that the water quality and pixel characteristics in the four areas were changed, and there may be deviations in the selection of training samples in the study. This exposed the problems of supervised classification in water body extraction: classification required extraordinary purity of training samples, and training samples should be typical and representative. When choosing this method, a priori knowledge of the research field was required, so it was difficult to achieve automatic extraction of water bodies.

4.2. Extraction of water bodies in urban areas

Extraction results of water bodies in urban areas based on different methods are shown in Figure 7, where the water body pixels are also marked in blue. Figure 7(a) is the real water body result of visual interpretation.

There were fewer types of ground features in urban areas than in mountainous areas, and there were fewer interference factors in the extraction process. The types of municipal water bodies were single, and the spectral curves of the water body pixels in areas of No. 5 and No. 6 were almost the same. The water quality of the two water bodies in the No. 5 area and the No. 6 area were both very good, and the boundaries of the water bodies were clearer.

In Figure 8, the water body extraction result obtained by the decision tree method based on multi-band combined operation (Figure 8(d)) was better than other methods. This method effectively extracted real water pixels, and avoided erroneous extraction of non-water pixels. In the extraction results of the supervised classification method based on SVM algorithm, although there were not many pixels extracted by mistake, the pixels of real water bodies were obviously omitted.

Among the results obtained by other methods, compared with the results of visual interpretation, the single-band threshold method (Figure 8(b) and 8(c)) and the unsupervised classification method (Figure 8(f) and 8(g)) had similar effects in extracting real water pixels. However, in these four results, there are more non-water pixels extracted by mistake than in Figure 8(d), especially, the large number of shadow pixels of buildings in Figure 8(f) and 8(g) were incorrectly classified as water bodies. In the extraction results based on the NDWI, many real water pixels in areas of No. 5 and No. 6 were omitted, and a large number of non-water pixels were also incorrectly extracted. The difference from the aforementioned four consequences was that the mis-mentioned pixel types in Figure 8(e) were mainly buildings and a small amount of building shadows. At this time, the NDWI index method could better remove the interference of the building shadow, but it could not avoid the influence of the building itself.

4.3. Accuracy analysis of different methods

Water body extraction results obtained based on different methods were compared and analyzed with the real water body manually interpreted, and confusion matrices were constructed respectively. Kappa coefficient, overall accuracy, misclassification error, and omission error could be calculated separately. The results are shown in Tables 2 and 3.

Table 2 shows the accuracy results of different methods for water body extraction in mountainous areas. The Kappa coefficient of the decision tree method based on multi-band combination was the highest, reaching 97.73%, and the Kappa coefficient of the supervised classification method based on SVM was the lowest, being 94.47%. Omission error values and misclassification error values of the decision tree based on multi-band combination operation were relatively low. Both were 1.89%. This method not only realized the effective extraction of water body pixels but also effectively avoided the wrong extraction of non-water body pixels. Although the omission error of water body extraction based on SVM was

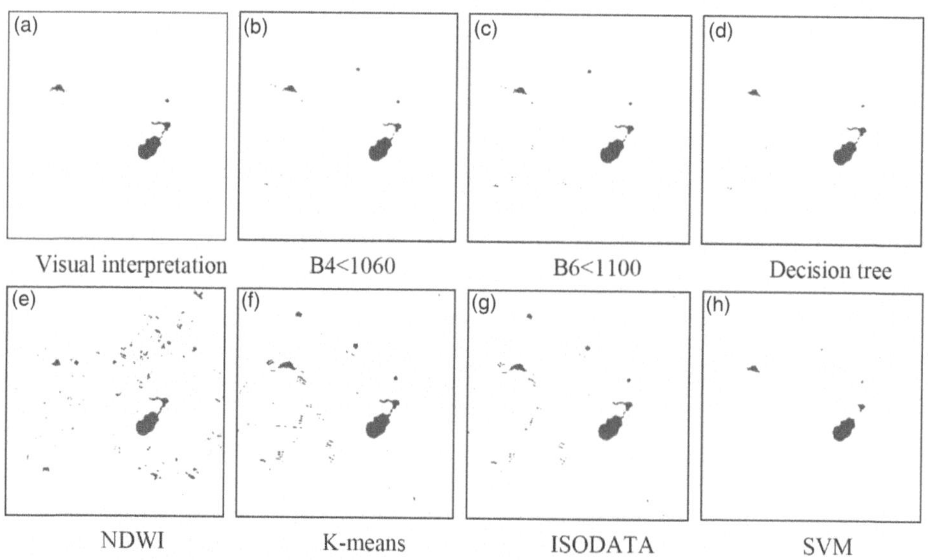

Figure 8 | Extraction results of water bodies in urban areas based on different methods.

Table 2 | The accuracy of different methods for water body extraction in mountainous areas

Serial number	Method	Kappa (%)	Overall accuracy (%)	Misclassification error (%)	Omission error (%)
1	Single-band threshold method based on B4 band	96.35	98.99	4.49	1.55
2	Single-band threshold method based on B6 band	96.18	98.95	5.01	1.26
3	Water index method based on NDWI	96.8	99.12	2.6	2.75
4	Unsupervised classification method based on ISODATA algorithm	96.61	99.07	2.88	2.79
5	Unsupervised classification method based on K-means algorithm	96.16	98.94	4.22	2.15
6	Supervised classification based on SVM	94.97	98.59	7.3	0.8
7	Decision tree model method based on multi-band combined operation	97.73	99.38	1.89	1.89

Table 3 | The accuracy of different methods for water body extraction in urban areas

Serial number	Method	Kappa (%)	Overall accuracy (%)	Misclassification error (%)	Omission error (%)
1	Single-band threshold method based on B4 band	94.06	99.89	6.51	5.23
2	Single-band threshold method based on B6 band	93.21	99.87	8.99	4.32
3	Water index method based on NDWI	65.66	99.14	48.29	8.54
4	Unsupervised classification method based on ISODATA algorithm	84.24	99.66	26.25	1.36
5	Unsupervised classification method based on K-means algorithm	76.79	99.46	37.1	0.56
6	Supervised classification based on SVM	91.63	99.85	1.31	14.35
7	Decision tree model method based on multi-band combined operation	94.41	99.90	4.42	6.6

the smallest of all methods, the error of misclassification was the highest. In the study, the training samples of the water pixels were carefully selected, but due to the complex water types in the mountainous areas, the extraction results were still not ideal.

In addition to the above two methods, Kappa coefficients of the extraction results obtained by other methods were about 96%, but there were differences in the error of misclassification and omission. The single-band threshold method based on the B4 band and the B6 band had a relatively low missed scoring error and a higher missed scoring. The single-band threshold method could better extract the real water pixels in the four regions, but many non-water pixels were mistakenly classified as water pixels, which was consistent with the result in Figure 7. Both extraction results based on NDWI and the unsupervised classification method had high error values for misclassification and omission. The disadvantages of the unsupervised classification method tended to be unable to effectively extract the water in the No. 1 area. The water body index law based on NDWI tends to extract non-water body pixels incorrectly.

Table 3 shows the accuracy results of different methods for water body extraction in urban areas. The Kappa coefficient of the decision tree method based on multi-band combination operation was the highest, reaching 94.41%, and the K-means algorithm-based unsupervised classification had the lowest Kappa coefficient of 76.79%. Kappa coefficients of the single-band threshold method extraction results based on B4 and B6 also reached 94.06% and 93.21%, respectively. The single-band threshold method had a better effect on water body extraction in urban areas, and the single-band threshold method based on the B6 band found and proposed in this study also showed very good classification performance. Both the unsupervised classification method and NDWI had relatively high misclassification errors in the extraction results, which were in accordance with the analysis results in Figure 8. It is worth noting that among all the methods, the omission error in the extraction results based on SVM was very high, reaching 14.35%. The reason may be that the water pixels at the boundary of the lake were not included in the training samples, and the pixel information of these pixels may be different from the pure water pixels in the center of the lake.

From the above analysis results, the decision tree method based on the combined operation of B2, B3, B4, B5, and B6 in this study has achieved very good results in surface water extraction in mountainous and urban areas.

4.4 Application value of red edge band

Improving the accuracy of remote sensing extraction of surface water is part of the core objectives of a large number of studies. In recent years, in order to improve the accuracy, researchers have put forward many new water indexes, such as MOWI (modified optimization water index) (Moradi *et al.* 2017), Terrain indices (MrVBF and HAND) (Huang *et al.* 2017), Sum 457 (sum of near infrared and two shortwave infrared bands) (Zeng *et al.* 2017), and so on. With the increasing popularity of machine learning algorithms, a large number of related methods based on machine learning have been applied to water remote sensing extraction, and some new classification algorithms have emerged (Isikdogan *et al.* 2017; Li *et al.* 2017). In addition, methods based on spectral mixture analysis are becoming ever more popular, because they can estimate the proportion of various spectra existing in pixels, which can better classify pixels (Bijeesh & Narasimhamurthy 2020). However, no matter which extraction method, it is inseparable from the spectral characteristics of remote sensing data. Therefore, it continues to be of great significance to explore the water extraction ability of different characteristic spectra, provided that remote sensing data containing more spectra can be obtained. The above research of Sections 4.1–4.3 showed that the red edge band itself has similar surface water extraction ability to the near infrared band. Whether in mountainous areas or urban areas, the single-band threshold method based on red edge 2 bands in GF-6 WFV data was similar to the single-band extraction accuracy based on near infrared bands, both of which were above 93%. The multi-band combination decision tree method proposed in this study was obtained through a great deal of constant debugging, and its extraction ability was the best among the methods adopted in this study. Compared with several typical conventional remote sensing extraction methods of surface water, the two newly discovered extraction methods using red sideband would not cause significant differences in extraction effects due to altered water environment. In sharp contrast, the extraction effect of NDWI was more easily affected by the environment of water. Kappa value of the index reached 96.8% in mountainous areas, but only 65.66% in cities.

Although the red edge band showed unforgettable extraction ability in this study, it still failed to achieve complete and correct extraction of the water body. The existence of shadows will still cause interference. Therefore, water extraction methods related to the red edge band and the water extraction methods based on GF-6 WFV data are still worth improving.

5. CONCLUSIONS

In this paper, the GF-6 WFV data were taken as the remote sensing data source, and the water bodies located in Hong Yashan mountainous area and county town in Minqin, Gansu Province were taken as the extraction objects, and the remote sensing extraction method was studied. The research further expands the application prospect of red edge band in the remote sensing field. Research results fully demonstrate that the red edge waveband extraction method we found and proposed had high utilization value in water extraction. The research drew the following conclusions.

Among the extraction methods based on different mechanisms employed in this research, the decision tree method based on multi-band combined operation obtained the best extraction results in both regions. Kappa coefficient of extraction result of water bodies in mountainous areas was 97.73%. Both the error of omission and the error of misclassification were 1.89%. Kappa coefficient of extraction consequence of water bodies in urban areas was 94.41%. The missing scoring error was 4.42%, and the misclassification error was 6.6%. The decision tree model used two red edge bands in the GF-6 WFV data, and successfully solved the problem of incorrect mention of shadows, vegetation, and buildings. It was necessary to extract and analyze the spectral information of the ground object pixels in advance in the process of building the model, so as to set specific thresholds for each layer. However, this decision tree model could obtain high-precision water body results in both mountainous and urban areas. The research results fully showed that this method could serve as a new method for water body extraction. In addition, the single-band threshold method based on the B6 band also achieved excellent water body extraction results in both mountainous and urban areas. The single-band threshold method based on the B6 band had a Kappa coefficient of 96.18% in mountain water bodies, and a Kappa coefficient of 93.21% in urban water bodies.

In order to be compared with the proposed method, four conventional methods were utilized to extract water from the study area by remote sensing. The pixel-based classification method used in the study had poor water extraction results. Regardless of the fact that Kappa coefficients of the two algorithms of the unsupervised classification method in mountain water body extraction results reached more than 96%, it was virtually impossible to achieve effective extraction of the water body pixels of the Shiyang River. The water body extraction results of these two algorithms in urban areas were even lower than 85%. Kappa coefficients of the water body extraction results obtained by using the supervised classification method of the SVM algorithm in the two areas were greater than 91%, but this method relied too much on the purity of the training samples. In areas with complex water features, the reliability of the method may become unstable. However, in the case that water information needs to be extracted quickly and the extraction accuracy is low, the supervised classification method still has good application value. NDWI showed good results in mountain water extraction, but it was very bad in urban water extraction, and the interference of buildings was very serious.

In this study, due to various reasons, it is impossible to measure more water bodies in different areas, which led to the study area being restricted to Minqin, China. In checking to ensure the reliability of the proposed method, water bodies located in mountainous areas and water bodies located in urban areas were selected as two independent research objects. In order to further verify the reliability of the water body method based on the red sideband, it is necessary to study more water bodies located in different regions in the future. At the same time, the application of the red sideband in remote sensing extraction of water resources needs further development, such as developing a general water resources index based on red sideband.

ACKNOWLEDGEMENTS

The authors deeply appreciate the anonymous reviewers and the editor's constructive comments and suggestions, all of which have led to great improvements in the presentation of this article.

AUTHOR CONTRIBUTIONS

Conceptualization, D.W. and Z.L.; Methodology, Z.L.; Software, Z.L.; Validation, Z.D. (Zhengdong Deng) and Z.D. (Zhibin Ding); Formal analysis, H.Z. and H.N.; Investigation, G.W. and H.X.; Data curation, H.Z. and X.Z.; Writing – original draft preparation, Z.L.; Writing – review and editing, Y.S. and J.Z.; Project administration, D.W. All authors have read and agreed to the published version of the manuscript.

CONFLICTS OF INTEREST

The authors declare no conflict of interest.

FUNDING

This research received no external funding.

DATA AVAILABILITY STATEMENT

All relevant data are included in the paper or its Supplementary Information.

REFERENCES

Acharya, T. D., Subedi, A. & Lee, D. H. 2018 Evaluation of water indices for surface water extraction in a Landsat 8 scene of Nepal. *Sensors* **18**, 2580.

Ali, M., Dirawan, G., Hasim, A. & Abidin, M. 2019 Detection of changes in surface water bodies urban area with NDWI and MNDWI methods. *Int. J. Adv. Sci. Eng. Inf. Technol.* **9**, 946–951.

Bijeesh, T. V. & Narasimha, M. 2019 A comparative study of spectral indices for surface water delineation using Landsat 8 Images. *2019 International Conference on Data Science and Communication (IconDSC)*, IEEE, Bangalore, India.

Bijeesh, T. V. & Narasimhamurthy, K. N. 2020 Surface water detection and delineation using remote sensing images: a review of methods and algorithms. *Sustain. Water Resour. Manag.* **6**, 68.

Blaschke, T. 2010 Object based image analysis for remote sensing. *ISPRS J. Photogramm. Remote Sens.* **65**, 2–16.

Câmara, G., Souza, R., Freitas, U. & Garrido, J. 1996 Integrating remote sensing and GIS by object-oriented data modelling. *Comput. Graphics* **20**, 395–403.

Chen, Q., Zheng, L., Li, X., Xu, C., Wu, Y., Xie, D. & Liu, L. 2019 Water body extraction from high-resolution satellite remote sensing image based on deep learning. *Geogr. Geo-Inf. Sci.* **35**, 43–49.

Deng, F., Zhang, X., Hua, L. & Li, Z. 2017 A surface water body extraction method based on domestic remote sensing imagery of high resolution. *Hydrogeol. Eng. Geol.* **44**, 43–50.

Deng, Z. D., Lu, Z., Wang, G. Y., Wang, D. Q., Ding, Z. B., Zhao, H. F., Xu, H. L., Shi, Y., Cheng, Z. J. & Zhao, X. N. 2021 Extraction of fractional vegetation cover in arid desert area based on Chinese GF-6 satellite. *Open Geosci.* **13**, 416–430.

Ding, F. 2009 Study on information extraction of water body with a new water index (NWI). *Sci. Surv. Mapp.* **34**, 155–157.

Du, Z. Q., Li, W. B., Zhou, D. B., Tian, L. Q., Ling, F. & Wang, H. L. 2014 Analysis of Landsat-8 OLI imagery for land surface water mapping. *Remote Sens. Lett.* **5**, 672–681.

Duan, Q., Meng, L., Fan, Z., Hu, W. & Xie, W. 2015 Applicability of the water information extraction method based on GF-1 image. *Remote Sens. Land Resour.* **27**, 79–84.

Feyisa, G., Meilby, H., Fensholt, R. & Proud, S. 2014 Automated water extraction index: a new technique for surface water map-ping using Landsat imagery. *Remote Sens. Environ.* **140**, 23–35.

Gong, W., Wang, P., Wang, S., Zhou, Y. & Cao, K. 2018 Methods of water body extraction in boundary river based onGF-2 satellite remote sensing image of high resolution. *J. Eng. Heilongjiang Univ.* **9**, 1–7.

Hao, Y. & Wan, L. 2019 Comparative analysis of water information extraction methods in Harbin area based on landsat8 OLI images. *Nat. Sci. J. Harbin Norm. Univ.* **35**, 87–92.

He, Y., Zhang, X. & Hua, L. 2016 Object-based distinction between building shadow and water in high-resolution imagery using fuzzy-rule classification and artificial bee colony optimization. *J. Sens.* 2385039. https://doi.org/10.1155/2016/2385039.

Huang, S., Ding, J. & Li, Y. 2016 Study of water information extraction based on domestic GF-1 remote sensing image by using object-oriented method. *Yangtze River* **47**, 23–28.

Huang, C., Nguyen, B. D., Zhang, S. Q., Cao, S. M. & Wagner, W. 2017 A comparison of terrain indices toward their ability in assisting surface water mapping from Sentinel-1 data. *ISPRS Int. J. Geo-Inf.* **6**, 140.

Huang, W., DeVries, B., Huang, C., Lang, M., Jones, J., Creed, I. & Carroll, M. 2018 Automated extraction of surface water extent from Sentinel-1 data. *Remote Sens.* **10**, 797.

Isikdogan, F., Bovik, A. C. & Passalacqua, P. 2017 Surface water mapping by deep learning. *IEEE J. Sel. Top. Appl. Earth Obs. Remote Sens.* **10**, 4909–4918.

Jason, Y. & Du, X. 2017 An enhanced water index in extracting water bodies from Landsat TM imagery. *ANN GIS* **23** (3), 141–148.

Jiang, X., Fang, S., Huang, X., Liu, Y. & Guo, L. 2021 Rice mapping and growth monitoring based on time series GF-6 images and red-edge bands. *Remote Sens.* **13**, 579.

Kaplan, G. & Avdan, U. 2017 Object-based water body extraction model using Sentinel-2 satellite imagery. *Eur. J. Remote Sens.* **50**, 137–143.

Kasvi, E., Salmela, J., Lotsari, E., Kumpula, T. & Lane, S. 2019 Comparison of remote sensing-based approaches for mapping bathymetry of shallow, clear water rivers. *Geomorphology* **333**, 180–197.

Kettig, R. & Landgrebe, D. 1976 Classification of multispectral image data by extraction and classification ofhomogeneous objects. *IEEE Trans. Geosci. Electron.* **14**, 19–26.

Li, H., Wang, X., Dai, S. & Tian, G. 2015 Flood monitoring in Hainan Island based on HJ-CCD data. *Trans. CSAE* **31** (17), 191–198.

Li, N., Martin, A. & Estival, R. 2017 An automatic water detection approach based on Dempster–Shafer theory for multi-spectral images. In: *2017 20th International Conference on Information Fusion*, Xi'an, China.

Li, L., Yan, Z., Shen, Q., Cheng, G., Gao, L. & Zhang, B. 2019 Water body extraction from very high spatial resolution remote sensing data based on fully convolutional networks. *Remote Sens.* **11**, 1162–1181.

Li, D., Wu, B., Chen, B., Qin, C., Wang, Y., Zhang, Y. & Xue, Y. 2020 Open-surface river extraction based on Sentinel-2 MSI imagery and DEM data: case study of the Upper Yellow River. *Remote Sens.* **12**, 2737.

Liang, S., Wang, Q., Zhang, C., Zhang, M., Sun, H. & Ren, H. 2018 Extraction method of high dive site mining area water-body based on GF-2 data. *Mine. Surv.* **46**, 5–9.

Liu, S., Wang, M., Yang, S., Yang, M. & Yang, L. 2019 Extraction accuracy and stability analysis of different waterbody index models in GF-2 images. *Bull. Surv. Mapp.* **8**, 135–139.

McFeeters, S. 1996 The use of the normalized difference water index (NDWI) in the delineation of open water features. *Int. J. Remote Sens.* **17**, 1425–1432.

Mondejar, J. & Tongco, A. 2019 Near infrared band of Landsat 8 as water index: a case study around Cordova and Lapu-Lapu City, Cebu, Philippines. *Sustain. Environ. Res.* **29**, 16.

Moradi, M., Sahebi, M. & Shokri, M. 2017 Modified optimization water index (MOWI) for Landsat-8 Oli/Tirs. *ISPRS-Int. Arch. Photogramm. Remote Sens. Spatial Inf. Sci.* **XLII**, 185–190.

Nie, Y., Yu, M. & Lan, T. 2019 Research on water information extraction based on MAWEI index. *J. Earth Environ.* **10**, 281–290.

Pierpaolo, S. 2018 Water for agriculture, irrigation management. *Appl. Soil Ecol.* **123**, 793–796.

Sarp, G. & Ozcelik, M. 2016 Water body extraction and change detection using time series: a case study of Lake Burdur, Turkey. *J. Taibah Univ. Sci.* **11** (3), 381–391.

Sekertekin, A. 2021 A survey on global thresholding methods for mapping open water body using Sentinel-2 satellite imagery and normalized difference water index. *Arch. Computat. Meth. Eng.* **28**, 1335–1347.

Sun, M. X., Yu, L. J., Zhang, P., Sun, Q. Q., Jiao, X., Sun, D. F. & Liu, F. 2021 Coastal water bathymetry for critical zone management using regression tree models from Gaofen-6 imagery. *Ocean Coast. Manage.* **204**, 105522.

Tong, L., Yan, Q., Luo, C. & Du, Y. 2017 Water body information extraction based on NDWI segmentation and object-oriented method. *Geo. Spat. Inf.* **15**, 57–59.

Wang, J., Ding, J., Zhang, C. & Chen, W. 2017 Method of water information extraction by improved SWI based on GF-1 satellite image. *Remote Sens. Land Resour.* **29**, 29–35.

Wang, H. & Qin, F. 2018 Summary of the research on water body extraction and application from remote sensing image. *Sci. Surv. Mapp.* **43** (5), 23–32.

Wang, J., Nuan, R., Chai, Y. & Lin, P. 2018 Object-oriented urban water information extraction based on GF-2. *Geo. Spa. Inf.* **16**, 34–37. + 40.

Wang, X., Xu, M., Li, K. & Ning, C. 2019 Effective water body extraction method for high resolution remote sensing images. *Comput. Eng. Appl.* **55**, 145–151 + 207.

Wu, X., Zheng, Y., Wu, B., Tian, Y., Han, F. & Zheng, C. 2016 Optimizing conjunctive use of surface water and groundwater for irrigation to address human-nature water conflicts: a surrogate modeling approach. *Agric. Water Manage.* **163**, 380–392.

Xi, X., Shen, N. & Li, X. 2009 Research of automatic extraction of water bodies on ETM+ images. *Comput. Eng. Des.* **30**, 993–996.

Xu, R., Zhang, Z. & Zhao, C. 2015 Different models used in extraction of lake water body based on MODIS date. *Remote Sens. Inf.* **30**, 111–118.

Yang, X., Zhao, S., Qin, X., Zhao, N. & Liang, L. 2017 Mapping of urban surface water bodies from sentinel-2 MSI imagery at 10 m resolution via NDWI-based image sharpening. *Remote Sens.* **9**, 596.

Yu, L., Wang, Z., Tian, S., Ye, F., Ding, J. & Kong, J. 2017 Convolutional neural networks for water body extraction from landsat imagery. *Int. J. Comput. Intell. Appl.* **16**, 1750001.

Yue, H., Li, Y., Qian, J. & Liu, Y. 2020 A new accuracy evaluation method for water body extraction. *Int. J. Remote Sens.* **41**, 1–32.

Zeng, L. F., Schimtt, M., Li, L. & Zhu, X. X. 2017 Analysing changes of the poyang lake water area using sentinel-1 synthetic. *Int. J. Remote Sens.* **38**, 7041–7069.

Zhang, M. 2008 Extracting water body information with improved model of spectral relationship in a higher montain area. *Geogr. Geo-Inf. Sci.* **24**, 14–16.

Zhang, D., Yang, S., Wang, Y. & Zheng, W. 2019 Refined water body information extraction of three gorges reservoir by using GF-1 satellite data. *Yangtze River* **50**, 233–239.

Zhao, Z. 2016 A new water index extracting water information automatically from landsat based on OSTU. *Geomatics Spat. Inf. Technol.* **39**, 57–60.

Zhao, X., Wang, P., Chen, C., Jiang, T., Yu, Z. & Guo, B. 2017 Waterbody information extraction from remote-sensing images after disasters based on spectral information and characteristic knowledge. *Int. J. Remote Sens.* **38** (5), 1404–1422.

Zhou, W., He, L., Ma, R., Liu, G. & Yuan, H. 2012 Extraction of water information based on LBV transformation of TM image using the object-oriented method. *J. Grad. Univ. Chin. Acad. Sci.* **29**, 775–779.

First received 12 May 2021; accepted in revised form 23 June 2021. Available online 29 July 2021

doi: 10.2166/nh.2022.111

Application of the artificial intelligence approach and remotely sensed imagery for soil moisture evaluation

Vahid Nourani [a,b]

[a] Center of Excellence in Hydroinformatics and Faculty of Civil Engineering, University of Tabriz, 29 Bahman Avenue, Tabriz 5166616421, Iran
[b] College of Engineering, Information Technology and Environment, Charles Darwin University, Ellengowan, Brinkin, NT 0810, Australia
E-mail: vnourani@yahoo.com, nourani@tabrizu.ac.ir

VN, 0000-0002-6931-7060

ABSTRACT

The current research attempts to present a modeling framework for determining soil moisture conditions by using remotely sensed imagery products. In this way, identifying various pixels with similar patterns from satellite images could be a reliable method to have an appropriate view over the soil moisture condition of a particular region. In this context, an artificial intelligence-based self-organizing map (SOM) method is employed to classify homogenous pixels over Phoenix, which is located in the south of Arizona, utilizing parameters extracted from satellite images. The central pixels of clusters are selected as the cluster indicator, with one from each cluster. Then, feed-forward neural networks (FFNNs) consisting of three layers of input, hidden, and output are trained by employing the extracted satellite images time series of the central pixels of the clusters. Finally, the soil moisture conditions of the representative pixels of the clusters are simulated by the trained models. The results reveal the suitability of SOM-based clustering to identify the specific points by which soil moisture can represent the soil moisture condition over the related regions. The proposed methodology and obtained results can be further used to provide a cost-effective method to determine the soil moisture condition of the region by reducing the costs of monitoring.

Key words: artificial intelligence, Phoenix, remote sensing, soil moisture, spatiotemporal modeling

HIGHLIGHTS

- An SOM is used to cluster homogenous pixels.
- The soil moisture conditions of the representative pixel for each cluster are simulated by using an ANN.
- The results reveal the suitability of the SOM clustering method to identify the specific points by which the soil moisture can represent the soil moisture condition.

GRAPHICAL ABSTRACT

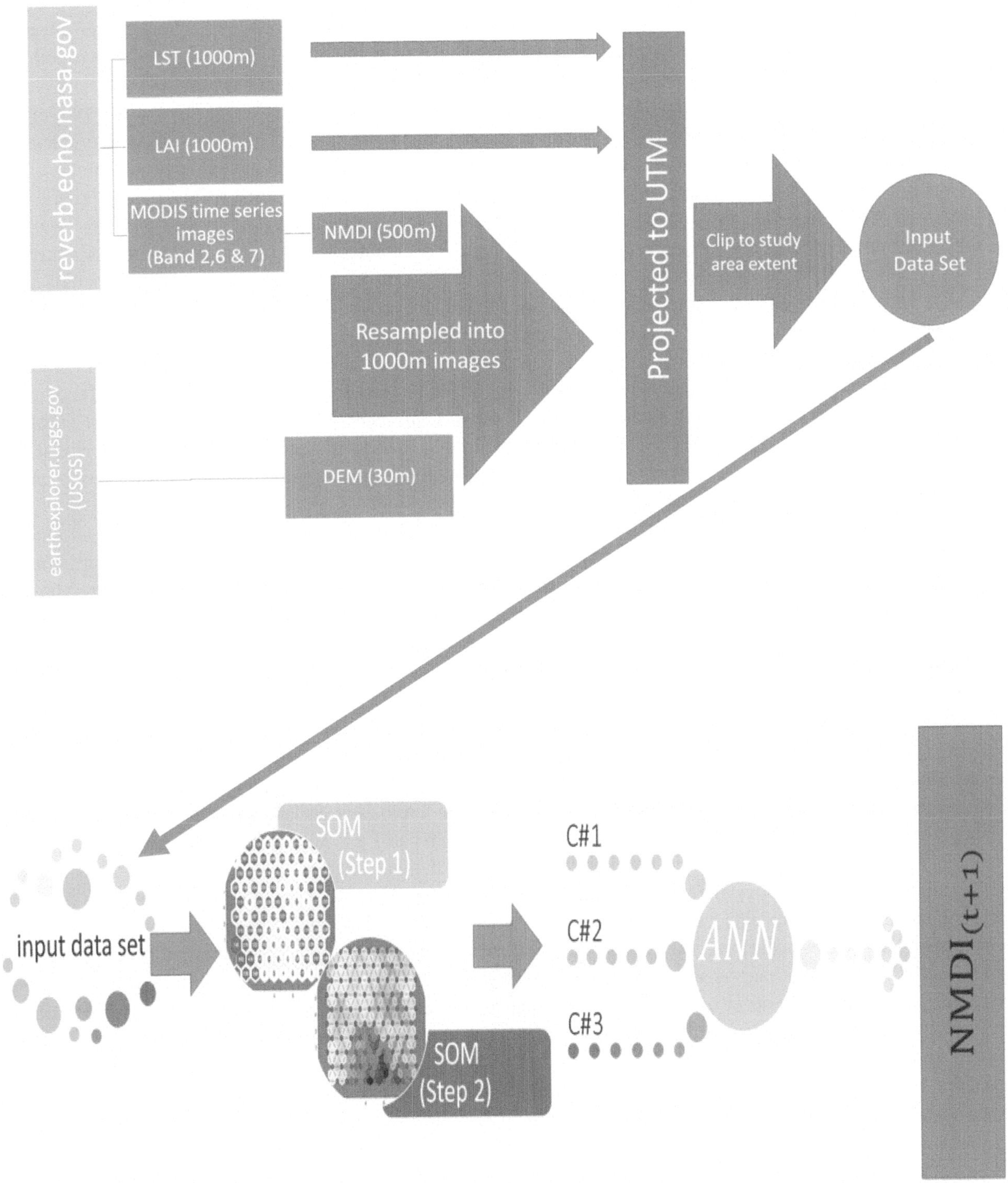

1. INTRODUCTION

Surface soil moisture (SM) is an important contributor to hydro-climatic processes, having critical roles to play in disaster prediction (e.g., flooding and drought), environmental monitoring (such as dust storms and erosion), and hydrological applications. Soil water content can affect the physio-biologic components of the biosphere and link the surface of the earth to the atmosphere via surface energy and moisture flux. SM is a source of water for the atmosphere through evapotranspiration from the land, which includes vegetable transpiration and bare soil evaporation. Moreover, SM conditions can affect the hydrological pattern of the land surface by controlling the infiltration capacity of the soil and portioning rainfall to runoff. The ecohydrology which focuses on the linkage between vegetation -water - climate relationships has been found to have a complicated dependency on the availability of the SM dynamics (Garcia-Estringana *et al.* 2013; Mulebeke *et al.* 2013). All these processes are highly characterized by the nonlinear behavior of SM and complex feedback mechanisms. Therefore, quantified conditions of SM are influential inputs for modeling agricultural, hydro-climatological, and meteorological attributes.

A set of components governs the dynamics of SM on the land surface with diverse time and spatial scales. Variations in both weather and climate are, therefore, influenced by the condition of SM. Reynolds (1970) classified the static (e.g., soil texture and topography) and dynamic (e.g., precipitation and vegetation) controlling elements of SM. An assessment of SM depends on the condition of the associated variables. Many of these elements are interrelated and range spatially and/or temporally, rendering the recognition of relationships among SM patterns and its riding variables complicated. Landscape elements, including topography, vegetation, and land use, are spatial and temporal governing elements of SM. SM's spatial variations are notably related to the features of the terrain (e.g., slope, elevation, and topographic wetness index). Therefore, the attributes of the terrain were utilized to estimate the parameters of SM patterns through regression, geospatial, and hydrological modeling in some previous research studies (e.g., see Western *et al.* 1999, 2004; Adab *et al.* 2020; Li *et al.* 2021). Additionally, the impacts of vegetation cover (e.g., type and distribution) on SM variation have been noted in various research studies. Also, the effects of spatial attributes on vegetation (normally interpreted from remote sensing images) have been utilized to generate SM patterns (Mohanty *et al.* 2000; Hupet & Vanclooster 2002).

Generally, long-term time series of SM can spatially detect changes in the water cycle associated with the weather or hydrological condition. In a large study area, the wide variety of networks and gauging SM remain restricted and furthermore, obtaining reliable approximations from field measurements is a challenging task due to excessive variability and the lack of correlations among the parameters. In several applications of SM, a wide variety of satellite-based products have the potential to help hydrologists to measure the SM condition in large regions. Since the 1970s, some remote-sensing strategies have been developed for analyzing and mapping SM by measuring specific regions of the electromagnetic spectrum ranging from optical to microwave areas (Musick & Pelletier 1988; Engman 1991; Wang & Qu 2009). As remote sensors cannot measure SM content directly, mathematical-based methods that can explain the relationship of measured signals and SM content need to be extracted. Microwave remote sensing strategies that include the Advanced Microwave Scanning Radiometer-Earth staring at system (AMSR-E) on-board the Aqua satellite for pc (from 2002), Soil moisture and ocean salinity satellite (SMOS from 2009), Multi-frequency Scanning Microwave Radiometer (MSMR from 1999), and Soil Moisture Active Passive (SMAP) (from January 2015) are currently operational, producing daily satellite records globally. Although these strategies provide many techniques to gauge SM on a large scale, they have almost low resolution (typically ~25 km) and are no longer suitable for small catchments or discipline scales. Optical/thermal infrared remote sensing records are known as the Surface Temperature/Vegetation Index Method and offer finer resolution (~1 km). Recently, Zhang & Zhou (2016) presented a new approach by which SM estimation could be made from optical/thermal remote sensing, and this method especially relies upon the affiliation among SM and the surface reflectance and the temperature or vegetation index. Retrieval strategies in this area, like thermal inertia, emphasize soil thermal traits or triangular dating techniques, indicating that the linkages among SM, the Normalized Difference Vegetation Index (NDVI), and the Land Surface Temperature (LST) of a given area are being utilized in different applications. However, their applications are limited due to the lack of sufficient spatial data including topography or low-density vegetation cover maps and data. The remotely sensed vegetation indices to estimate SM (e.g., NDVI, the Normalized Difference Water Index (NDWI), and the Normalized Multi-Band Drought Index (NMDI)) are proper alternatives; however, the distribution of SM cannot be anticipated by a single parameter and by figuring out parameter modifications among specific land-surface-aspect intensities. Extensive efforts have been made to estimate SM by satellite images by establishing a connection between remotely sensed LST and vegetation indices (e.g., Dari *et al.* 2021; Zhu *et al.* 2021). One of the practical advantages of remotely sensed imagery is that vegetation and LST parameters are available with a high spatial resolution (from 30 m to 1 km) via imagery, in addition to topography data. Predicting the SM condition by utilizing the structured landscape factors extracted from remotely sensed imagery instead of *in situ* measurement makes it viable to obtain quick and real-time tracking of the SM condition.

As stated above, because of the complicated nature of the SM distribution process, a wide variety of ambiguous land covers (especially vegetation cover) and the effect of different functions, which include topography and terrain, make SM studies more complicated. The growing availability of remotely sensed imagery, linked to advances in modern computers, effectively contributes to low-priced broad-scale monitoring of hydro-environmental processes. In this regard, black box (lumped) methods may be superior to physical-based methods for a spatial-temporal estimation of SM. Unlike the data-driven methods, physical-based models require sequential weather elements and parameters of soil and vegetation types. Therefore, physical-based models are not so flexible for real-world practices because of the requirements of spatial parameters and high-quality field data. Neural networks is a kind of data-driven approach that describes nonlinear and complicated relationships to map inputs into a desirable output based on a sequence of complicated mathematical functions without any mechanistic information of the system. Several research studies have already described the application of neural network–based models to estimate SM over watersheds (Jiang & Cotton 2004; Pandey *et al.* 2010; Kornelsen & Coulibaly 2014).

Moreover, to handle massive data and records provided by remotely sensed satellites (for every pixel, there are numerous numbers or indices) that are to be incorporated into the model that uses remotely sensed land surface parameter data (e.g., SM), some data preprocessing algorithms are required. Clustering is one such statistical preprocessing technique of classifying data to homogeneous groups to make an operational assessment of the records through a regionalization of the study area. A cluster analysis can be employed for classifying multivariate data into subgroups to enhance the overall performance of the modeling. Clustering strategies recognize the shape of unlabeled data sets by dividing the members with minimum within-group-item dissimilarity and maximum between-group-item dissimilarity into homogeneous groups. There are several methods to cluster the data. A self-organizing map (SOM) is a neural-based model permitting unsupervised classification based on similarity among separable data sets. Therefore, the SOM approach for clustering an evaluation of the complicated spatiotemporal data sets and dominant patterns has gained the attention of hydrologists. For instance, Hsu & Li (2010) employed the SOM to recognize the most homogenous patterns of precipitation stations at the high-dimensional wavelet-transformed space. For an evaluation of the drought frequency, Chen *et al.* (2011) used the SOM-based clustering method by employing the standardized precipitation index (SPI) and the geographic characters of the gauges to group the rain gauges into different clusters. For groundwater data clustering, Nourani *et al.* (2015) used the SOM to recognize the patterns of groundwater level to be fed into a feed-forward neural network (FFNN) for one- and multi-step-ahead groundwater level forecasting. The evaluation, simulation, and analysis of the SOM in order to solve hydrologic problems have been reviewed by Kalteh *et al.* (2008) for the fields of meteorology and oceanography, respectively. In the field of ecohydrology, Olden *et al.* (2012) used the SOM to regionalize the ecohydrological function and provided an explanation, because the final result given by the SOM ought to replicate the ecohydrological traits of regionalization. Due to the aforementioned advantages, the SOM as a robust method to recognize the spatial patterns of remotely sensed SM conditions, was used in the present study.

On the other hand, for the subsequent section of ANN-based modeling, the FFNN model with specific properties for nonlinear modeling was used in this study to predict the SM condition in the future. The capacity to recognize existing patterns from a given data set makes the FFNN to ferret out complicated relationships among the inputs and outputs. The conjunction of the FFNN with the mathematical tool of the clustering technique can be applied for the prediction of physical processes.

The main objective of the current study was to present a method to estimate the SM condition over the land surface. To this end, the time series of Moderate Resolution Imaging Spectroradiometer (MODIS) soil and vegetation indices and the digital elevation model (DEM) were used to obtain information about the SM condition over the study area. The method employs remotely sensed imagery to assist SM mapping, particularly for in situ data-poor areas. In order to fulfill this aim, two approaches in the field of artificial intelligence (AI) were applied. First, the SOM was used as a clustering technique to recognize homogeneous and comparable areas and essential pixels, after which for every cluster, a three-layer FFNN model for each centroid pixel of each cluster was designed as the method to discern the underlying nonlinear pattern within the data, spatially and temporally. Based on the literature survey and to the best of the author's knowledge, this is the first application of SOM-based clustering analysis to improve AI-based modeling of the SM to enjoy the advantages of both methods within a robust unique framework.

In the subsequent sections of the paper, remotely sensed data sources and their processing and the study area are presented. Thereafter, the proposed and used methods (i.e., SOM and FFNN) are reviewed and the technique for model development and evaluation criteria are explained. The outcomes acquired by the proposed technique are then presented and discussed, followed by the conclusions.

2. MATERIALS AND METHODS

2.1. The study area and data analysis

Located in the southwestern region of the USA (situated between latitudes of 32 °30′ and 34 °01′N and longitudes of 110 °18′ to 113 °20′W), in the south-central part of Arizona, Phoenix is a semi-arid desert region (Figure 1). It is situated at a mean elevation of 331 m, in the northeastern part of the Sonoran Desert. The topography of Phoenix is largely flat, except for highlands located around the city, including scattered and low mountains surrounding the valley: in the northeast, McDowell Mountain; in the west, White Tank Mountain; in the south/southwest, South Mountain and Sierra Estrella; and in the west, Camelback Mountain. North Mountain, Sunnyslope Mountain, and Piestewa Peak are located at the midpoint of the valley. The borders of Phoenix consist of large fields of irrigated cropland and Native American reservation lands. The Salt River flows westward through the city, but owing to massive diversions for irrigation purposes, the riverbed often goes dry.

Typical of the Sonoran Desert, the Phoenix climate (Koppen climate classification BWh) consists of subtropical desert. In the Phoenix metropolitan area and the surrounding agricultural land (in the center of Phoenix), the mean annual precipitation and evapotranspiration potentials are 180 and 2,286 mm, respectively. The summers of Phoenix are long and extremely hot and the winters are short and mild to warm. In USA, the average high temperatures in summer are the hottest among any main city. From late May to early October, there are 107 days in a year, on average, and on most of the days, the temperature is at least 38 °C. But even in peak summer when there is extreme heat, the highest average annual temperature is not reached.

In Sonoran, a hot desert, the average annual precipitation is 76–500 mm, which is dependent on the location, with considerable inter- and intra-annual changeability in timing and quantity. The air temperature in summer exceeds 40 °C and frequently reaches 48 °C. Severe thunderstorms of the summer monsoons occur due to the interaction of this high near-surface temperature with cool, moist air in the atmosphere. Subsequently, moisture on the soil surface and near-surface air evaporates and temperatures may drop by 10 °C or more, and this drop occurs following storms within a matter of minutes. In winter, the temperature is mild, with valley bottoms typically without frost, while the surrounding mountains may have compressed snow cover at high elevations and in the northern and eastern parts. During any season, daytime temperature swings of 15 °C or more are common, because the dry atmosphere and relatively low vegetation cover simplify the process of re-radiation of daylight heat into the atmosphere overnight.

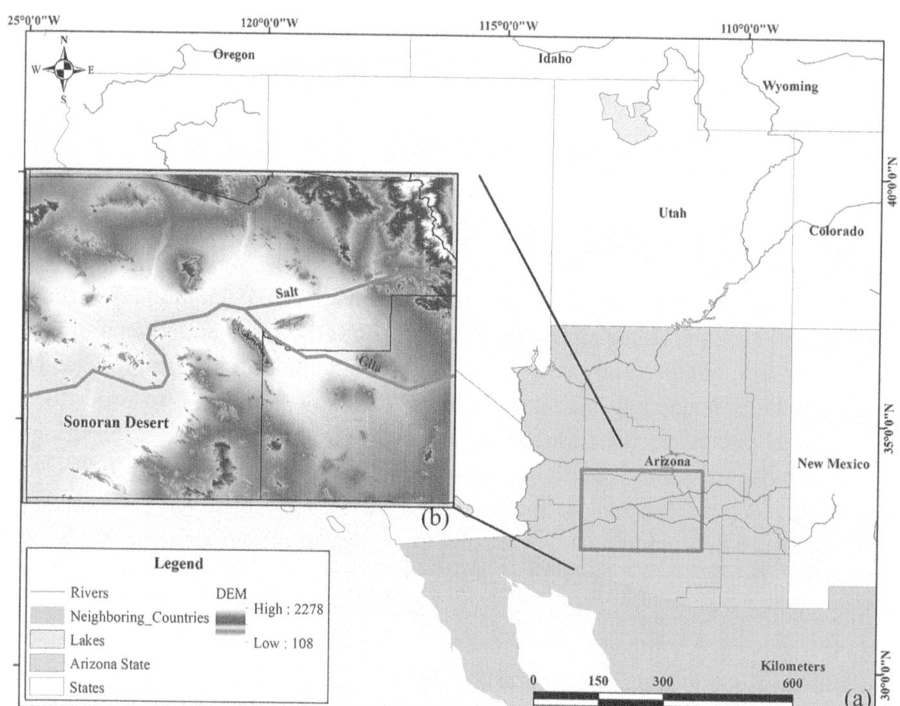

Figure 1 | Study area: (a) location map and (b) the DEM.

In hydro-environmental studies, the MODIS sensor is used, and it can be beneficial at global and regional scales (>250 m). This sensor has several cutting-edge features of high temporal resolution, wide spectral range, real time, and low cost (Zhang *et al.* 2014). The temperature of the land surface and vegetation indices are obtained from the Terra/Aqua satellite in the observation cycle comprising a comparatively short period (https://reverb.echo.nasa.gov/reverb/). The LST is one of the parameters for monitoring the physical and biological procedures occurring on the land surface. In areas with bare soil, the LST indicates the soil surface temperature, and in areas with dense vegetation, the LST shows the vegetation covering temperature under the assumption of energy balance. Consequently, the LST can be used to display the SM form in bare soil statues. To quantify the disparity in vegetation during the study period from January 2001 to December 2004, the leaf area index (LAI) was used. The LAI is defined as the one-sided green leaf zone per unit of the ground surface zone. For the LST product (MODIS/TERRA Land Surface Temperature/Emissivity 13 global v005) and the LAI product (the level-4 MODIS global Leaf Index and Fraction of Photosynthetically Active Radiation (FPAR)), the 8-day merged images were shaped by mending together images from a period of ± 4 days, so that the number of clouds could be reduced to the minimum in the sinusoidal map projection system covering an area of approximately 1,000 m$\times$1,000 m. Similarly, the data input to the satellite related to the derivation of the NMDI is the atmospherically modified MODIS surface reflectance product (MOD09A1) in the years 2001–2004, which is an 8-day composite of products related to the surface reflectance, used for the calculation of the NMDI. A gridded level-3 product made of 8-day composites ends at bands 1–7 at 500 m resolution in the sinusoidal projection. In space distant sensing of soil and vegetation moisture capacity, the NMDI is calculated on the basis of three bands, as

$$\mathrm{NMDI} = \frac{R_{860\,\mathrm{nm}} - (R_{1640\,\mathrm{nm}} - R_{2130\,\mathrm{nm}})}{R_{860\,\mathrm{nm}} + (R_{1640\,\mathrm{nm}} - R_{2130\,\mathrm{nm}})} \tag{1}$$

R stands for the reflectivity of each band. In MODIS sensors, 860 nm represents the Channel 2 band, 1640 nm represents the Channel 6 band, and 2130 nm represents the Channel 7 band (Wang & Qu 2007). This is an index exhibiting good sensitivity to drought severity, and, therefore, Zhang *et al.* (2009) used this index for Henan drought supervision and ascertain whether the NMDI can be used in areas with modest vegetation coverage. Mobasheri & Bidkhan (2013) studied the relationship between two SM indices: NMDI and PDI, and SM content (%) on the basis of *in situ* spectral measurements. They showed that the mentioned indices are well suited for monitoring SM, particularly for bare soil or thinly vegetated areas.

A robust absorption of near-infrared (860 nm) light in the vegetation canopy is used in this index, as well as the sensitivity of short-wave infrared 1,640 and 2,130 nm bands for the absorption of variance between soil and vegetation water contents. The SM condition includes three classes (on the basis of volumetric SM range): dry: 0–0.1; intermediate: 0.1–0.2 and wet: >0.2 (Miller *et al.* 2004); consequently, a pixel will be mapped as a dry soil condition, related to bare soil or areas with weak vegetation, if the NMDI is <0.7, intermediate condition if the NMDI is in the range of 0.6–0.7, and wet if the NMDI is <0.6. The DEM extracted from ASTER, a sensor mounted on Terra satellite, as part of NASA's EOS (Earth Observing System), was used in the current study with a 30 -m resolution retrieved via https://earthexplorer.usgs.gov/.

The raw data collected from MODIS images are in the form of sinusoidal projection and could be transferred to the UTM coordinate system via the MODIS Conversion Toolkit (MCTK). The images from the MODIS surface reflectance and DEM have 500- and 30-m spatial resolutions, respectively; hence, they were resampled to match the spatial degree of the MODIS LAI and LST products with a 1- km spatial resolution. Using vector data related to the study region boundaries, a water mask was applied, and finally the clouds were removed from the images. Each of the satellite images was cropped to the geographic extent of the study area. For each of MODIS product (e.g., LST, LAI, and NMDI) in order to create 8-day time series matrix for the study period (2001–2004), a three-dimensional matrix 184$\times$(168 $\times$ 193) was formed, including 184 records (with 8-day interval) for 168 $\times$ 193 spatial cells. After fully removing unvalued data, the size of the input matrix for SOM classification included 553 (184 LST images+184 LAI image+184 NMDI image+1 DEM) rows and 21,022 (168$\times$193 unvalued data) columns.

The 8-day MODIS products (LST, LAI, and NMDI) of the Phoenix for 2001–2004 were used for FFNN modeling of the centroid pixels of clusters: the first 3 years for the training phase and the rest for the verification phase. Extreme values of the inputs were separated for the training phase to make the learning process more reliable.

2.2. AI-based methods

It is necessary to combine appropriate information about SM conditions in different pixels of the region and recognize the forceful pixels to predict SM in future. Therefore, the SOM as an ANN-based clustering approach, a common machine learning method for pattern recognition, is able to discover the structure of relationships among groups of unlabeled data without considering the physically inherent natural phenomenon. One simple description of the SOM in this study is that it is able to separate the images into meaningful regions with homogeneous behavior. The proposed model consists of two phases. After preprocessing satellite data, in phase 1, a two-step SOM was employed for classifying pixels into classes with close patterns. Generally, such a two-step SOM clustering is proposed to gain an understanding of homogeneous regions and estimate the number of clusters with regard to the topology of the plain. Then, for determining the predominant (center) pixel of each cluster, the Euclidean distance criterion was used. The predominant pixel was determined in order to extract the best sample to represent the pattern of the whole cluster.

2.2.1. Self-organizing map

The SOM is one of the high-performance tools for analyzing data with high dimensions. It orderly maps a distribution with high dimensions to a grid with regular low dimensions. Therefore, it can convert statistically complex and nonlinear relationships of data with high dimensions to simple geometric relationships with low dimensions. Moreover, the SOM is proven to effectively preserve the topological structure of the data during the conversion process (Kohonen 1990). The application of the SOM network in image clustering is to place the pixels with similar attributes in the same cluster by reducing the dimension and creating a map of commonly one or two dimensions. Therefore, the SOM achieves two goals of distinguishing similarities and dissimilarities and reducing the dimension of the data. Commonly, SOM architectures are double-layered, consisting of the input layer and the Kohonen layer. The described two-level SOM neural network (see Figure 2) is shown to be an efficient method to obtain an initial understanding of the complicated data sets. The classical SOM network can be extended by another one-dimensional Kohonen layer, where neurons are interconnected to the neurons of the previous Kohonen layer.

Steps of iterative learning of an SOM network are described as follows:

- The weights of each node are randomly assigned for initializing the network.
- A vector from the input vector (x) is selected randomly.
- All nodes are inspected to find the weights most similar to the input vector. The Euclidean distance is used as distance measurement (similarity and dissimilarity).
- The Best Matching Unit (BMU) is the winning node with the closest weights to the related input. The BMU and the nodes in the neighborhood are calculated and permitted to learn by changing weights through iterations. The iterations are processed to obtain the closest distance between the input and the weights.
- The training process continues until obtaining the convergence results.
- Finally, the clusters (homogeneous region) will be recognized on the basis of the structure of the SOM.

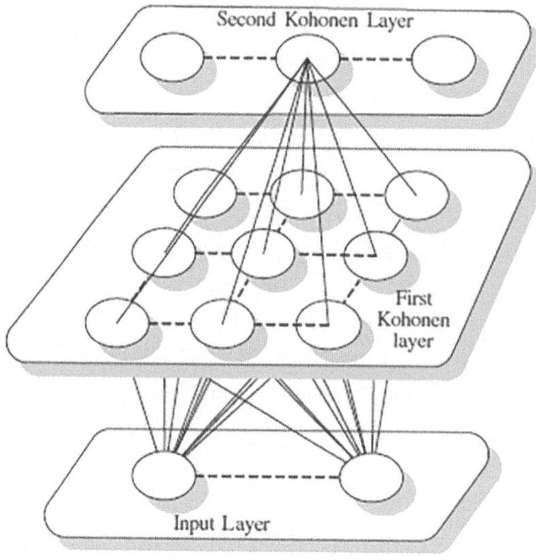

Figure 2 | Schematic of the proposed two-level SOM model.

2.2.2. Feed-forward neural network

An ANN is an extensively employed tool in studies for modeling environmental phenomena. The back-propagated (BP) feed-forward neural network is a common ANN approach. It has been shown that in the majority of issues related to engineering, BP network models have performed satisfactorily (ASCE 2000). A three-layered FFNN consists of a linear combination of the input vector, going through a nonlinear activation function. The hyperbolic Tangent-sigmoid (Tansig) was chosen as the activation function in this study. As shown in Figure 3, the three-layered FFNN can present an applicable network for understanding the nonlinear relationships between input and output data sets and a flexible method for predicting environmental parameters. The output value of a three-layered FFNN is given by (Kim & Valdés 2003):

$$\hat{y_k} = f_o\left[\sum_{j=1}^{A} w_{kj}.f_h\left(\sum_{i=1}^{B} w_{ji}x_i + w_{jo}\right) + w_{ko}\right] \tag{2}$$

is the connection between the ith neuron in the input layer and the jth neuron in the hidden layer denoted by a weight w_{ji}, the bias of the jth hidden neuron is w_{jo}, f_h is the nonlinear activation function (e.g., sigmoid or tangent-sigmoid) of the hidden neuron, w_{kj} is a weight in the output layer connecting the jth neuron in the hidden layer and the kth neuron in the output layer, w_{ko} is the bias for the kth output neuron, f_o is the activation function for the output neuron, x_i is the ith input variable for the input layer, and y_k and y are the calculated and observed output variables, respectively. A and B represent the number of neurons of the input and hidden layers, respectively. The weights that can be altered and adapted in the training phase are not the same in the hidden and output layers.

The Levenberg–Marquardt was selected as an optimization algorithm for a three-layer FF–BP network model in this study. It is worth mentioning that the efficiency of the developed FFNN-based models is associated with the appropriate selection of the neuron number in the input and hidden layers and the number of training iterations (epoch). Additionally, the model may display overfitting in the training phase due to a high epoch number and poor quality or quantity of data, thus limiting the model's performance by limiting the ability to predict new data outside the training set.

2.3. Modeling performance criteria

All the data sets were first normalized before FFNN-based SM modeling to reduce the dimension of the input and output variables. Then, the training and verification sets were selected from normalized data. In this study, the root mean square error (RMSE) and the determination coefficient (DC), also known as Nash Sutcliffe efficiency, could be employed to evaluate

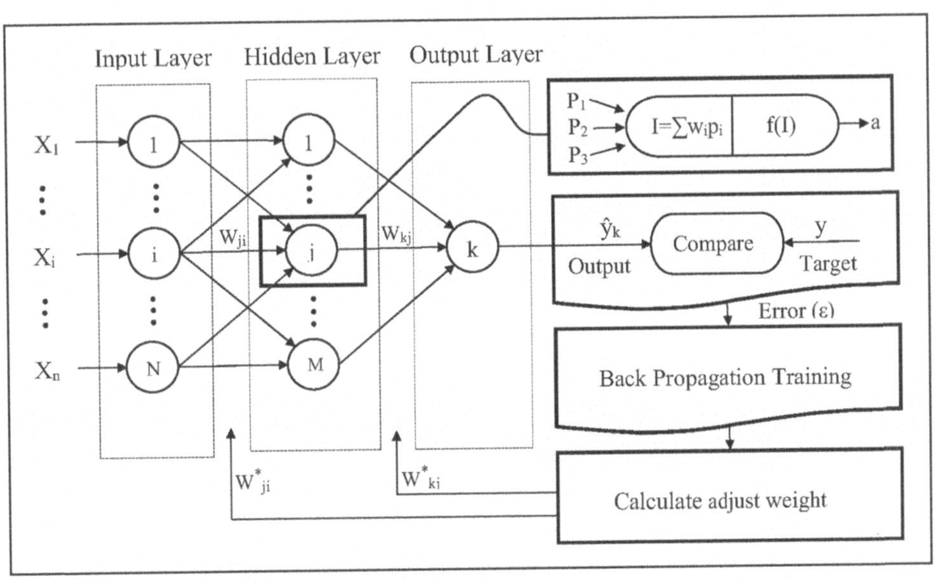

Figure 3 | Three-layered feed-forward neural network with the BP-training algorithm.

the performance of the proposed methodology (Nourani & Kalantari 2010). The RMSE and DC reveal inconsistencies between predictions and observations as

$$\mathrm{RMSE} = \sqrt{\frac{\sum_{i=1}^{M}(Z_i - \hat{Z}_i)^2}{M}} \tag{3}$$

$$\mathrm{DC} = 1 - \frac{\sum_{i=1}^{M}(Z_i - \hat{Z}_i)^2}{\sum_{i=1}^{M}(Z_i - \underline{Z})^2} \tag{4}$$

In Equations (3) and (4), M denotes the number of data, Z_i is the observed target value of the data and $\hat{Z}_i$ is the calculated value for intercepting the observed target, and $\underline{Z}$ is the average value of the observed data. In a high-performance model, the DC is closer to one, and the RMSE is closer to zero.

The silhouette coefficient was used to examine the competence of spatial clustering by the SOM. The silhouette coefficient is calculated for each cluster representing the likeness degree of pixels for the cluster as (Nourani *et al.* 2015):

$$S(a) = \frac{b(x) - a(x)}{\max[a(x), b(x)]} \tag{5}$$

where $S(a)$ shows the silhouette coefficient of pixel x; $a(x)$ is the Euclidean distance showing the average difference of pixel x with other pixels of cluster A; $b(x)$ is the minimum average dissimilarity of pixel x to all other pixels in a cluster other than the considered cluster. A small silhouette coefficient denotes a better similarity among the pixels in the same cluster. The comprehensive efficiency of the SOM approach can be assessed by calculating the mean of silhouette widths for the whole data set.

3. RESULTS AND DISCUSSION

The satellite-based data (as introduced in Table 1) were used for SOM-based clustering of the area. A total of 553 normalized preprocessed satellite images in the form of (193 × 168) matrixes with a pixel resolution of 1,000 m were considered as inputs of the SOM in order to classify the region of the case study. Studies conducted on the NMDI illustrate that it does not respond to moisture variation of the soil, starting from LAI equal to 2, which means no soil background impact is found on the NMDI for any SM domain (Wang & Qu 2007). Therefore, for a vegetation canopy with an LAI equal or greater than 2, which means heavily vegetated areas, the NMDI becomes a complete index for predicting vegetation water content, compared with the SM index. In order to achieve the overall view of the situation of SM over the study region, pixels with LAI values higher than 2 were removed from the data.

Since selecting the appropriate number of neurons is based on a topology of inputs in the first step to form the two-step SOM, the size of the Kohonen layer could be determined as a 10-by-10 lattice by a trial and error process. Figure 4(a) illustrates the hits plan of the output layer for the applied data. The SOM output layer is explained by the hits plan in which every single neuron represents the number of arranged input vectors. Also, the size of the colored patch reveals the proportionate number of vectors. Figure 4(b) presents the neighbor weight distances in which the darker colors represent large distances and the lighter colors exhibit small distances. After forming a 10-by-10 Kohonen layer, the associated zones of the clusters

Table 1 | Characteristics of the used satellite-based data

Product	Data type	Spatial resolution (m)	Temporal resolution (day) from January 2001 to December 2004	Data source
LAI	Raster	1,000	8	https://reverb.echo.nasa.gov/reverb
LST	Raster	1,000	8	https://reverb.echo.nasa.gov/reverb/
NMDI	Raster	500	8	https://reverb.echo.nasa.gov/reverb
DEM	Raster	30	–	http://earthexplorer.usgs.gov

on the map could be recognized on the basis of the number of zones allotted to the neurons (see Figure 4(a)). A one-dimensional SOM with neurons arranged within a 1-by-5 Kohonen layer was then organized in the second step of the clustering to group the pixels into five distinct classes, in which the number 5 for the final classes of numbers would be determined in the first step of clustering (two-dimensional map), visually.

Since SM is known as a dynamic component of the underlying process happening on the surface of the earth, it is important to consider the effects of dynamic temporal variability in the classification parameters. In order to investigate the impact of time series as the input of SOM classification parameters for the study area, the time series and the average of time series for 4 years were imposed to the SOM separately. The silhouette coefficient as a cluster validity measure was utilized to assess the homogeneity of the clusters obtained by the proposed two-step SOM.

Figure 5 illustrates the performance of each SOM clustering for two types of selected input data. Horizontal and vertical axes show the number of clusters and silhouette value, respectively. Although an averaging of time series produces higher silhouette values for all numbers of clusters, and the average of the data reduces the complexity and the costs of calculation, averaging can conceal the temporal fluctuation within the data. Therefore, the whole time series imposed to the SOM was preferred for the classification. Since the maximum silhouette value was obtained when using clusters 4 and 5, these are the eligible cluster numbers and a strong-enough reason to choose 5 as the best cluster number in this study.

It is clear (as seen in Figure 5) that by increasing the number of clusters, the Silhouette coefficient decreases, which denotes a higher homogeneity of the clusters and efficiency of the clustering. On the other hand, more clusters mean more FFNN models are needed to model the SM over the region. It is seen in Figure 5 that there is a significant decrease in the Silhouette coefficient till the cluster number 5, and thereafter, by increasing the cluster number, only a lower homogeneity can be seen in the clusters. Therefore, and to have a tradeoff between the modeling cost and accuracy, the clustering size of 5 was considered in this study to classify the region.

The mentioned SOM clustering technique applied to the set of remotely sensed data clustered the region into five clusters. Figure 6 shows a spatial distribution of the clustered pixels and the SM condition in the study region with regard to the input data used for the clustering (LST, DEM, and LAI). Considering the DEM map of the area (Figure 6(b)), the diversity of the clusters in the higher elevations located in the northwest part of the map indicates that the rate of changes in parameter comparison with other areas is remarkable. Also, it is clear by considering Figures 6(a)–(d) that the spatial distribution of clusters is correlated more to the mean LAI and LST data than to the DEM, where such data include temporal variations and can handle the seasonality of the process, but the DEM does not have temporal variations. To explore the capability of the SOM, the main statistical features of each attribute are described and discussed in the following paragraphs to characterize the spatiotemporal variability of the process.

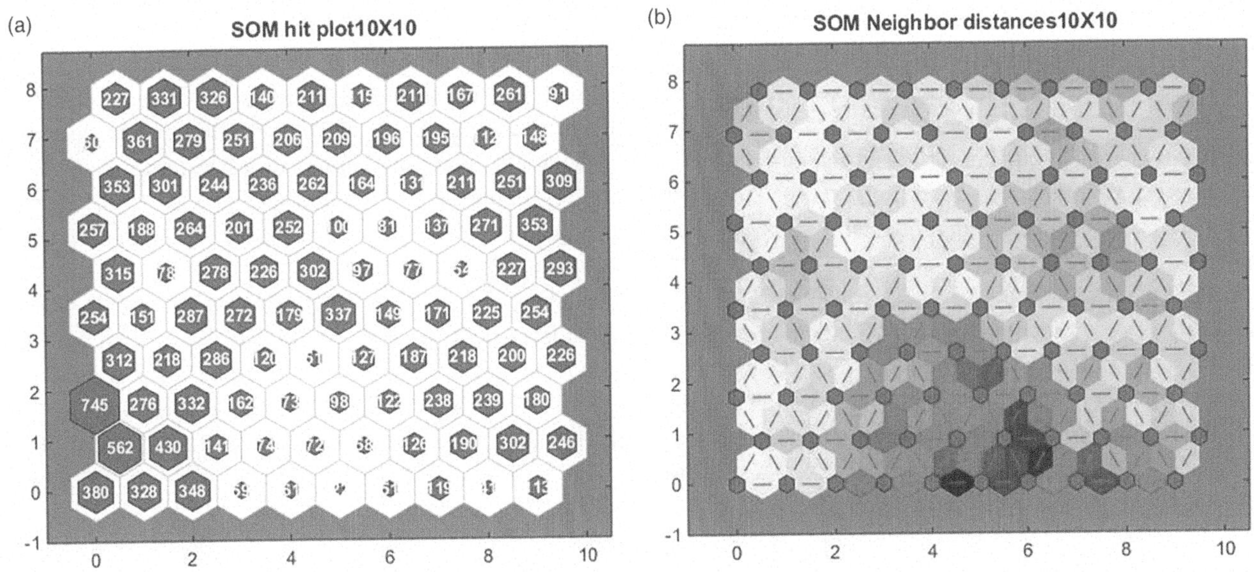

Figure 4 | Two-dimensional SOM clustering of satellite data: (a) SOM hits and (b) SOM neighbor weight distances plan. Please refer to the online version of this paper to see this figure in colour: http://dx.doi.org/10.2166/nh.2022.111.

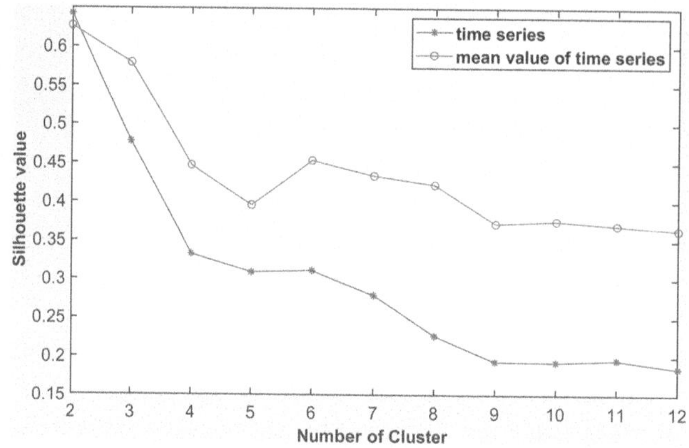

Figure 5 | Silhouette coefficient comparison for time series and mean value of time series as two input alternatives for the SOM.

Figure 7 shows the distribution of elevation and average time series of the LST for each cluster sorted on the basis of increasing the mean value of each cluster. As expected, the elevation and LST have a negative relationship, which means clustering with high elevation typically produces lower average temperatures. Clusters 1 and 2 have a wide variation range of both parameters among other clusters. The downstream areas with low elevation were allocated to clusters 4 and 5, and gradually with increasing the elevation, the pixels located in other clusters, which illustrates the critical role of elevation in the clustering outcome (see Figure 7(a)).

LST histogram presented by Figure 8 shows the high non-homogeneity of LST values in the clusters. The histograms of the NMDI for the clustered areas demonstrate a regular distribution of this variable for each cluster (Figure 9). Most of the percentages of the NMDI values of the pixels of clusters 1, 2, and 3 are lower than 0.6 (based on the NMDI description in the data set section), so these clustered areas are mapped having a wet soil condition, and cluster 4 has a normal soil condition; cluster 5, which tends to the right of the diagram, denotes the dominant nature of the dry soil condition.

According to the range of LAI variation in Figure 10, clusters 1 and 2 are included in a wide range of LAI values, so this indicates that the non-homogeneity of these clusters is remarkable. But the remaining clusters have a low range of LAI values.

As is illustrated in Figure 7, clusters 4 and 5 are located in low elevation lands and high-temperature areas, which cover the Sonoran Desert. During dry conditions in these clusters, a low SM (see Figure 8(d) and (e)) can also strongly limit evapotranspiration, and, in turn, can reduce the potential variability caused by vegetation covers (see Figure 10(d) and (e)).

The downhill of the northeastern mountains is allocated to cluster 3. The high evapotranspiration rates in summer, as well as poor vegetation cover (Figure 10(c)), reduce the SM in this area. But the impact of topography on water redistribution, which is caused by rainfall at high altitudes, will increase the SM during rainy days and improve vegetation cover in wet conditions.

Clusters 1 and 2, which include the mountainous region and agricultural land surrounding the Phoenix metropolitan area, illustrates the wet soil condition. During the growth season in these areas, decreasing temperatures and maximum rainfall could improve the vegetation cover, and consequently, the evapotranspiration rate of the SM declines.

In the last step of the modeling (prediction step), an FFNN was trained for each of the representative pixels of each cluster to predict the NMDI one time-step ahead. The LAI and LST data sets of the present step and NMDI values of the previous step were used as the inputs for the training phase. These input orderlies, especially the NMDI lags, were selected by sensitivity analysis to consider the seasonality and autoregressive characteristics of the process. The results showed that the model efficiency was not strongly dependent on cases where larger lags were considered. Therefore, to obtain satisfying results, the minimum numbers of the input data were used to reduce the error of imposing redundant information and noise. The verification data were used to test the efficiency of the trained models. The obtained results are presented in Table 2. According to these results, it is clear that by imposing the NMDI value at the previous time step (NMDI($t-1$))) as input, the performance of the model improves, which indicates the autoregressive property of the process.

When each of the LAI or LST parameters was individually imposed (as the sole input) to the FFNNs, the modeling performance (in terms of verification DC) dipped, and the DC was less than 0.4 in.all clusters, which indicates poor modeling by these input sets. But when both LAI and LST were used as inputs of FFNNs simultaneously, the verification DC was

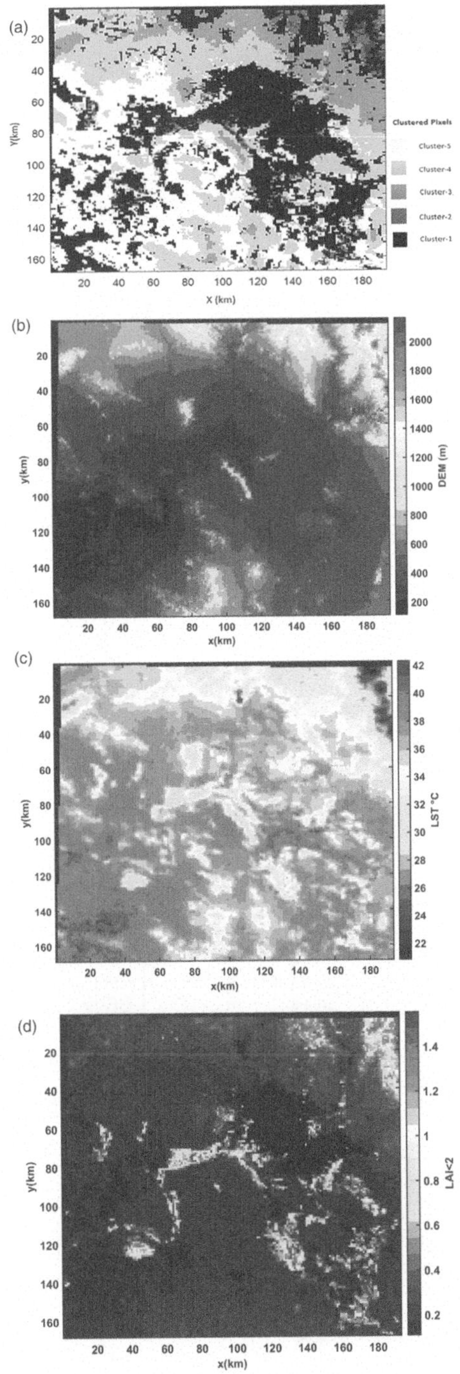

Figure 6 | Spatial distribution map of the study area for (a) clustered pixels, (b) DEM, (c) mean LST, and (d) mean LAI.

at least 0.6. Adding the NMDI (t − 1) to the input layer resulted in a DC verification of at least 0.76, which shows the autoregressive property of the process.

It should be noted that the temporal modeling size of the study was 8 days, but when the modeling lag (horizon) was increased to have two- or three-time-step-ahead modeling (as a multi-step-ahead prediction scheme), the verification DC dropped below 0.5 in most cases, and because of this, the results of the multi-step-ahead modeling are not provided in the table. The reason for this poor multi-step-ahead modeling performance is that, after 8 days, climatic and soil conditions

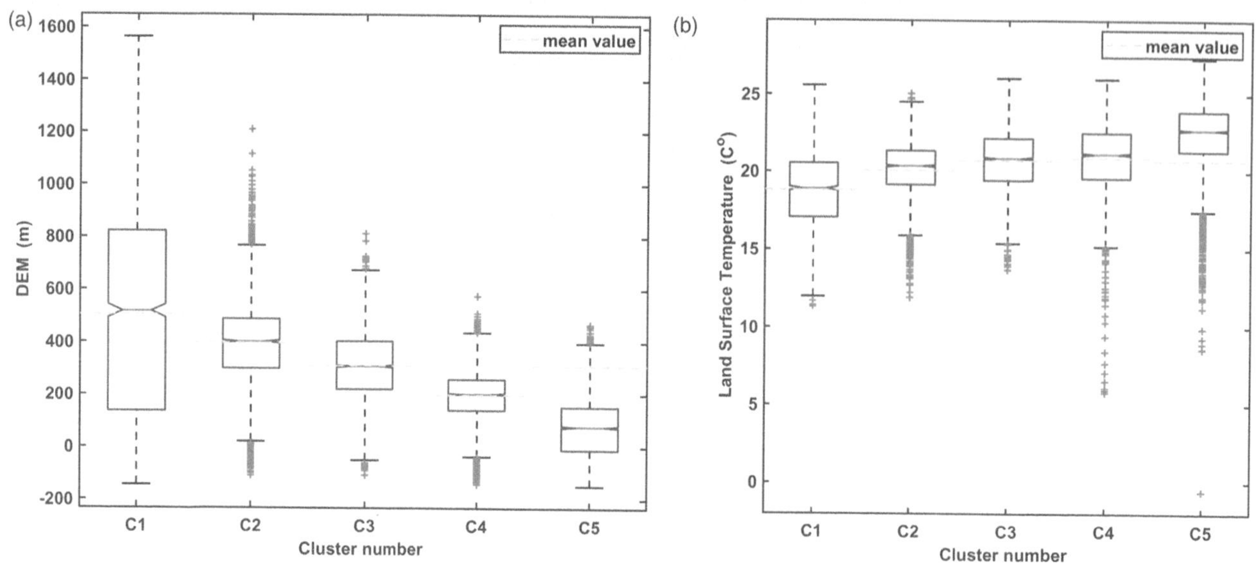

Figure 7 | Box plot of the spatial distribution of (a) DEM and (b) average time series of LST values for each cluster.

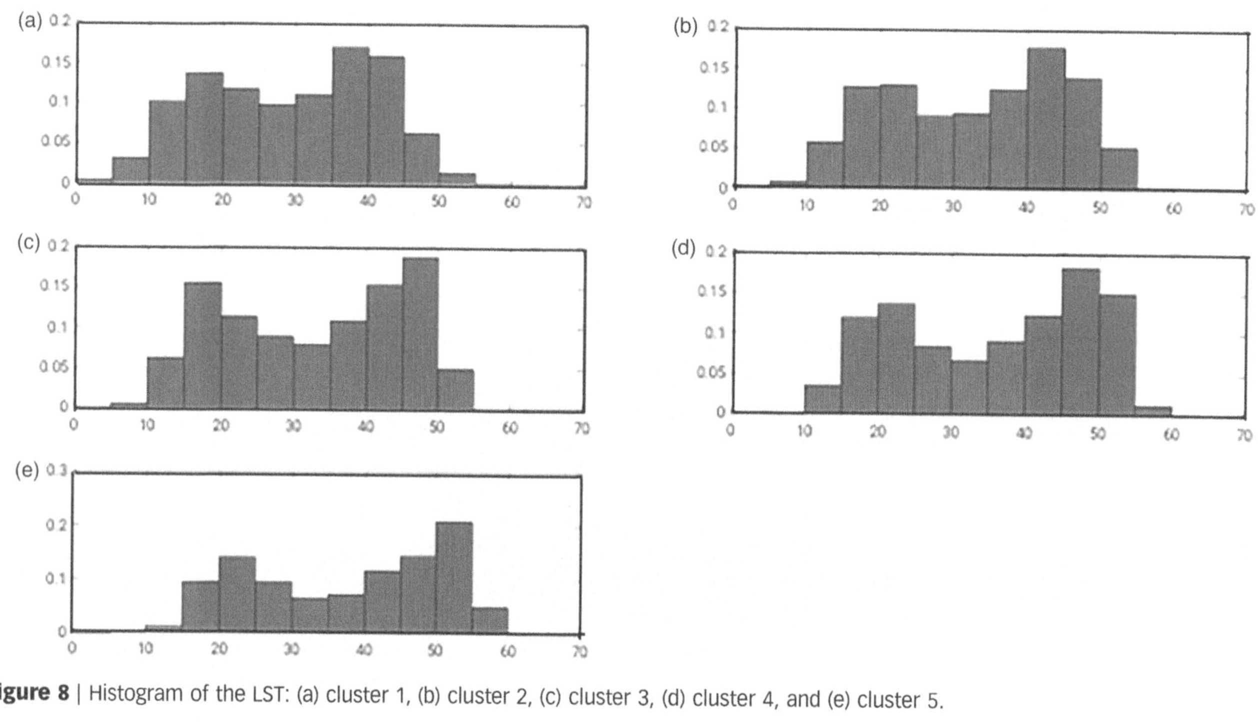

Figure 8 | Histogram of the LST: (a) cluster 1, (b) cluster 2, (c) cluster 3, (d) cluster 4, and (e) cluster 5.

can undergo large changes, which make it more difficult to find a mathematical relationship between the present conditions and those that prevailed 16 (or 24) days back.

4. CONCLUSIONS

In the present research, a combined SOM clustering method and FFNN approach was applied to devise an AI-based black box methodology for the multivariate SM condition modeling of the study region located in the southwestern part of the USA. In the first step, the SOM was utilized to recognize the pixel data with high similarity and to determine the dominant pixel in

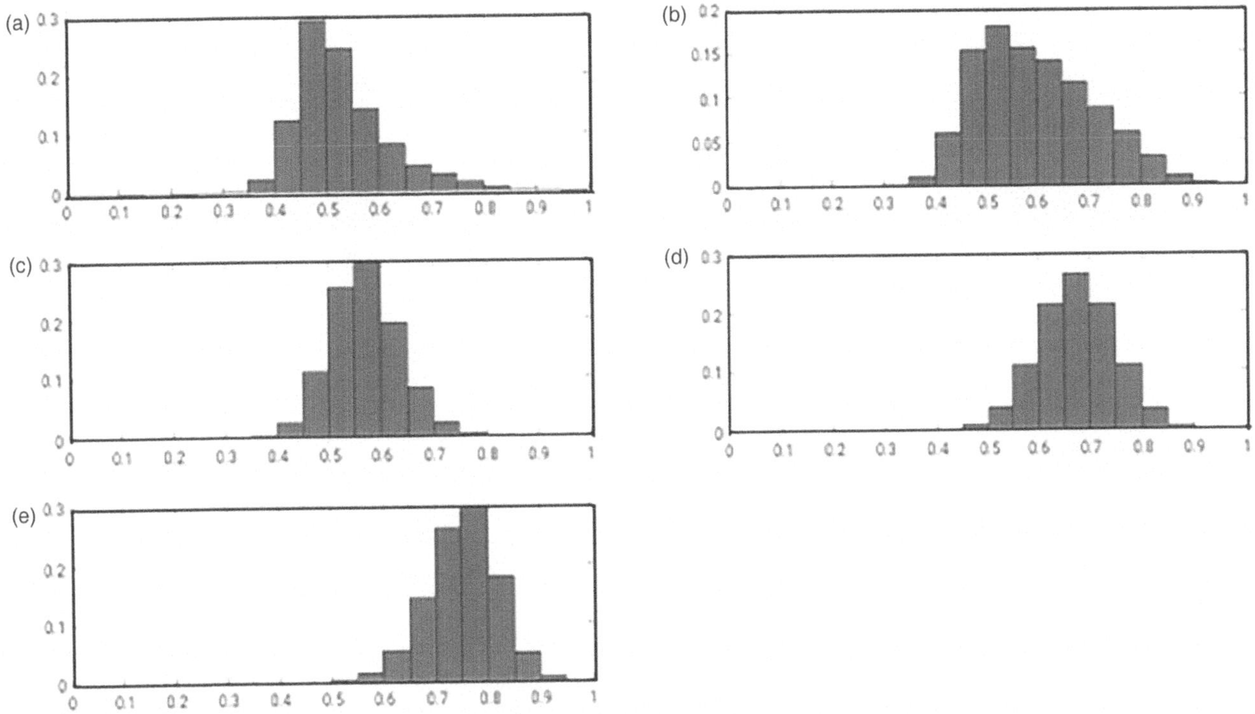

Figure 9 | Histogram of the NMDI: (a) cluster 1, (b) cluster 2, (c) cluster 3, (d) cluster 4, and (e) cluster 5.

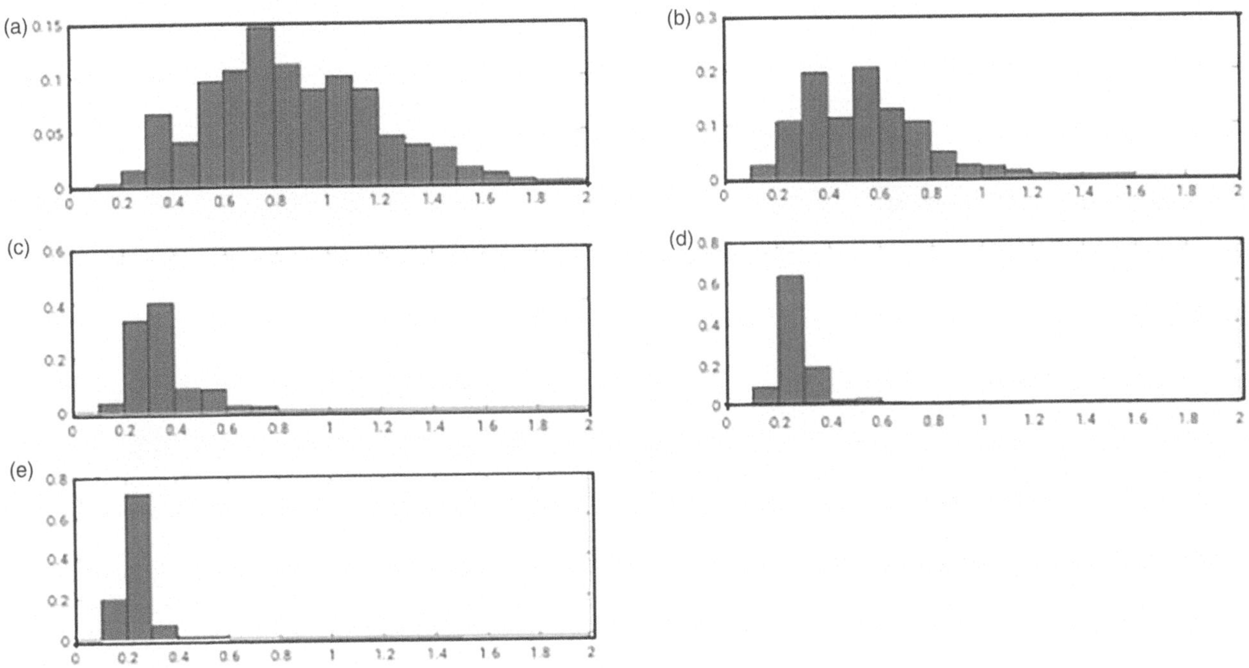

Figure 10 | Histogram of the LAI: (a) cluster 1, (b) cluster 2, (c) cluster 3, (d) cluster 4, and (e) cluster 5.

each cluster to identify the best patterns as a cluster indicator. The selection of suitable pixels by the SOM resulted in the definition of regions with similar SM conditions. Also, it could be concluded that sufficient *in situ* SM measurements can be done in the future. By selecting such representative and dominant pixels, more tools and investments on these dominant points for *in situ* measurement may be employed to improve the data-recording performance. In this study, nonlinear

Table 2 | Results and structures of the trained FFNN models

Cluster no.	Input variable	Output variable	Network structure[a]	DC train	DC verify	RMSE train (normalized)	RMSE verify (normalized)
Cluster1	LST(t),LAI(t) < 2	NMDI(t)	2-5-1	0.69	0.74	0.0039	0.0047
	LST(t),LAI(t) < 2,NMDI(t – 1)	NMDI(t)	3-8-1	0.87	0.88	0.0036	0.0035
Cluster2	LST(t),LAI(t) < 2	NMDI(t)	2-7-1	0.59	0.63	0.0060	0.0066
	LST(t),LAI(t) < 2,NMDI(t – 1)	NMDI(t)	3-8-1	0.80	0.76	0.0048	0.0051
Cluster3	LST(t),LAI(t) < 2	NMDI(t)	2-6-1	0.64	0.64	0.0070	0.0067
	LST(t),LAI(t) < 2,NMDI(t – 1)	NMDI(t)	3-7-1	0.76	0.77	0.0060	0.0060
Cluster4	LST(t),LAI(t) < 2	NMDI(t)	2-6-1	0.62	0.65	0.0100	0.0110
	LST(t),LAI(t) < 2,NMDI(t – 1)	NMDI(t)	3-9-1	0.83	0.89	0.0070	0.0050
Cluster5	LST(t),LAI(t) < 2	NMDI(t)	2-6-1	0.63	0.70	0.0083	0.0086
	LST(t),LAI(t) < 2,NMDI(t – 1)	NMDI(t)	3-8-1	0.82	0.82	0.0063	0.0053

[a]Structure of a-b-c denotes a, b, and c neurons, respectively, in the input, hidden, and output layers.

relationship between LAI, LST and previous (lagged) values of NMDI were discovered by the FFNN for each of the representative pixel of each cluster to predict the SM condition. The obtained results indicated encouraging evidence when comparing the observed and predicted NMDI values.

As with any kind of black box (data-driven) method, the numerical results obtained through the current modeling are case sensitive for the study area data. But this methodology can be applied similarly to any other area, and it is expected that the same regression will be seen between the same data but with different order of strengths. As a suggestion for future studies, the proposed methodology can be further developed using other hydrological time series and variables as additional inputs for predicting the NMDI (e.g., precipitation or/and evapotranspiration). The combination of the SOM and FFNN models can be improved and updated by more recent remotely sensed images and field analysis. Due to the fact that the presented method of the current research mostly relies on data, which is a limitation of this study, using a high volume of data may increase the noise and produce abundant information, resulting in a time-consuming training phase with inefficient results. To overcome this limitation, an SOM-based clustering approach that employs wavelet-based entropies of time series as inputs of clustering is suggested to optimize the SOM input layer (see Nourani *et al.* 2015).

DATA AVAILABILITY STATEMENT

Data cannot be made publicly available; readers should contact the corresponding author for details.

REFERENCES

Adab, H., Morbidelli, R., Saltalippi, C., Moradian, M. & Ghalhari, G. A. F. 2020 Machine learning to estimate surface soil moisture from remote sensing data. *Water* **12** (11), 3223.

ASCE task committee on application of Artificial Neural Networks in hydrology 2000 Artificial neural networks in hydrology 1: hydrologic applications. *Journal of Hydrologic Engineering* **5** (2), 124–137.

Chen, L. H., Hong, Y. T. & Hsu, C. W. 2011 A study on regional drought frequency analysis using self-organizing map and l-moments. *Journal of Taiwan Agricultural Engineering* **57** (2), 57–77.

Dari, J., Quintana-Seguí, P., Escorihuela, M. J., Stefan, V., Brocca, L. & Morbidelli, R. 2021 Detecting and mapping irrigated areas in a Mediterranean environment by using remote sensing soil moisture and a land surface model. *Journal of Hydrology* **596**, 126129.

Engman, E. T. 1991 Applications of microwave remote sensing of soil moisture for water resources agriculture. *Remote Sensing of Environment* **35** (2), 213–226.

Garcia-Estringana, P., Latron, J., Llorens, P. & Gallart, F. 2013 Spatial and temporal dynamics of soil moisture in a Mediterranean mountain area (Vallcebre, NE Spain). *Ecohydrology* **6** (5), 741–753.

Hsu, K. C. & Li, S. T. 2010 Clustering spatial–temporal precipitation data using wavelet transform self-organizing map neural network. *Advances in Water Resources* **33** (2), 190–200.

Hupet, F. & Vanclooster, M. 2002 Intraseasonal dynamics of soil moisture variability within a small agricultural maize cropped field. *Journal of Hydrology* **261** (1), 86–101.

Jiang, H. & Cotton, W. R. 2004 SM estimation using an artificial neural network: a feasibility study. *Canadian Journal of Remote Sensing* **30** (5), 827–839.

Kalteh, A. M., Hjorth, P. & Berndtsson, R. 2008 Review of the self-organizing map (SOM) approach in water resources: analysis, modelling and application. *Environmental Modelling and Software* **23**, 835–845.

Kim, T. W. & Valdés, J. B. 2003 Nonlinear model for drought forecasting based on a conjunction of wavelet transforms neural networks. *Journal of Hydrologic Engineering* **8** (6), 319–328.

Kohonen, T. 1990 The self-organizing map. *Proceedings of the IEEE* **78** (9), 1464–1480.

Kornelsen, K. C. & Coulibaly, P. 2014 Root-zone soil moisture estimation using data-driven methods. *Water Resources Research* **50**, 2946–2962.

Li, Z.-L., Leng, P., Zhou, C., Chen, K.-S., Zhou, F.-C. & Shang, G.-F. 2021 Soil moisture retrieval from remote sensing measurements: current knowledge and directions for the future. *Earth-Science Reviews* **218**, 103673.

Miller, D., Mohanty, B. P., Jacobs, J. M. & Hsu, E. C. 2004 SMEX02: field scale variability, time stability and similarity of soil moisture. *Remote Sensing of Environment* **92**, 436–446.

Mobasheri, M. R. & Bidkhan, N. G. 2013 'Development of new hyperspectral angle index for estimation of soil moisture using in situ spectral measurments', ISPRS-international archives of the photogrammetry. *Remote Sensing and Spatial Information Sciences* **40** (3), 481–486.

Mohanty, B. P., Famiglietti, J. S. & Skaggs, T. H. 2000 Evolution of soil moisture spatial structure in a mixed vegetation pixel during the southern great plains 1997 (SGP97) hydrology experiment. *Water Resources Research* **36** (12), 3675–3686.

Mulebeke, R., Kironchi, G. & Tenywa, M. M. 2013 Soil moisture dynamics under different tillage practices in cassava-sorghum based cropping systems in eastern Uganda. *Ecohydrology & Hydrobiology* **13** (1), 22–30.

Musick, H. B. & Pelletier, R. E. 1988 Response to soil moisture of spectral indexes derived from bidirectional reflectance in thematic mapper wavebands. *Remote Sensing of Environment* **25** (2), 167–184.

Nourani, V. & Kalantari, O. 2010 Integrated artificial neural network for spatiotemporal modeling of rainfall-runoff-sediment processes. *Environmental Engineering Science* **27** (5), 411–422.

Nourani, V., Alami, M. T. & Vousoughi, F. D. 2015 Wavelet-entropy data preprocessing approach for ANN-based groundwater level modeling. *Journal of Hydrology.* **524**, 43–52.

Olden, J. D., Kennard, M. J. & Pusey, B. J. 2012 A framework for hydrologic classification with a review of methodologies and applications in ecohydrology. *Ecohydrology* **5** (4), 503–518.

Pandey, A., Jha, S. K., Srivastava, J. K. & Prasad, R. 2010 Artificial neural network for the estimation of SM and surface roughness. *Russian Agricultural Sciences* **36** (6), 428–432.

Reynolds, S. G. 1970 The gravimetric method of soil moisture determination part iii an examination of factors influencing soil moisture variability. *Journal of Hydrology* **11** (3), 288–300.

Wang, L. & Qu, J. J. 2007 NMDI: A normalized multi-band drought index for monitoring soil and vegetation moisture with satellite remote sensing. *Geophysical Research Letters* **34**, 20.

Wang, L. & Qu, J. J. 2009 Satellite remote sensing applications for surface soil moisture monitoring: a review. *Frontiers of Earth Science in China* **3** (2), 237–247.

Western, A. W., Grayson, R. B., Blöschl, G., Willgoose, G. & McMahon, T. A. 1999 Observed spatial organization of soil moisture and its relation to terrain indices. *Water Resources Research* **35** (3), 797–810.

Western, A. W., Zhou, S. L., Grayson, R. B., McMahon, T. A., Blöschl, G. & Wilson, D. J. 2004 Spatial correlation of soil moisture in small catchments its relationship to dominant spatial hydrological processes. *Journal of Hydrology* **286** (1), 113–134.

Zhang, D. & Zhou, G. 2016 Estimation of soil moisture from optical thermal remote sensing: a review. *Sensors* **16** (8), 1308.

Zhang, H. W., Chen, H. L. & Shen, S. 2009 The application of normalized multi-band drought index (NMDI) method in cropland drought monitoring. In *'SPIE Europe Remote Sensing', International Society for Optics Photonics.* pp. 74721Q–74721Q.

Zhang, F., Zhang, L. W., Jing-Jing, S. H. I. & Huang, J. F. 2014 Soil moisture monitoring based on land surface temperature-vegetation index space derived from MODIS data. *Pedosphere* **24** (4), 450–460.

Zhu, Q., Wang, Y. & Luo, Y. 2021 Improvement of multi-layer soil moisture prediction using support vector machines and ensemble Kalman filter coupled with remote sensing soil moisture datasets over an agriculture dominant basin in China. *Hydrological Process* **35**, 14154.

First received 28 October 2021; accepted in revised form 4 April 2022. Available online 15 April 2022

doi: 10.2166/nh.2021.028

Hybrid point and interval prediction approaches for drought modeling using ground-based and remote sensing data

Kiyoumars Roushangar[a,b,*], Roghayeh Ghasempour[a], V. S. Ozgur Kirca[c] and Mehmet Cüneyd Demirel[c]

[a] Department of Water Resource Engineering, Faculty of Civil Engineering, University of Tabriz, Tabriz, Iran
[b] Center of Excellence in Hydroinformatics, University of Tabriz, Tabriz, Iran
[c] Division of Hydraulics, Department of Civil Engineering, Istanbul Technical University, Istanbul, Turkey
*Corresponding author. E-mail: kroshangar@yahoo.com

ABSTRACT

Drought as a severe natural disaster has devastating effects on the environment; therefore, reliable drought prediction is an important issue. In the current study, based on lower upper bound estimation, hybrid models including data preprocessing, permutation entropy, and artificial intelligence (AI) methods were used for point and interval predictions of short- to long-term series of Standardized Precipitation Index in the Northwest of Iran. Ground-based and remote sensing precipitation data were used covering the period of 1983–2017. In the modeling process, first, the data processing capability via variational mode decomposition (VMD), ensemble empirical mode decomposition, and permutation entropy (PE) was investigated in drought point prediction. Then, interval prediction was applied for tolerating increased uncertainty and providing more details for practical operation decisions. The simulation results demonstrated that the proposed integrated models could achieve significantly better performance compared to single models. Hybrid PE models increased the modeling accuracy up to 40 and 55%. Finally, the efficiency of developed models was verified for Normalized Difference Vegetation Index prediction. Results demonstrated that the proposed methodology based on remote sensing data and VMD–PE–AI approaches could be successfully used for drought modeling, especially in limited or non-gauged areas.

Key words: artificial intelligence, data preprocessing, drought, permutation entropy, remote sensing, variational mode decomposition

HIGHLIGHTS

- Apply ground- and remote sensing-based data to point and interval predictions of drought series.
- Propose new model structures via ensemble empirical mode decomposition, variational mode decomposition, permutation entropy, and artificial intelligence approaches for drought modeling.
- The possibility of using the proposed hybrid models for the Normalized Difference Vegetation Index series forecasting was assessed.
- The obtained results showed the high ability of developed models in drought indices modeling.

GRAPHICAL ABSTRACT

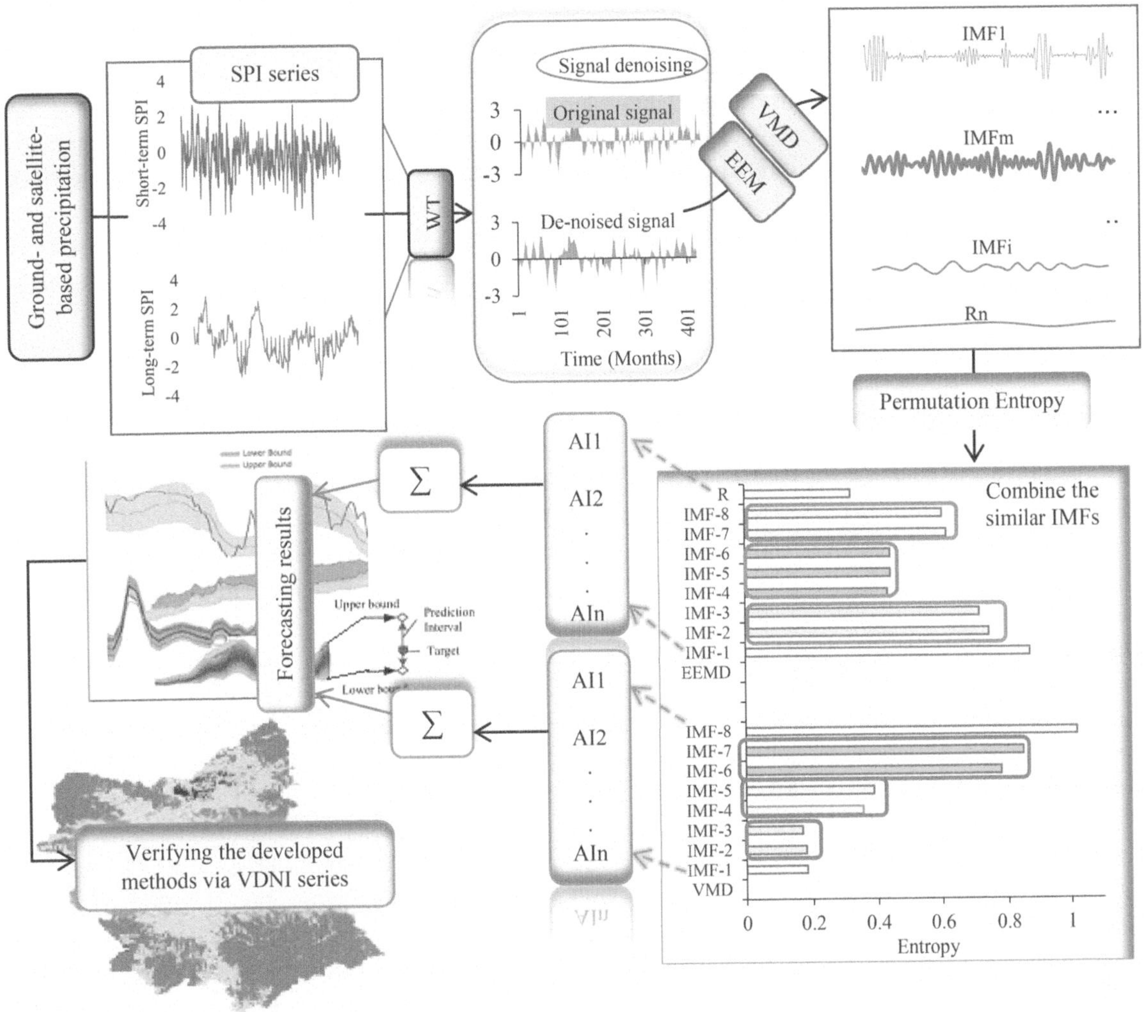

INTRODUCTION

One of the most complex phenomena among all extreme climate events is drought (Ghulam *et al.* 2007). Drought has serious negative effects on different parts of human life (Hayes *et al.* 1999). Some of these effects are water supply shortage, low agricultural produce, reduced soil moisture, economic losses, migration, and famine (Von Hardenberg *et al.* 2001; Zhou *et al.* 2018). These widespread negative effects can be seen anywhere in the world (Mann & Gleick 2015; Marvel *et al.* 2019). An accurate understanding of spatial and temporal variation in rainfall and drought will provide a good insight into the arranging and management of drought subordinate exercises (Basalirwa *et al.* 1999; Nischitha *et al.* 2014). Among hydrometeorological variables (i.e. precipitation, evapotranspiration, soil moisture, runoff, reservoir level, streamflow, and temperature), precipitation is the most appropriate variable to monitor the meteorological drought. The use of drought index is a common practical method to monitor the drought phenomenon. Therefore, different indices have been developed for precise applications (Kwak *et al.* 2016; Mukherjee *et al.* 2018) such as Surface Water Supply Index, Palmer Drought Severity Index, Standardized Runoff Index, Evapotranspiration Deficit Index, Standardized Precipitation Evapotranspiration Index, and Standardized Precipitation Index (SPI). Li *et al.* (2015) suggested the SPI as an ideal tool in drought severity characterization, since it can show both short- and long-term effects of drought. Different time scales of drought index makes it possible to investigate the impacts of rainfall shortages on different parts of water resources. At shorter timescales, the SPI is related to soil moisture, whereas at longer timescales, it shows the groundwater and reservoir storage conditions.

On the other hand, the traditional drought indices are usually obtained from hydro-meteorological data from stations, the spatial resolution of which does not necessarily meet the drought monitoring requirements in large-scale areas. According to Yilmaz *et al.* (2008), spatially and temporally continuous information can be provided by data obtained from meteorological satellites. Therefore, the use of satellite data can compensate the limitation of point-based observation data for hydro-meteorological variables and makes it possible to monitor drought conditions in non-gauged regions. Meteorological satellites can acquire multitemporal, multispectral, continuous, and complete data. Sensors used to measure light reflections from surfaces are mounted to an aircraft, drone, or even a satellite. This positioning of the sensors allows them to cover a large area in a short time, which renders remote sensing a fast process. Remote sensors are also able to scan and create maps of inaccessible areas, and they provide more homogeneous data quality compared to ground observations. However, raw satellite datasets have some disadvantages; for example, the amount of cloudiness affects the accuracy of the results. Therefore, satellite-based datasets need to be examined and verified by comparison with ground-based data.

The Normalized Difference Vegetation Index (NDVI) is an alternative remote sensing drought index, which is successfully used for monitoring the dynamic variations in vegetation cover (Chen *et al.* 2010). In recent years, many researchers have tried to use satellite data as an efficient tool to measure and monitor drought-related variables on a global scale. Although the use of satellite products has many benefits in drought analysis, there are still significant challenges to boost the reliability and efficiency of various approach outcomes (Brown *et al.* 2015). So far, numerous models have been developed to enhance the efficiency of drought predictions (Mishra *et al.* 2007; Mehr *et al.* 2014; Geng *et al.* 2016; Sakizadeh *et al.* 2019). Time-series, regression, physical-based, and artificial intelligence (AI) methods are some examples. In recent decades, AI methods [e.g., support vector machine (SVM), artificial neural networks (ANNs), genetic programming, adaptive neuro-fuzzy inference system (ANFIS), feed-forward neural network (FFNN), and Gaussian process regression (GPR)] have been widely used to investigate the complex hydraulic and hydrologic phenomena (Yin *et al.* 2017). Side weir discharge coefficient modeling (Azamathulla *et al.* 2016), rainfall–runoff modeling (Chadalawada *et al.* 2017), pan evaporation estimation (Wu *et al.* 2019), groundwater-level prediction (Sakizadeh *et al.* 2019), forecasting different types of drought simultaneously (Aghelpour *et al.* 2020a), and river discharge relationship modeling (Roushangar *et al.* 2021) are some examples. Deo *et al.* (2017) investigated the potential of several AI models in the SPI series forecasting. The obtained results highlighted the importance of periodicity in drought forecasting. Kisi *et al.* (2019) investigated the accuracy of novel heuristic methods in forecasting various time scales of the SPI series in a semi-arid environment and compared the obtained results with the classical ANFIS method. Mohamadi *et al.* (2020) used integrated machine learning models with a nomadic people optimization (NPA) algorithm to forecast meteorological droughts in Iran. The general results indicated that the NPA and wavelet coherence analysis are useful tools for modeling drought indices. Aghelpour *et al.* (2020b) used meta-heuristic models to model different drought indices. The applied models led to desirable predictions in arid, semi-arid, and Mediterranean climates, while humid climate provided the weakest predictions. Aghelpour *et al.* (2020c) investigated the snow cover of the Northern and Southern slopes of Alborz Mountains in Iran by considering two issues: (1) estimating the snow cover area and (2) investigating the effects of droughts on snow cover. The results showed that the effects of a drought event on the snow cover area would remain up to 5 (or 6) months in the region.

Since most of the hydrological time series are nonstationary, trendy, or contain seasonal fluctuations, the signal preprocessing methods are commonly used by researchers especially in the past three decades (Kumar & Foufoula 1993; Adamowski *et al.* 2009; Roushangar & Alizadeh 2018). These methods have been used for breaking down and excavating complex, periodic, and irregular hydrological time series (Roushangar & Alizadeh 2018). Wavelet transform (WT) analysis is a common method for time-series analysis, and many researchers have used it to decompose signals in different fields. Beside the WT, the empirical mode decomposition (EMD) and variational mode decomposition (VMD) methods have also been used recently for signal decomposition. These methods are particularly suitable for decomposition of nonlinear and nonstationary signals (Huang *et al.* 1998). According to Abdoos (2016), the correlation frequency band and modal components can be found by the use of the variational mode; so, the VMD has better anti-noise performance compared to the EMD.

In the present study, point and interval predictions of short- to long-term drought were performed by using ground- and remote sensing-based datasets. In this regard, the FFNN, SVM, and GPR models were applied for point and prediction intervals (PIs) of the SPI series at 1-month (short-term), 9-month (mid-term), and 24-month (long-term) timescales. SVM and GPR are relatively new and promising methods based on different kernel types, which are based on the statistical learning theory initiated. Such methods can model nonlinear decision boundaries, and moreover, they offer many alternative kernels to choose from. A kernel-based (KB) approach is capable of adapting itself to predict any variable of interest via sufficient inputs and is fairly robust against over-fitting, especially in high-dimensional space. FFNN is a commonly encountered classical type of ANN that has been applied to many diverse fields. In this type of network, the information moves in only one direction, forward from the input nodes to the output nodes, through the hidden nodes (if any). The training of this method is fast, it has high accuracy, and the probability of occurrence of data over-training in this method is relatively less. PIs are prevalent measures to investigate the uncertainty of prediction. PIs assimilate the accuracy of predicted values vs. the measured values (Khosravi *et al.* 2011). Since the uncertainties are considered in PIs, a bound (envelope) for the variables of concern can be obtained, which makes PIs a useful instrument to address decision-making challenges and the management of critical issues arising from design and adaptation plans in the context of climate change. In the present research, lower upper bound estimation (LUBE) approach was used for constructing PIs for GPR-based modeling of drought series. The precipitation data from meteorological stations and two PERSIANN-CDR and CMAP satellites were used for the SPI series modeling. During modeling processes, data of 12 stations located in the Northwest of Iran were used. Iran, located in the West of Asia, has diverse climatic characteristics. However, in most regions, arid and semi-arid climates are predominant, and therefore, drought is a major concern in the country. For calculating the SPI time series, the rainfall data in the period of 1983–2017 was used. The accuracy and homogeneity of data were checked and confirmed. In the proposed modeling, first, the EEMD and VMD methods were utilized for decomposition of the original time series into several subseries named intrinsic mode functions (IMFs). Then, the arrangement entropy of each IMF was calculated. Based on the closeness of these entropy values, the IMFs were combined to achieve a recombinant subseries. The recombinant subseries were inserted into AI methods to the point prediction of the drought time series. However, due to the difficulties in accurate point prediction, the LUBE method, which can tolerate increased uncertainty, was applied to construct PIs for drought time series. As such, the present study aims at proposing a novel hybrid system for short- to long-term SPI time-series point and interval predictions by combining data preprocessing, permutation entropy (PE), and AI approaches using ground- and satellite-based data. Also, for verifying the proposed hybrid model performance, the capability of the proposed methods was investigated for the NDVI time-series prediction.

MATERIALS AND METHODS

Characteristics of study area and data used

In the current study, East Azerbaijan province in the Northwest of Iran was selected as the study area. The annual precipitation variability is high in the Northwest regions of Iran; therefore, drought is one of the most frequently encountered climate-related disasters. The average annual rainfall in the selected area varies from 198 to 845 mm. Most of this precipitation is received by the mountainous areas. The mean monthly precipitation in the studied 35-year statistical period varies between 0 and 159.4 mm. Based on this monthly precipitation data, May, April, March, and October come out to be the rainiest months of the year, whereas August, July, and September are the driest months. Also, in summer, the prevailing concentration of rain gauge stations is strongly irregular in the selected area. In this study, as seen from Figure 1, 12 synoptic stations were used to assess drought modeling. The datasets used were obtained from the meteorology portal of Iran (http://www.weather.ir). The considered stations had records of 35 years, and they showed a desirable spatial distribution over the study area. The Thiessen polygon method was applied for calculating the spatial average precipitation of the region; then, these spatially averaged precipitation values were used for calculating the region SPI time series.

In addition to the ground-based data, two remote sensing observation products were used to model the SPI time series: (1) the Precipitation Estimation from Remotely Sensed Information using Artificial Neural Networks-Climate Data Record (PERSIANN-CDR) and (2) CPC Merged Analysis of Precipitation (CMAP). According to Xie & Arkin (1996), the CMAP is a group of rainfall dataset, which can be generated via gauge data analysis and satellite-based rainfall estimations. The PERSIANN-CDR is a valuable precipitation dataset, which is based on rainfall prediction via ANN algorithm and infrared satellite images (Ashouri *et al.* 2015). Table 1 shows the characteristics of spatio-temporal remotely sensed products used in this study. Additionally, the NDVI series, which can be obtained from the Moderate Resolution Imaging Spectroradiometer (MODIS) on the National Aeronautics and Space Administration (NASA) Earth Observation Satellite (EOS) Terra (https://modis.gsfc.nasa.gov/), was used for predicting the vegetation patterns.

Standard Precipitation Index (SPI)

The impacts of a water deficit are complex functions of water source and water use. The time scale over which precipitation deficits accumulate becomes extremely important and separates different types of drought. As described by McKee *et al.*

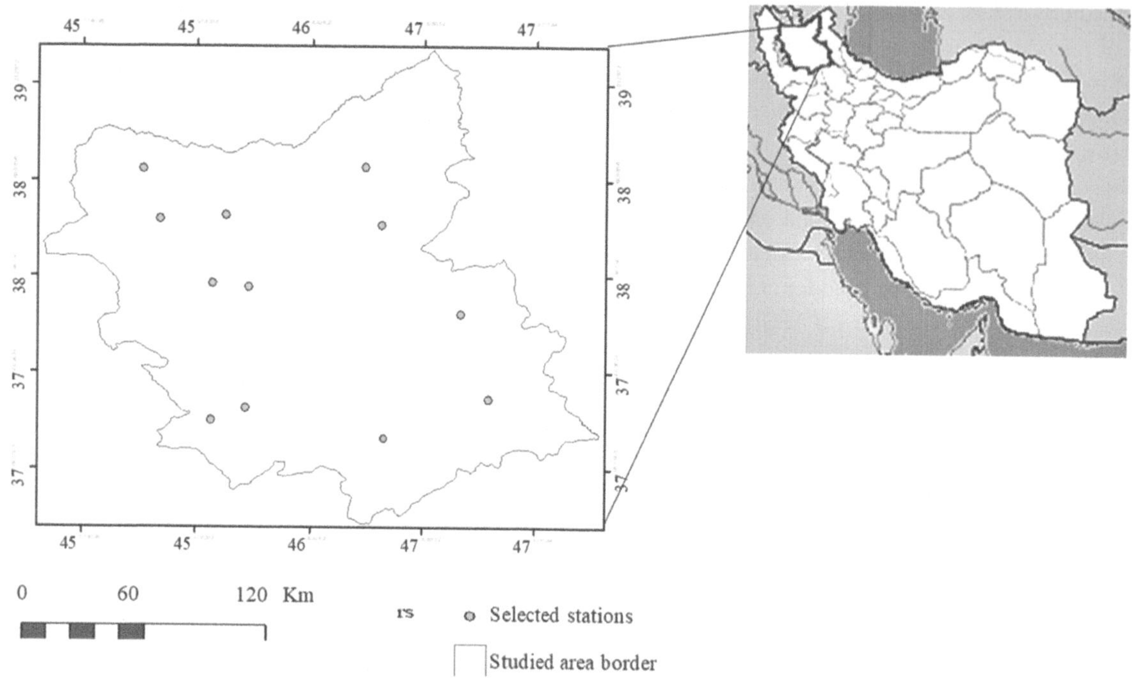

Figure 1 | Location of the selected stations for drought modeling.

Table 1 | Information of the used satellite products

Satellite product	Date range	Spatial resolutions	Temporal resolutions	Input data
PERSIANN-CDR	1983–2017	0.25° × 0.25°	Daily	Multiple satellite microwave sensors
CMAP	1979–2017	2.5° × 2.5°	Monthly	Gauge analysis; microwave and infrared observations from polar orbiting and geostationary satellites
MODIS	2000–2017	1 km	16 days	

(1993), some of the practical issues that are important in drought analysis are time scale, probability (the expected frequency of an event), and precipitation deficit. Addressing these issues, McKee *et al.* (1993) developed a parameter termed as SPI, where the standardization is based on probability. The SPI quantifies observed precipitation as a standardized departure from a selected probability distribution function that models the raw precipitation data. The raw precipitation data are typically fitted to a gamma or a Pearson Type-III distribution and then transformed to a normal distribution. The SPI values can be interpreted as the number of standard deviations by which the observed anomaly deviates from the long-term mean (Mirabbasi *et al.* 2013).

The SPI index has statistical consistency and gives a representation of abnormal wetness or dryness. The SPI was designed to quantify the precipitation deficit for multiple timescales, and it is capable of showing both short- and long-term drought effects in variable timescales of precipitation anomalies. The use of different time scales allows the effects of a precipitation deficit on different water resources components (groundwater, reservoir storage, soil moisture, and streamflow) to be investigated (Tsakiris & Vangelis 2004; Mirabbasi *et al.* 2013). For example, soil moisture conditions respond to precipitation anomalies on a relatively short scale, whereas groundwater, streamflow, and reservoir storage reflect the longer-term precipitation anomalies. The SPI values can be positive or negative: positive SPI values show wet conditions, and negative SPI values show dry conditions. The descriptive features of SPI are listed in Table 2.

Table 2 | Classification of drought based on the SPI

SPI values	Class	SPI values	Class
>2	Extremely wet	−1 to −1.49	Moderately dry
1.5–1.99	Very wet	−1.5 to −1.99	Severely dry
1.0–1.49	Moderately wet	<−2	Extremely dry
−0.99 to 0.99	Near normal		

Methods used in data preprocessing

EMD is a popular and widely used method for time-series processing (Wang *et al.* 2015). This approach can be followed for processing the nonlinear and nonstationary signals, and for decomposition of these signals into several components terms as inherent mode functions (IMFs). EMD is capable of determining the signal instantaneous frequency. According to Lei *et al.* (2009), in the process of obtaining the frequency components of a signal by decomposition, first, the components with higher frequency shall be separated, and then the process will continue until the lowest frequency component remains. The ensemble EMD (EEMD) solves the mode mixing issue of EMD via the noise-added data analysis approach (Wu & Huang 2009). This method can be summarized as follows:

1. A random white noise signal is added to the original signal $x(t)$, and then, the new signal is decomposed into several IMF series.
2. If the number of added white noises is less than the number of trials, step 1 should be repeated; otherwise, step 3 should be performed.
3. For obtaining the ensemble IMF, the average of the sum of all IMFs is computed ($I_j(t)$).
4. Finally, the original signal can be obtained as $x(t) = \sum_{j=1}^{n} I_j(t)$.

VMD is another relatively new time-series processing method developed to decompose the time series into a sequence of discrete sub-modules with a special frequency band (He *et al.* 2019). In the VMD method, each IMF is assumed as a finite bandwidth with a different center frequency, and the aim is the minimization of the sum of the estimated bandwidths for each IMF. The detailed description is as follows:

$$\min_{u_k(t),w_k} \left\{ \sum_{k=1}^{K} \left\| \partial_t \left[\left(\delta(t) + \frac{j}{\pi t} \right) \otimes u_k(t) \right] e^{-jw_k t} \right\|_2^2 \right\} \tag{1}$$

$$\text{subject to } \sum_{k=1}^{K} u_k(t) = f(t)$$

where K, ∂_t, $\delta(t)$, j, $\otimes$, $u_k(t)$, w_k, and $f(t)$ are the number of modes, partial derivative for time t, Dirac distribution, imaginary unit, convolution operator, tth data in the kth decomposed mode, center frequency of the kth decomposed mode, and tth data in the original signal, respectively. In general, due to complex features (such as convexity or concavity) in objective and constraint functions in Equation (1), directly detecting the solution of constrained optimization problem is difficult. For reducing the difficulty of solution, He *et al.* (2019) combined the quadratic penalty function guaranteeing the reconstruction fidelity and the Lagrange multiplier, strictly enforcing constraints for transforming the problem to an unconstrained optimization problem, as shown below:

$$\begin{aligned} Z(u_k(t), w_k, \lambda) = a\left\{ \sum_{k=1}^{K} \left\| \partial_t \left[\left(\delta(t) + \frac{j}{\pi t} \right) \otimes u_k(t) \right] e^{-jw_k t} \right\|_2^2 \right\} + \left\| \sum_{K=1}^{R} u_k(t) - f(t) \right\|_2^2 \\ + \prec \lambda(t), f(t) - \sum_{K=1}^{K} u_k(t) \succ \end{aligned} \tag{2}$$

where Z, a, and λ are augmented Lagrange function, penalty parameter, and Lagrange multiplier, respectively. In Equation

(2), for optimizing the modified unconstrained problem, the alternate direction method of multiplier method can be used (He *et al.* 2019). The corresponding solution can be described as below:

$$\overline{u}_k^{n+1}(w) = \frac{\overline{f}(w) - \sum_{i=1, i\neq k}^{K} \overline{u}_i(w) + 0.5\overline{\lambda}^n(w)}{1 + 2a(w - w_k^n)^2}; \quad w_k^n = \frac{\int_0^\infty w|\overline{u}_k^{n+1}(w)|^2 dw}{\int_0^\infty |\overline{u}_k^{n+1}(w)|^2 dw} \tag{3}$$

where $\overline{f}(w)$, $\overline{u}_i(w)$, $\overline{\lambda}(w)$, and $\overline{u}_k^{n+1}(w)$ are Fourier transforms of $f(t)$, $u_i(t)$, $\lambda(t)$, and $u_k^{n+1}(t)$, respectively; and n is the number of iterations.

After data decomposition with EEMD and VMD, too many IMF components are obtained, which are needed to be reorganized or regrouped. In this study, PE algorithm was used to carry out IMF reorganization. Based on the Shannon entropy form, the PE $H_p(m)$ of the kth different symbol sequence of time series $X(i)$ can be expressed as follows:

$$H_p(m) = -\sum_{j=1}^{k} p_j \ln p_j \tag{4}$$

Based on the above equation, when P_j is $1/m!$, $H_p(m)$ has the maximum value $\ln(m!)$. Standardize $H_p(m)$ for convenience.

$$H_{\mathrm{PE}}(m) = \frac{H_{\mathrm{P}}(m)}{\ln(m!)} \tag{5}$$

where $0 \leq H_{\mathrm{PE}}(m) \leq 1$. The value of the H_{PE} indicates the time-series randomness and complexity. The larger the H_{PE} value, the stronger the randomness will be and vice versa.

Data-driven methods for the prediction of drought indices

Feed-forward neural network

The ANN, as a data-driven method, aims at finding a nonlinear relationship between data series of inputs and output. An ANN is based on a connected node collection called neurons (Najafi *et al.* 2018). These neurons are linked to certain nodes in their neighborhood with varying connectivity coefficients which show the connection strengths. The three-layer FFNN (i.e. input, hidden, and target layers) is one of the most common ANN algorithms, and it has been widely applied for problems related to water resources engineering. The FFNN is considered as a classical type of ANN. In this method, the information transfer is only in one direction: from the input nodes and output nodes, through hidden nodes (if any). The hidden neuron sums up the weights of input connections. For selecting the appropriate information to be passed on to the next neuron, the weighted summation is passed through an activation function in the hidden layer (for more details, see Tayfur 2012).

KB methods

Among data-driven techniques, KB methods such as GPR and SVM are considered as relatively innovative and significant techniques in terms of various kernel types and the statistical learning theory (Roushangar *et al.* 2019). These models can adapt themselves for predicting any parameter of interest by adequate inputs, and numerous kernel-type alternatives exist in this regard (Babovic 2009). Nevertheless, the proper selection of the kernel type is the most crucial step in the GPR and SVM methods because of its direct effect on the precision of classification and training. According to Smola (1996), there are various kernel functions for selecting. For instance, in the SVM, the linear, polynomial, sigmoid functions, and radial basis function (RBF) and in the GPR, the polynomial, normalized polynomial, RBF, and Pearson kernels can be selected based on the nature of the phenomenon to be dealt with.

For assessing the performance and capability of the utilized AI methods, three criteria were used, including determination coefficient (DC), correlation coefficient (R), and root-mean-square errors (RMSE), which can be formulated, respectively, as follows:

$$DC = 1 - \frac{\sum_{i=1}^{N} (SPI_o - SPI_p)^2}{\sum_{i=1}^{N} (SPI_o - \overline{SPI_p})^2} \tag{6}$$

$$R = \frac{\sum_{i=1}^{N} (SPI_o - \overline{SPI_o}) \times (SPI_p - \overline{SPI_p})}{\sqrt{\sum_{i=1}^{N} (SPI_o - \overline{SPI_o})^2 \times (SPI_p - \overline{SPI_p})^2}} \tag{7}$$

$$RMSE = \sqrt{\sum_{i=1}^{N} \frac{(SPI_o - SPI_p)^2}{ND}} \tag{8}$$

where SPI_o, SPI_p, $\overline{SPI_o}$, and $\overline{SPI_p}$ are the observed, predicted, mean observed, and mean predicted values, respectively. ND is the number of data.

Lower bound and upper bound estimation

According to Krupnick *et al.* (2006), the uncertainty of any prediction model contains both uncertainties stemming from the modeling itself and those stemming from prediction. The uncertainty of modeling is due to the inputs, parameters, and structures that generally generate the uncertainty of estimation (Kasiviswanathan *et al.* 2013). One of the most commonly used methods in quantifying the uncertainty of a model prediction is PI. Generally, the traditional interval prediction tries to make the PI based on point forecasting. The lower and upper bounds can be computed based on the forecasting values and the level of confidence. The point forecasting accuracy plays a key role in the PI accuracy. The lower bound and upper bound estimation (LUBE) attempts to directly approximate the PI upper and lower bounds based on the set of inputs. Based on Meade & Islam (1995), a PI shows the estimation uncertainty of a random parameter future realization and accounts for more uncertainty sources.

Two measures for the quantitative assessment of constructed PIs are PI coverage probability (PICP) and mean PI width (MPIW). PICP measures the percentage of observed values that fall within the prediction band. If the prediction band is wider, it covers most of the observed values. The closer PICP is to its nominal value, the more reliable it is. PICP, described in the following equation, is the measure that shows how good the PIs are (Kasiviswanathan *et al.* 2013):

$$PICP = \frac{1}{ND} \sum_{i=1}^{ND} c_i \tag{9}$$

if $L(X_i) < x_i < U(X_i)$, $c_i = 1$; else $c_i = 0$

in which $L(X_i)$ and $U(X_i)$ show the PI lower bound and upper bound corresponding to the ith sample, respectively. The ideal value of PICP could be obtained if the upper and lower bounds of PIs had extremely large values. Actually, too wide PIs are not valuable, because PIs carry no report around the observed values. Therefore, an additional criterion should be used for quantification of the width of prediction bounds as follows (Kasiviswanathan *et al.* 2013):

$$MPIW = \frac{1}{ND} \sum_{i=1}^{ND} L(X_i) - U(X_i) \tag{10}$$

For comparing the achieved PIs via various methods, the MPIW normalization is needed:

$$NMPIW = \frac{MPIW}{R} \tag{11}$$

where R shows the underlying target range (i.e. difference between the maximum and the minimum of the target amounts). The NMPIW is a dimensionless parameter and demonstrates how narrow PIs are. The consideration of these two measures

(NMPIW and PICP) independently may not lead to accurate results, because they have a contrary relationship. Therefore, both measures should be simultaneously evaluated in the procedure of PI construction.

Overall description of the proposed methodology

In this paper, hybrid models were developed for point and interval predictions of drought time series based on data preprocessing, PE, AI, and LUBE approaches. Three precipitation datasets were used in the calculation of SPI times series. The first of these was ground-based data (precipitation data from 12 synoptic stations), and the other two datasets were based on satellite precipitation data. Given that at least 30-year long data should be used in the SPI series calculations (Mishra & Singh 2010; Sahoo *et al.* 2015); the PERSIANN-CDR and CMAP satellites datasets were utilized for this aim. Figure 2 shows the applied methodology for data multi preprocessing. Since in the collection process, many internal and external factors affect the raw data, the integration of unrelated information in the data may increase the difficulty of accurate prediction of the intended parameter. Therefore, at first, the original data were refined via WT. WT was used for the signals de-noising via specifying a threshold for the detail coefficients, which represent noise (Donoho 1995). Figure 3 shows de-noised signals of SPI-9 time series for three selected datasets.

Two scenarios were considered in the modeling process. According to previous studies (e.g. Mokhtarzad *et al.* 2017; Xu *et al.* 2018; Zhang *et al.* 2019; Li *et al.* 2020; Khan *et al.* 2020), the previous values of the SPIs, precipitation (P), relative humidity (R), and temperature (T) series are effective in drought modeling. Therefore, in the first scenario, all effective variables were considered as inputs in predicting the SPI time series in the scales of 1-, 9-, and 24-month timescales. FFNN, GPR,

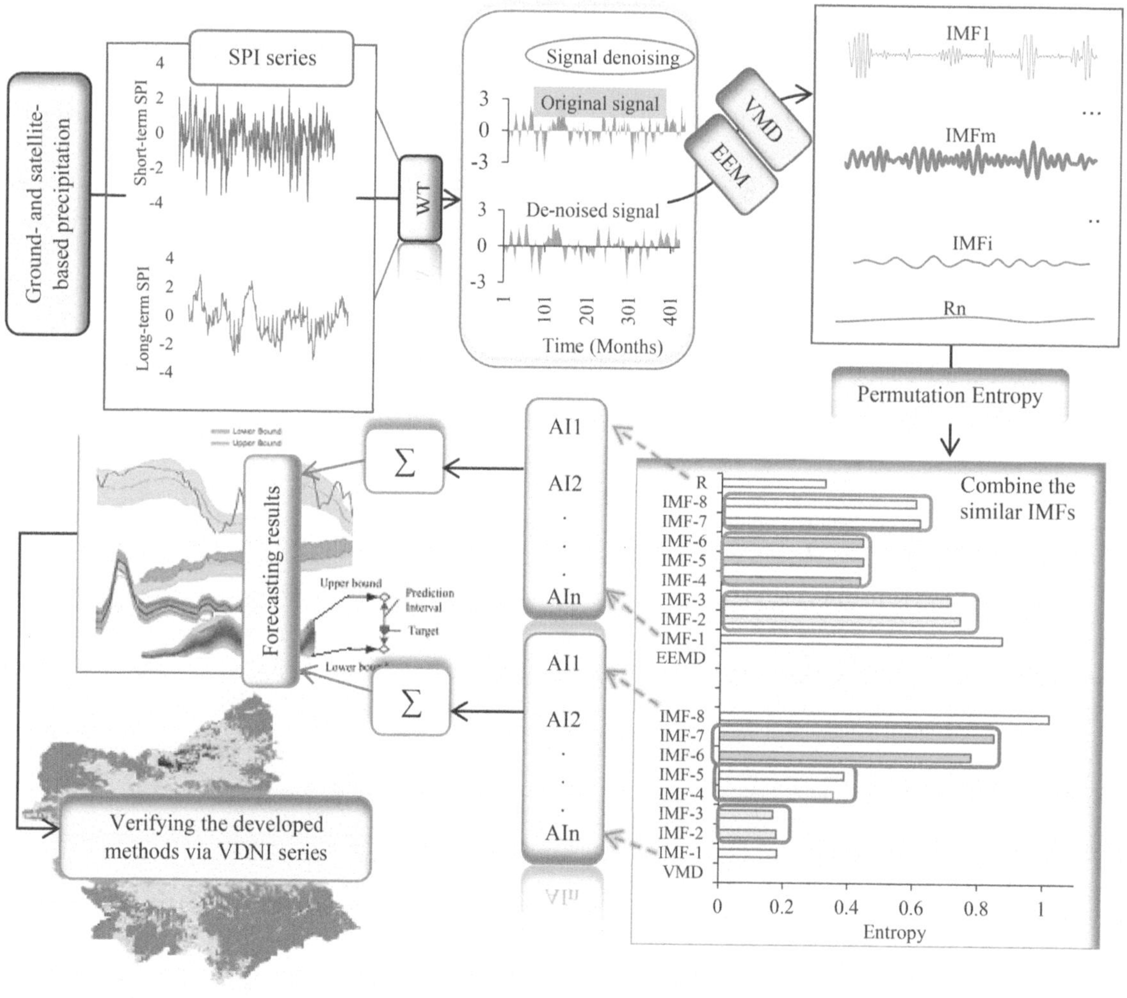

Figure 2 | Considered methodology for data multi preprocessing.

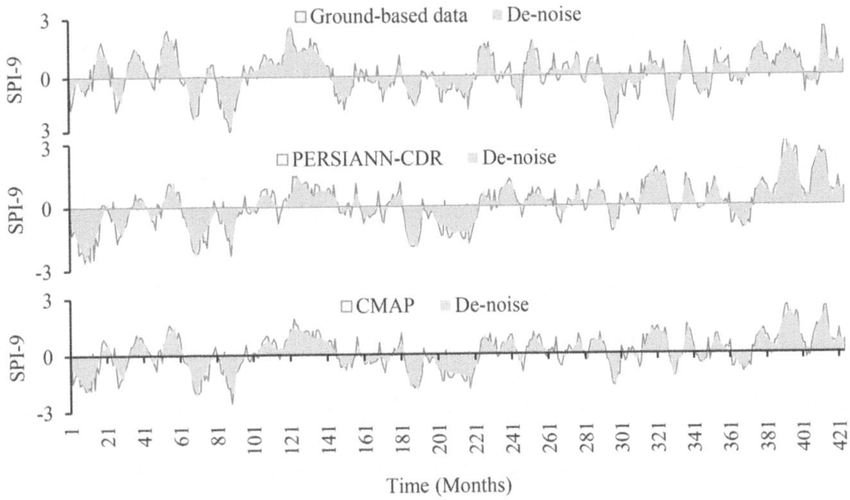

Figure 3 | The de-noised SPI signals via WT.

and SVM approaches were used for point prediction purpose. In the second scenario, only the decomposed subseries of the SPI time series were used as inputs. Directly selecting of the original time series as input parameter may affect the capability of the single predicting models. Different types of factors affect the hydrological time series and contribute to nonlinearity and uncertainty properties of time series. This causes the time series to have different resolution components. Therefore, in this study, an attempt was made to develop practical hybrid preprocessing-based AI models to improve the drought prediction efficiency. In this regard, two data processing models, including EEMD and VMD, as well as PE, were applied for original drought series decomposition to increase the efficiency of single AI methods. For analyzing the complexity of each IMFs, component PE was used. Then, for solving the computational burden and over-decomposition issues, the IMFs with similar entropy amounts were recombined. In the next step, hybrid models were used for interval prediction based on the LUBE method. The proposed hybrid models could be described step by step as follows:

1. Calculation of short- to long-term SPI series from the selected datasets.
2. *Decomposition*: the VMD and EEMD approaches were used to decompose time series into several subseries (IMFs).
3. *Calculation of the PE of IMFs*: entropy values of each IMF were computed. Based on the closeness of entropy values, the IMFs were mixed for obtaining a recombinant subseries.
4. *Model development*: the AI models were applied to construct an approximate network.
5. *Model prediction*: combined predicted result of the developed AI models for all the subseries was considered as the final predicted result of the original drought time series. For assessing the efficiency of the models used, the datasets were divided into training, validation, and testing parts. In total, 70% of the datasets from the beginning was utilized to train the developed models, and the rest 30% of data was used for validation (15%) and testing (15%) purposes. It should be noted that in this study, all used variables (input and output parameters) were scaled between 0 and 1 (i.e. normalization) in order to eliminate the dimensions of input and output variables. Data normalization eliminates the influence of the variables with different absolute magnitudes and increases the training speed. The used normalization only rescales the data to another range, which could be de-normalized to real values after modeling. The following equation was used to normalize the utilized data in this study:

$$x_n = \left(\frac{x - x_{\min}}{x_{\max} - x_{\min}}\right) \tag{12}$$

where x_n, x, $x_{\max}$, and $x_{\min}$, respectively, are the normalized value of variable x, the original value, the maximum of variable x, and the minimum of variable x.

6. *Interval prediction*: the LUBE method was used to determine the uncertainty of predictions performed by developed methods.

7. *Verifying the efficiency of developed methods*: in the final step, for verifying the developed hybrid models, the capability of the proposed methods was assessed for the NDVI series modeling.

RESULTS AND DISCUSSION

A comparison of the monthly precipitation time series with the long-term average value revealed that the selected region had suffered from frequent droughts. The correlation between ground-based and satellite precipitation data is shown in Figure 4(a).

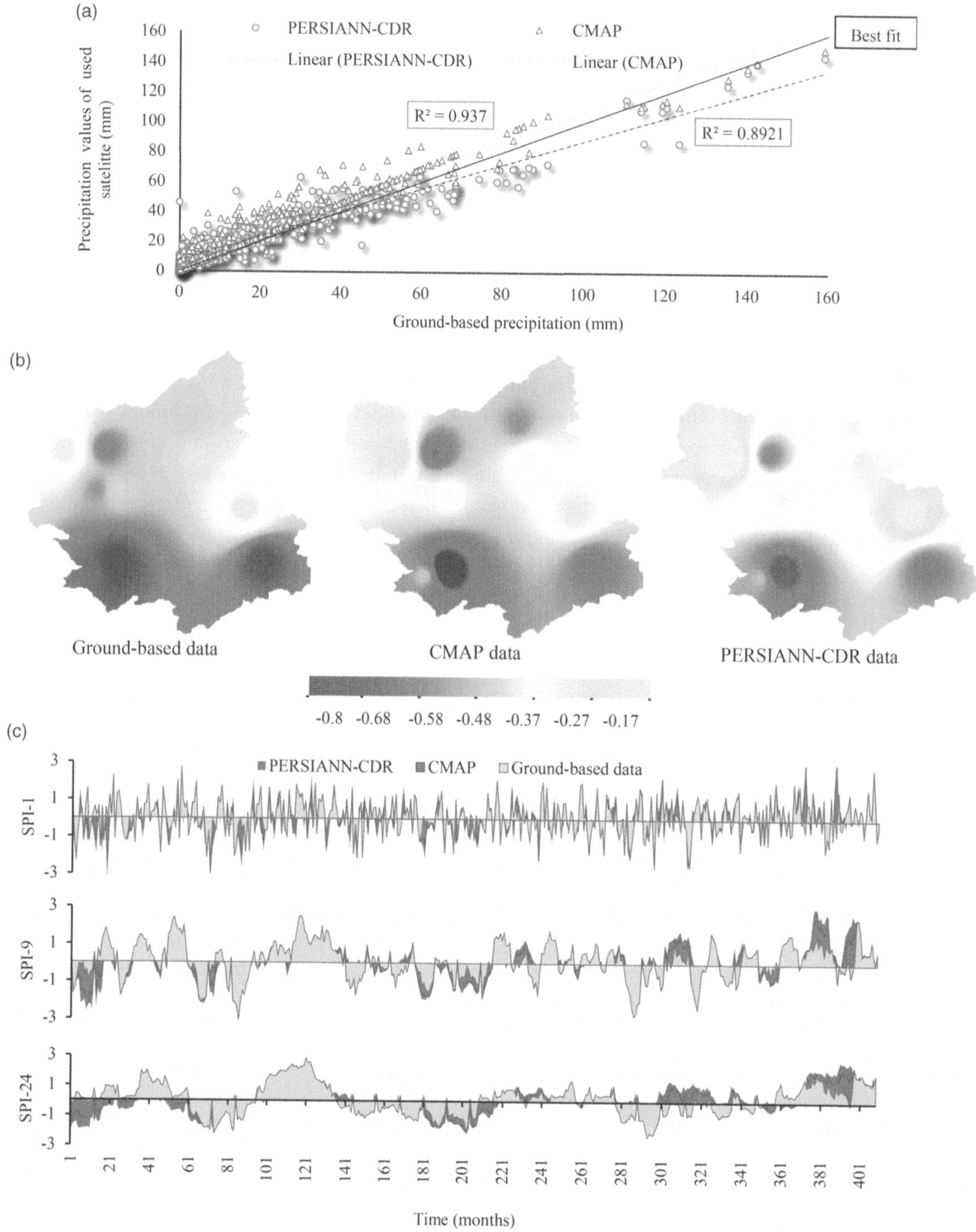

Figure 4 | (a) The correlation between ground-based and satellite precipitation data, (b) zoning map of the selected area based on the mean yearly SPI series, and (c) the SPI series obtained from three datasets.

As it could be seen from this figure, there was a good correlation between ground-based and satellite data. Precipitation data obtained from three datasets were used to calculate the SPI time series. Figure 4(b) shows the zoning map of the mean yearly SPI values during the selected statistical period for these datasets. Based on this figure, the results of PERSIANN-CDR and CMAP satellite datasets were similar to those of ground-based stations. For all datasets, drought was more severe in the Southern parts of the region, and slight drought had occurred in the Northwest of the selected area. Therefore, the SPI time series obtained from the precipitation products of the PERSIANN-CDR and CMAP satellites were successful in drought monitoring in the selected region. Also, the results of CMAP satellite data were closer to the ground-based station results. Figure 4(c) shows the SPI time series obtained from three datasets at 1-, 9-, and 24-month timescales (SPI-1, SPI-9, and SPI-24). From these SPI time series, it was found that different drought levels occurred in the area, but in most of the cases, the drought events were slight droughts (i.e. $-1 < \text{SPI} \leq 0$). The satellite datasets were used in order to investigate whether it is possible to model drought conditions using the satellite precipitation products and obtain desirable prediction accuracy via the hybrid methodology proposed in this study, especially in the areas without any stations. In the next part of the study, the SPI time series were applied for the prediction of the short-term, mid-term, and long-term droughts.

Results of AI models for point predictions (Scenario 1)

In this part, the capability of AI methods was investigated in modeling the SPIs via the original series (i.e. without using data processing methods). The previous values in the SPI time series, relative humidity, average temperature, and precipitation were used as inputs. For obtaining desirable modeling results, setting specific parameters of each AI approach is required, and their optimal values should be determined. For the design of KB approaches, the appropriate kind of kernel function selection should be done. According to Gill *et al.* (2006), there are various kernel functions for selection; however, among them, the RBF has led to better prediction results in the study of various hydrological phenomena. Also, for the FFNN modeling, various networks were tested to determine the node numbers of hidden layer, since the network topology has direct impacts on the capability of FFNN computational complexity and generalization. Appropriate activation functions for the hidden and output nodes were selected as the tangent sigmoid and pure linear functions, respectively. The backpropagation approach scheme was applied to train the model. The results of the single AI models are shown in Table 3. From the obtained results of statistical parameters (i.e. R, DC, RMSE), it can be seen that the prediction accuracy of single methods was almost the same. However, among them, the GPR approach yielded slightly better prediction results. It could be stated that modeling accuracy was deteriorated as the SPI degree reduced. The single AI methods did not provide desirable accuracy in modeling the SPI-1, while their prediction accuracy was satisfactory for the SPI-24. The reason is that the SPI-24 is the normalization of rainfall data over the last 24 months; therefore, unlike the SPI-1, it is not sensitive to data changes from 1 month to the next. In Figure 5, the scatter plots of observed vs. predicted SPI-24 data are shown for the superior model, i.e. GPR.

Results of hybrid EEMD-AI and VMD-AI for point predictions (Scenario 2)

The capability of preprocessing methods in enhancing the accuracy of the models was investigated in this part. In this regard, at first, the de-noised SPI time series were decomposed using EEMD and VMD methods. According to Wu & Huang (2009), the principle of EEMD is decomposition of signal to different IMFs and one residual signal. The sum of these signals will be the same original signal. In the VMD method, signal is decomposed to several IMFs. Via EEMD and VMD, the original drought time series were decomposed into a series of disjoint subseries; then, each subseries was predicted via an appropriate GPR model construction. Since the prediction results of the single GPR model were somewhat better than those of the FFNN and SVM models, this method was used in hybrid models for prediction aims. The aggregated output of all predicted subseries was considered as the final simulated output. Figure 6(a) shows the decomposition results of two preprocessing techniques for the SPI-9 obtained from ground-based data. From the results, it can be indicated that the decomposed high-frequency IMFs via the VMD method had a relatively stable trend, which was conducive for prediction. Since the final forecast result of the VMD method was an accumulation of the forecast result of each IMFs, the characteristics of VMD result helped to enhance the prediction accuracy. However, end-point impacts (large fluctuations) occurred during the EEMD decomposition process and affected all IMFs continuously. Therefore, the EEMD prediction result led to a large error.

In the second step, the EEMD, VMD, and PE (H_{PE}) were used for decomposing the original drought series into several subseries to alleviate the modeling complexity. For analyzing each IMF complexity and recombination of IMFs that have similar entropy amounts, the PE was used. The PE values for SPI-9 via all used datasets are shown in Figure 6(b). For avoiding

Table 3 | Statistical parameters of the single AI models for three datasets

	Validation			Testing			Validation			Testing		
Model	***R***	**DC**	**RMSE**	***R***	**DC**	**RMSE**	***R***	**DC**	**RMSE**	***R***	**DC**	**RMSE**
SPI 1	Ground-based data						PERSIANN-CDR data					
GPR	0.631	0.491	0.821	0.588	0.473	0.841	0.657	0.579	0.705	0.597	0.506	0.726
SVM	0.547	0.426	0.893	0.539	0.428	0.965	0.562	0.494	0.792	0.538	0.478	0.795
FFNN	0.562	0.437	0.875	0.525	0.419	0.973	0.592	0.515	0.778	0.558	0.475	0.797
	CMAP data											
GPR	0.548	0.483	0.886	0.523	0.457	0.894						
SVM	0.473	0.431	0.921	0.518	0.409	0.971						
FFNN	0.490	0.427	0.954	0.501	0.412	0.969						
SPI 9	Ground-based data						PERSIANN-CDR data					
GPR	0.837	0.766	0.701	0.817	0.720	0.746	0.829	0.801	0.504	0.802	0.727	0.578
SVM	0.808	0.739	0.775	0.779	0.685	0.780	0.789	0.706	0.584	0.772	0.679	0.642
FFNN	0.811	0.723	0.805	0.779	0.666	0.821	0.779	0.686	0.628	0.734	0.638	0.651
	CMAP data											
GPR	0.833	0.750	0.730	0.812	0.718	0.758						
SVM	0.821	0.699	0.793	0.777	0.667	0.821						
FFNN	0.805	0.674	0.814	0.791	0.653	0.832						
SPI 24	Ground-based data						PERSIANN-CDR data					
GPR	0.918	0.854	0.488	0.849	0.815	0.503	0.906	0.871	0.395	0.876	0.816	0.421
SVM	0.841	0.793	0.526	0.811	0.782	0.658	0.891	0.806	0.412	0.864	0.781	0.514
FFNN	0.829	0.784	0.530	0.795	0.769	0.679	0.875	0.791	0.511	0.856	0.773	0.523
	CMAP data											
GPR	0.903	0.842	0.492	0.847	0.807	0.529						
SVM	0.863	0.790	0.591	0.823	0.768	0.673						
FFNN	0.844	0.781	0.615	0.815	0.746	0.684						

excessive decomposition and decreasing the scale of calculation, the IMFs with similar entropy were combined (see Figure 6(b)).

Table 4 and Figure 7 show the results of ensemble data processing models. By comparing the presented results in Tables 3 and 4, it was found that the signal processing enhanced the prediction accuracy of drought series. The ensemble methods were more efficient than single AI approaches. As it could be seen, the RMSE values of the GPR model for the test series of PERSIANN-CDR data were 0.726 for the SPI-1, 0.578 for the SPI-9, and 0.421 for the SPI-24; the error criterion values of the PE–VMD–GPR reduced to 0.393 for the SPI-1, 0.314 for the SPI-9, and 0.283 for the SPI-24. Therefore, using the ensemble methods, the short- to long-term SPI modeling was performed more accurately. Also, the predictive effect of the PE–VMD–GPR method was significantly better than others. Generally, the data preprocessing (VMD and EEMD) enhanced the modeling efficiency from 25 to 40% in testing sets. For hybrid PE methods, the efficiency improvement of the models was between 40 and 55%.

Constructed PIs using the LUBE method for the SPI time series

For assessing the capability of the LUBE method in constructing the PIs of short- to long-term SPI time series, the IMFs were used as the GPR model inputs to project the SPI values (point predictions) and related PIs. Due to the better prediction results of single GPR model, this method was used for PI prediction aim. The associated confidence level for all PIs was considered as 95%. Two criteria, PICP and NMPIW, were used to determine the optimum structure of GPR in the PI construction (with the target to maximize PICP and minimize NMPIW). Table 5 and Figure 8 show the obtained results. As can be seen, the

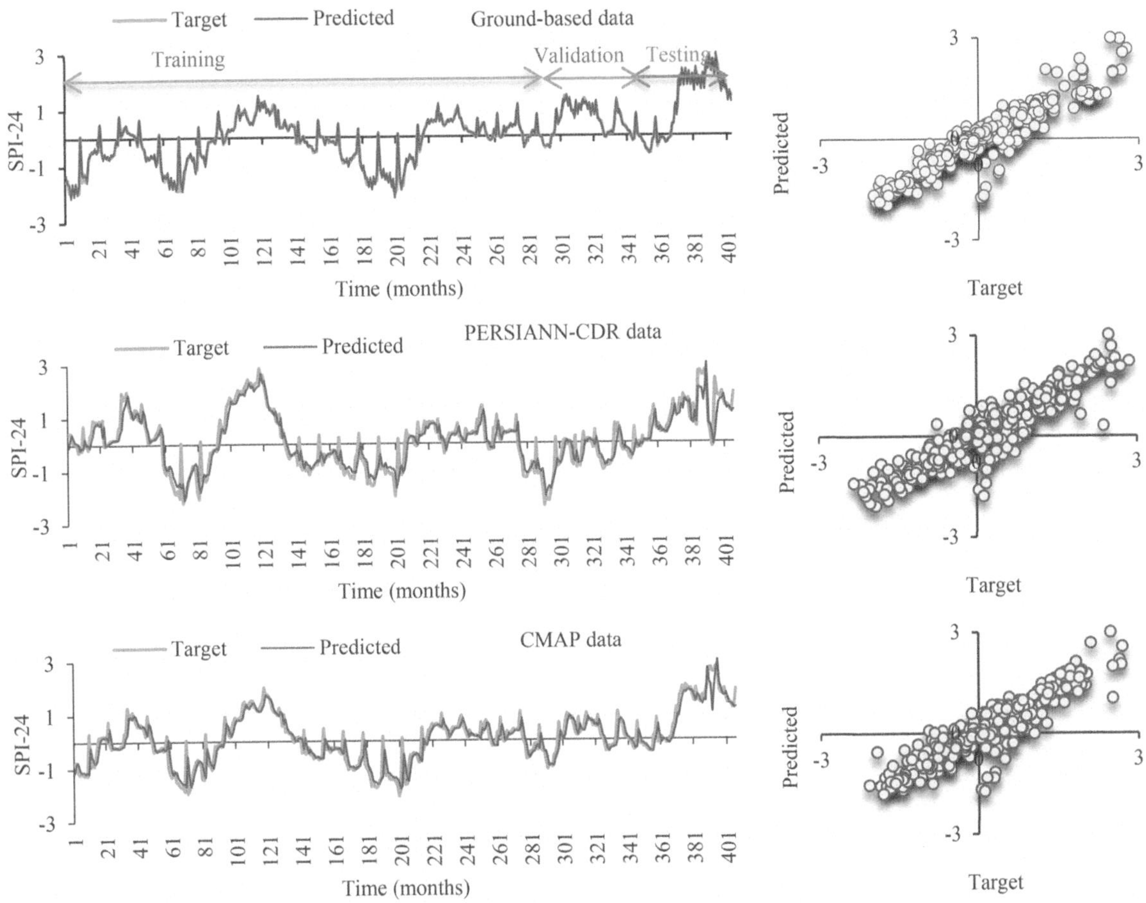

Figure 5 | The observed vs. predicted SPI-24 scatter plots for the GPR model.

PICP values for the hybrid PE–VMD–GPR model were up to 24, 19, and 12% higher than the single GPR, while the NMPIW values were up to 30, 26, and 32% less than the single GPR, for the SPI-1, SPI-9, and SPI-24, respectively, in comparison with other models. Therefore, the hybrid PE–VMD–GPR model led to significantly better outcomes both in point and interval predictions.

Verifying the capability of developed methods for modeling NDVI time series

According to Tucker (1979), satellite products make it possible to monitor vegetation density via an index called vegetation index, which makes the use of the difference in spectral reflection of green vegetation in the near-infrared (NIR) and the red parts of the spectrum (RED) (Fensholt *et al.* 2009). The data provided by weather satellites like Terra can be used for obtaining the vegetation index time series over long periods. Many researchers (e.g. Nicholson *et al.* 1990; Liu & Juárez 2001; Karabulut 2003; Fensholt *et al.* 2009) assessed the relationship between precipitation and NDVI in different areas and indicated that the NDVI can be successfully used for drought monitoring. Therefore, in this study, NDVI series was further used for verifying the proposed methods. The NDVI values vary from −1 to +1, and according to Tucker (1979), higher NDVI values show the greater photosynthetic capacity of vegetation canopy. The NDVI series is defined below:

$$NDVI = \frac{NIR - RED}{NIR + RED} \quad (13)$$

In this study, the NDVI time series were obtained for the period of February 2000 to December 2017 from MODIS. The 16-day composite NDVI images were used for obtaining the monthly average NDVI values. In Figure 9(a), the NDVI images of one dry-year and one wet-year are shown. As seen, the NDVI successfully showed vegetation conditions in these 2 years, and

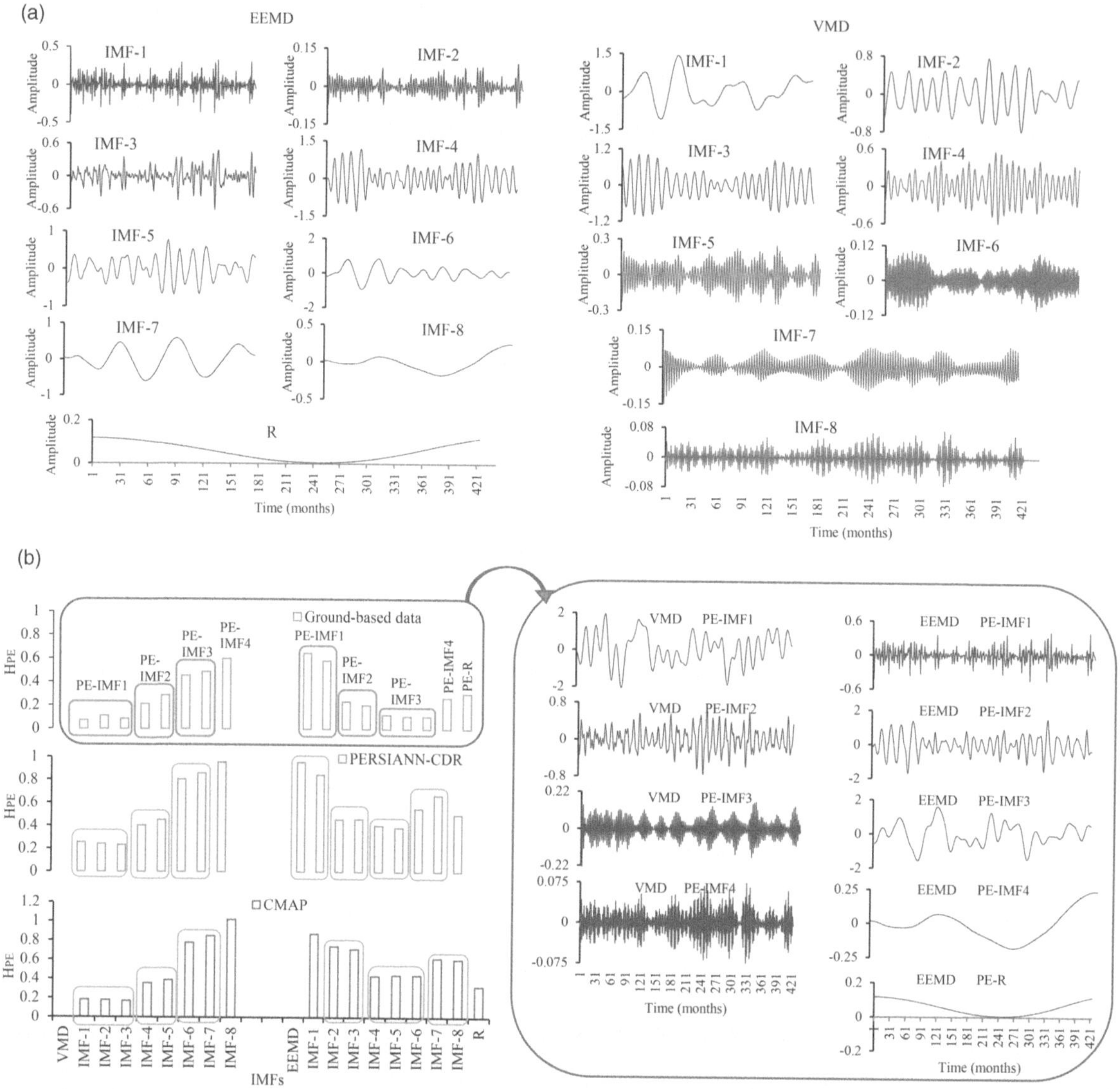

Figure 6 | (a) Decomposition subseries of the SPI-9 obtained from ground-based data and (b) the H_{PE} values for SPI-9 via all used datasets and recombined IMFs for the ground-based dataset.

it could be used for effectively assessing and monitoring the changes in vegetation cover. In the next step, the developed methods were applied for point and interval predictions of the monthly NDVI time series. The obtained results are listed in Table 6 and shown in Figure 9(b). From the analysis, it can be seen that the hybrid models were successful in the modeling process and the PE–VMD–GPR approach resulted in more reliable outcomes in both point and interval predictions of the monthly NDVI time series. The PICP value for the hybrid PE–VMD–GPR model was up to 15% higher than the single GPR, whereas the NMPIW value was approximately 37% less than GPR.

CONCLUSIONS

The accurate prediction of drought is crucial in understanding the drought process and improving water management practices. The current study investigated the capability of single and integrated AI methods in point and interval drought predictions, considering short-, mid-, and long-term timescales. The ground-, PERSIANN-CDR-, and CMAP-based precipitation datasets were used for the prediction of 1-, 9-, and 24-month timescale SPI time series. First, unprocessed time series were applied in the modeling process; then, these time series were decomposed into IMF subseries using EEMD and VMD. The PE was used to select the IMFs with similar entropy values and recombine them. According to the results,

Table 4 | Statistical parameters of the hybrid GPR models for three datasets

Model	Validation			Testing			Validation			Testing		
	R	DC	RMSE	*R*	DC	RMSE	*R*	DC	RMSE	*R*	DC	RMSE
SPI 1	Ground-based data						PERSIANN-CDR data					
EEMD–GPR	0.757	0.623	0.657	0.706	0.618	0.673	0.788	0.695	0.564	0.716	0.607	0.581
PE–EEMD–GPR	0.902	0.752	0.398	0.877	0.725	0.412	0.939	0.887	0.316	0.890	0.775	0.422
VMD–GPR	0.856	0.666	0.578	0.797	0.641	0.592	0.891	0.785	0.496	0.810	0.686	0.511
PE–VMD–GPR	0.952	0.873	0.376	0.927	0.841	0.389	0.948	0.887	0.299	0.940	0.823	0.393
	CMAP data											
EEMD–GPR	0.658	0.580	0.687	0.628	0.548	0.694						
PE–EEMD–GPR	0.783	0.740	0.444	0.780	0.700	0.449						
VMD–GPR	0.743	0.655	0.624	0.709	0.620	0.629						
PE–VMD–GPR	0.833	0.859	0.419	0.830	0.812	0.424						
SPI 9	Ground-based data						PERSIANN-CDR data					
EEMD–GPR	0.870	0.881	0.561	0.850	0.839	0.597	0.914	0.921	0.308	0.862	0.836	0.384
PE–EEMD–GPR	0.928	0.939	0.314	0.906	0.894	0.345	0.975	0.962	0.223	0.919	0.891	0.318
VMD–GPR	0.897	0.907	0.494	0.875	0.864	0.525	0.942	0.949	0.293	0.888	0.861	0.374
PE–VMD–GPR	0.948	0.944	0.296	0.926	0.902	0.326	0.985	0.971	0.211	0.939	0.896	0.314
	CMAP data											
EEMD–GPR	0.866	0.863	0.584	0.926	0.826	0.606						
PE–EEMD–GPR	0.924	0.919	0.334	0.968	0.880	0.354						
VMD–GPR	0.892	0.888	0.514	0.953	0.850	0.534						
PE–VMD–GPR	0.944	0.924	0.315	0.970	0.885	0.334						
SPI 24	Ground-based data						PERSIANN-CDR data					
EEMD–GPR	0.968	0.937	0.415	0.920	0.897	0.428	0.973	0.938	0.301	0.950	0.915	0.304
PE–EEMD–GPR	0.979	0.973	0.285	0.941	0.931	0.296	0.985	0.974	0.215	0.971	0.949	0.291
VMD–GPR	0.969	0.940	0.365	0.931	0.899	0.376	0.975	0.941	0.295	0.961	0.917	0.315
PE–VMD–GPR	0.980	0.978	0.269	0.946	0.936	0.280	0.990	0.979	0.203	0.976	0.954	0.283
	CMAP data											
EEMD–GPR	0.970	0.935	0.418	0.918	0.920	0.450						
PE–EEMD–GPR	0.981	0.951	0.288	0.939	0.937	0.316						
VMD–GPR	0.971	0.937	0.368	0.929	0.923	0.396						
PE–VMD–GPR	0.986	0.956	0.272	0.944	0.941	0.298						

the single AI methods did not lead to desirable efficiency in short-term drought modeling in the selected area. For longer time-scale SPI time series, the model efficiency was improved, and standalone methods were slightly better in SPI-24 modeling. It was shown that hybrid preprocessing methods enhanced the performance of standalone AI approaches approximately 25–40%, whereas hybrid PE methods enhanced the model performance between 40 and 55%. On the other hand, since point predictions via AI methods do not convey any details about the uncertainty of prediction, PI can be an essential index to quantify the reliability of AI-based modeling of hydrological time series. In this study, PIs of the GPR-based model were obtained using the LUBE method. The NMPIW and PICP criteria were used to quantitatively evaluate the efficiency of the constructed PIs. Obtained results indicated that the VMD–PE–GPR led to more desirable results. The PICP of this hybrid model was approximately 24, 19, and 12% higher, and the NMPIW was approximately 30, 26, and 32% lower than the GPR method for SPI-1, SPI-9, and SPI-24, respectively. Furthermore, the performance of the developed models was evaluated for the NDVI series modeling. It was found that the proposed hybrid models were successful in the modeling process

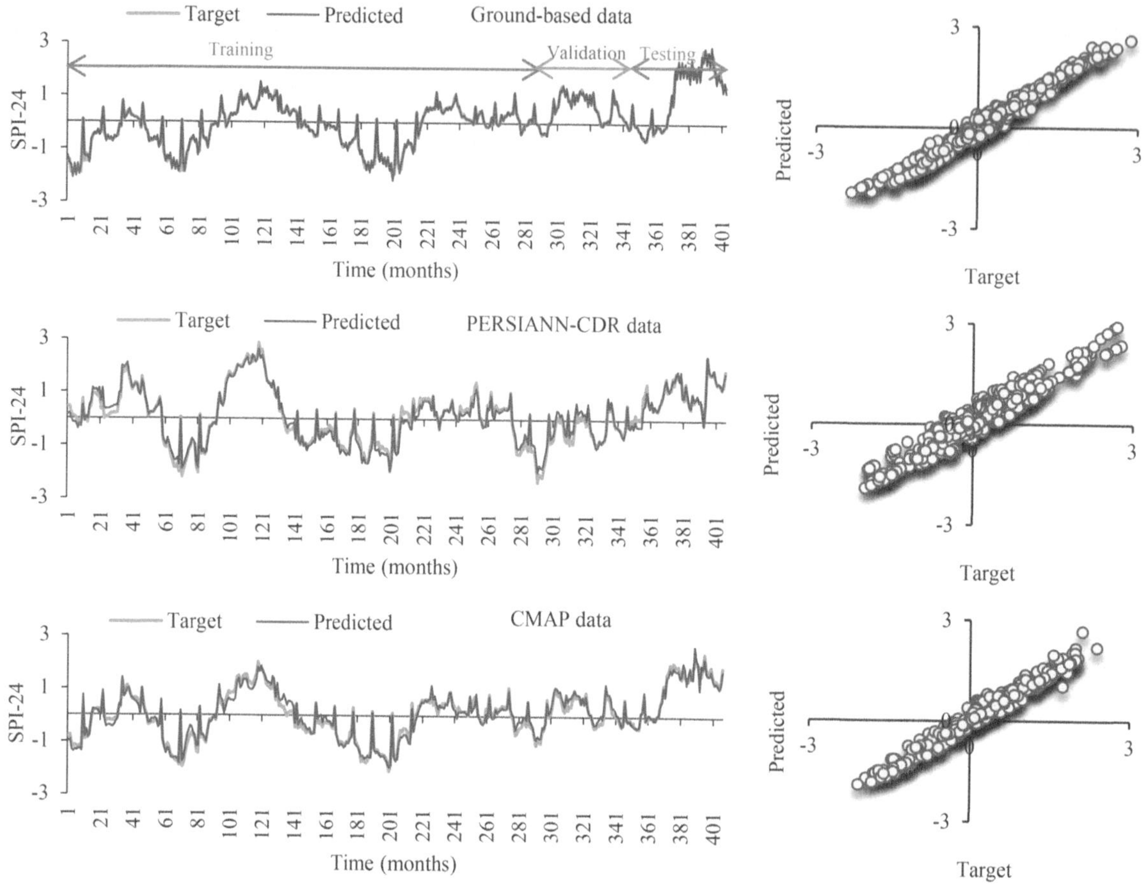

Figure 7 | The observed vs. predicted SPI-24 scatter plots for the PE–VMD–GPR model.

Table 5 | Constructed PI results for three datasets of LUBE methods

Model	Ground-based data		PERSIANN-CDR data		CMAP data	
	PICP	NMPIW	PICP	NMPIW	PICP	NMPIW
SPI 1						
GPR	0.77	2.49	0.75	2.57	0.78	2.55
EEMD–GPR	0.91	2.21	0.90	2.22	0.89	2.23
PE–EEMD–GPR	0.94	1.78	0.94	1.82	0.93	1.64
VMD–GPR	0.94	1.79	0.95	1.68	0.95	1.08
PE–VMD–GPR	0.96	0.80	0.95	1.03	0.96	1.02
SPI 9						
GPR	0.82	0.71	0.84	0.81	0.86	0.77
EEMD–GPR	0.93	0.69	0.91	0.79	0.91	0.73
PE–EEMD–GPR	0.96	0.55	0.97	0.51	0.96	0.54
VMD–GPR	0.96	0.62	0.93	0.65	0.95	0.63
PE–VMD–GPR	0.98	0.56	0.95	0.54	0.98	0.55
SPI 24						
GPR	0.88	0.64	0.86	0.65	0.90	0.67
EEMD–GPR	0.95	0.55	0.94	0.59	0.92	0.61
PE–EEMD–GPR	0.96	0.49	0.94	0.48	0.96	0.49
VMD–GPR	0.97	0.48	0.95	0.52	0.97	0.51
PE–VMD–GPR	0.99	0.48	0.97	0.51	0.98	0.47

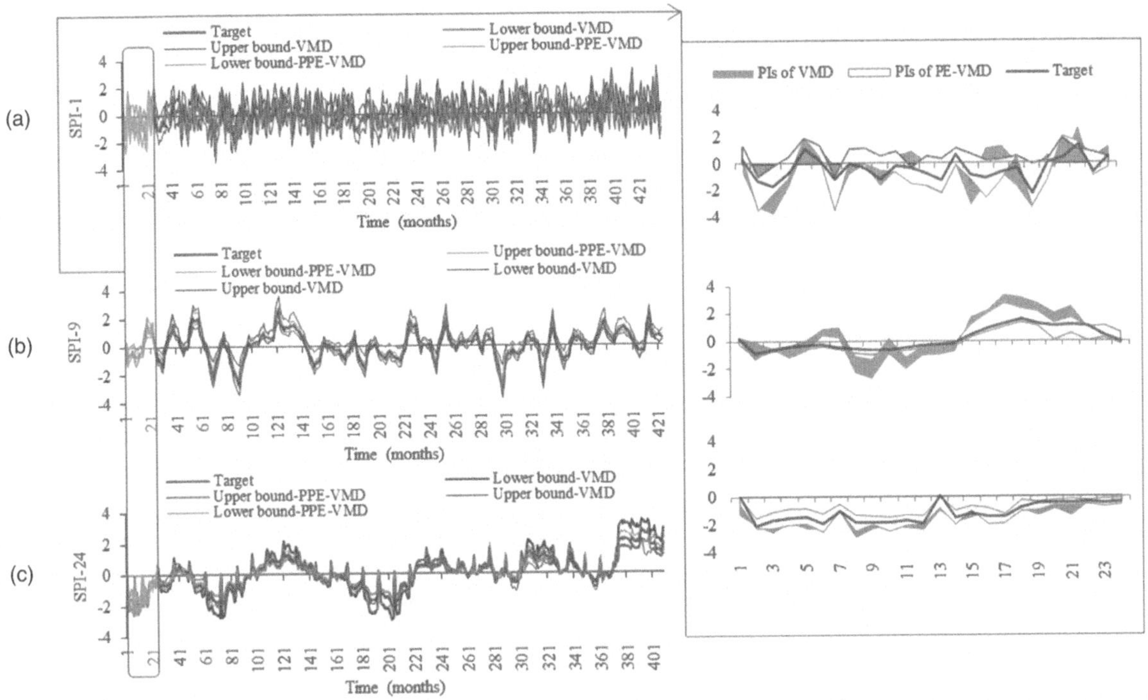

Figure 8 | Results of the PI construction by the LUBE method for (a) CMAP data, (b) ground-based data, and (c) PERSIANN-CDR data.

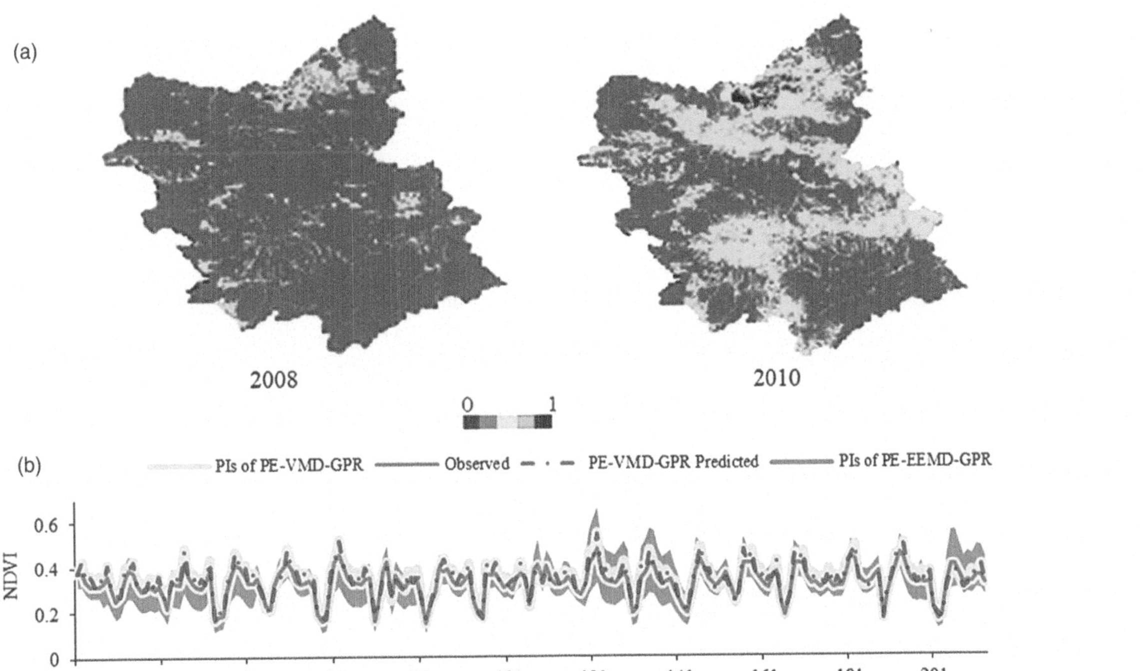

Figure 9 | (a) Comparison of the NDVI images of one dry-year (2008) and one wet-year (2010) for the selected area and (b) results of the PIs construction by the LUBE method for NDVI modeling.

Table 6 | Statistical parameters and constructed PI results of the hybrid GPR models for the NDVI modeling

	Validation			Testing				
Model	**R**	**DC**	**RMSE**	***R***	**DC**	**RMSE**	**PICP**	**NMPIW**
GPR	0.814	0.802	0.038	0.814	0.794	0.041	0.82	0.29
EEMD–GPR	0.833	0.855	0.029	0.832	0.818	0.033	0.88	0.21
PE–EEMD–GPR	0.924	0.913	0.016	0.901	0.901	0.021	0.96	0.13
VMD–GPR	0.901	0.905	0.018	0.912	0.892	0.024	0.91	0.14
PE–VMD–GPR	0.948	0.931	0.012	0.922	0.908	0.017	0.95	0.11

and provided desirable accuracy. The results obtained in this study supported the premise that hybrid data processing models had considerable potential to be used as an alternative approach for drought prediction based on ground-based and remote sensing data. For the prediction of complex short- to long-term hydrological time series, the developed hybrid models can be used. It should, however, be noted that the AI models are data-driven models, and thus, they are data sensitive. That is why studies on regions with different climatic characteristics are required to conclusively prove the advantages of the proposed models in predicting the drought time series. In addition, different types and lengths of meteorological data and satellite datasets used in this paper may affect the fitting and prediction accuracy of meteorological and remote sensing data models, which should be further discussed in the future.

DATA AVAILABILITY STATEMENT

All relevant data are included in the paper or its Supplementary Information.

REFERENCES

Abdoos, A. A. 2016 A new intelligent method based on combination of VMD and ELM for short term wind power forecasting. *Neurocomputing* **203**, 111–120.

Adamowski, K., Prokoph, A. & Adamowski, J. 2009 Development of a new method of wavelet aided trend detection and estimation. *Hydrological Processes: An International Journal* **23** (18), 2686–2696.

Aghelpour, P., Mohammadi, B., Biazar, S. M., Kisi, O. & Sourmirinezhad, Z. 2020a A theoretical approach for forecasting different types of drought simultaneously, using entropy theory and machine-learning methods. *ISPRS International Journal of Geo-Information* **9** (12), 701.

Aghelpour, P., Bahrami-Pichaghchi, H. & Kisi, O. 2020b Comparison of three different bio-inspired algorithms to improve ability of neuro fuzzy approach in prediction of agricultural drought, based on three different indexes. *Computers and Electronics in Agriculture* **170**, 105279.

Aghelpour, P., Guan, Y., Bahrami-Pichaghchi, H., Mohammadi, B., Kisi, O. & Zhang, D. 2020c Using the MODIS sensor for snow cover modeling and the assessment of drought effects on snow cover in a mountainous area. *Remote Sensing* **12** (20), 3437.

Ashouri, H., Hsu, K. L., Sorooshian, S., Braithwaite, D. K., Knapp, K. R., Cecil, L. D., Nelson, B. R. & Prat, O. P. 2015 PERSIANN-CDR: daily precipitation climate data record from multisatellite observations for hydrological and climate studies. *Bulletin of the American Meteorological Society* **96** (1), 69–83.

Azamathulla, H. M., Haghiabi, A. H. & Parsaie, A. 2016 Prediction of side weir discharge coefficient by support vector machine technique. *Water Science and Technology: Water Supply* **16** (4), 1002–1016.

Babovic, V. 2009 Introducing knowledge into learning based on genetic programming. *Journal of Hydroinformatics* **11** (3–4), 181–193.

Basalirwa, C., Odiyo, J., Mngodo, R. & Mpeta, E. 1999 The climatological regions of Tanzania based on the rainfall characteristics. *International Journal of Climatology* **19** (1), 69–80.

Brown, S. C., Versace, V. L., Lester, R. E. & Walter, M. T. 2015 Assessing the impact of drought and forestry on streamflows in south-eastern Australia using a physically based hydrological model. *Environmental Earth Sciences* **74**, 6047–6063.

Chadalawada, J., Havlicek, V. & Babovic, V. 2017 A genetic programming approach to system identification of rainfall-runoff models. *Water Resources Management* **31** (12), 3975–3992.

Chen, T., De, J. R. A. M., Liu, Y., Van, D. W. G. R. & Dolman, A. J. 2010 Using satellite based soil moisture to quantify the water driven variability in NDVI: a case study over mainland Australia. *Ecological Informatics* **5** (5), 400–409.

Deo, R. C., Kisi, O. & Singh, V. P. 2017 Drought forecasting in eastern Australia using multivariate adaptive regression spline, least square support vector machine and m5tree model. *Atmospheric Research* **184**, 149–175.

Donoho, D. L. 1995 De-noising by soft-thresholding. *IEEE Transactions on Information Theory* **41** (3), 613–627.

Fensholt, R., Rasmussen, K., Nielsen, T. T. & Mbow, C. 2009 Evaluation of earth observation based long term vegetation trends-intercomparing NDVI time series trend analysis consistency of Sahel from AVHRR GIMMS, Terra MODIS and SPOT VGT data. *Remote Sensing of Environment* **113** (9), 1886–1898.

Geng, G., Wu, J., Wang, Q., Lei, T., He, B., Li, X., Mo, X., Luo, H., Zhou, H. & Liu, D. 2016 Agricultural drought hazard analysis during 1980–2008: a global perspective. *International Journal of Climatology* **36** (1), 389–399.

Ghulam, A., Qin, Q., Teyip, T. & Li, Z. L. 2007 Modified perpendicular drought index (MPDI): a real-time drought monitoring method. *ISPRS Journal of Photogrammetry and Remote Sensing* **62** (2), 150–164.

Gill, M. K., Asefa, T., Kemblowski, M. W. & McKee, M. 2006 Soil moisture prediction using support vector machines 1. *JAWRA Journal of the American Water Resources Association* **42** (4), 1033–1046.

Hayes, M. J., Svoboda, M. D., Wilhite, D. A. & Vanyarkho, O. V. 1999 Monitoring the 1996 drought using the standardized precipitation index. *Bulletin of the American Meteorological Society* **80** (3), 429–438.

He, F., Zhou, J., Feng, Z. K., Liu, G. & Yang, Y. 2019 A hybrid short-term load forecasting model based on variational mode decomposition and long short-term memory networks considering relevant factors with Bayesian optimization algorithm. *Applied Energy* **237**, 103–116.

Huang, N. E., Shen, Z., Long, S. R., Wu, M. C., Shih, H. H., Zheng, Q., Yen, N. C., Tung, C. C. & Liu, H. H. 1998 The empirical mode decomposition and the Hilbert spectrum for nonlinear and non-stationary time series analysis. *Proceedings of the Royal Society A* **454** (1971), 903–995.

Karabulut, M. 2003 An examination of relationships between vegetation and rainfall using maximum value composite AVHRR-NDVI data. *Turkish Journal of Botany* **27** (2), 93–101.

Kasiviswanathan, K. S., Cibin, R., Sudheer, K. & Chaubey, I. 2013 Constructing prediction interval for artificial neural network rainfall runoff models based on ensemble simulations. *Journal of Hydrology* **499**, 275–288.

Khan, M. M. H., Muhammad, N. S. & El-Shafie, A. 2020 Wavelet based hybrid ANN-ARIMA models for meteorological drought forecasting. *Hydrology* **590**, 125380.

Khosravi, A., Nahavandi, S., Creighton, D. & Atiya, A. F. 2011 Comprehensive review of neural network-based prediction intervals and new advances. *IEEE Transactions on Neural Networks* **22** (9), 1341–1356.

Kisi, O., Gorgij, A. D., Zounemat-Kermani, M., Mahdavi-Meymand, A. & Kim, S. 2019 Drought forecasting using novel heuristic methods in a semi-arid environment. *Journal of Hydrology* **578**, 124053.

Krupnick, A., Morgenstern, R., Batz, M., Nelson, P., Burtraw, D., Shih, J. S. & McWilliams, M. 2006 *Not a Sure Thing: Making Regulatory Choices under Uncertainty*. Resources for the Future, Washington, DC.

Kumar, P. & Foufoula, G. E. 1993 Multicomponent decomposition of spatial rainfall fields, 1. Segregation of large- and small-scale features using wavelet transforms. *Water Resources Research* **29** (8), 2515–2532.

Kwak, J., Kim, S., Jung, J., Singh, V. P., Lee, D. R. & Kim, H. S. 2016 Assessment of meteorological drought in Korea under climate change. *Advances in Meteorology* **2016**, 1–13.

Lei, Y., He, Z. & Zi, Y. 2009 Application of the EEMD method to rotor fault diagnosis of rotating machinery. *Mechanical Systems and Signal Processing* **23** (4), 1327–1338.

Li, X., He, B., Quan, X., Liao, Z. & Bai, X. 2015 Use of the standardized precipitation evapotranspiration index (SPEI) to characterize the drying trend in southwest China from 1982–2012. *Remote Sensing* **7** (8), 10917–10937.

Li, Z., Chen, T., Wu, Q., Xia, G. & Chi, D. 2020 Application of penalized linear regression and ensemble methods for drought forecasting in Northeast China. *Meteorology and Atmospheric Physics* **132** (1), 113–130.

Liu, W. T. & Juárez, R. N. 2001 ENSO drought onset prediction in northeast Brazil using NDVI. *International Journal of Remote Sensing* **22** (17), 3483–3501.

Mann, M. E. & Gleick, P. H. 2015 Climate change and California drought in the 21st century. *Proceedings of the National Academy of Sciences* **112** (13), 3858–3859.

Marvel, K., Cook, B. I., Bonfils, C. J., Durack, P. J., Smerdon, J. E. & Williams, A. P. 2019 Twentieth-century hydroclimate changes consistent with human influence. *Nature* **569**, 59–65.

McKee, T. B., Doesken, N. J. & Kleist, J. 1993 The relationship of drought frequency and duration to time scales. In: *Proceedings of the 8th Conference on Applied Climatology*, pp. 179–184.

Meade, N. & Islam, T. 1995 Prediction intervals for growth curve forecasts. *Journal of Forecasting* **14**, 413–430.

Mehr, A. D., Kahya, E. & Özger, M. 2014 A gene–wavelet model for long lead time drought forecasting. *Hydrology* **517**, 691–699.

Mirabbasi, R., Anagnostou, E. N., Fakheri-Fard, A., Dinpashoh, Y. & Eslamian, S. 2013 Analysis of meteorological drought in northwest Iran using the Joint Deficit Index. *Journal of Hydrology* **7** (492), 35–48.

Mishra, A. K. & Singh, V. P. 2010 A review of drought concepts. *Hydrology* **391** (1–2), 202–216.

Mishra, A., Desai, V. & Singh, V. 2007 Drought forecasting using a hybrid stochastic and neural network model. *Journal of Hydrologic Engineering* **12** (6), 626–638.

Mohamadi, S., Sammen, S. S., Panahi, F., Ehteram, M., Kisi, O., Mosavi, A., Ahmed, A. N., El-Shafie, A. & Al-Ansari, N. 2020 Zoning map for drought prediction using integrated machine learning models with a nomadic people optimization algorithm. *Natural Hazards* **104** (1), 537–579.

Mokhtarzad, M., Eskandari, F., Vanjani, N. J. & Arabasadi, A. 2017 Drought forecasting by ANN, ANFIS, and SVM and comparison of the models. *Environmental Earth Sciences* **76** (21), 1–10.

Mukherjee, S., Mishra, A. & Trenberth, K. E. 2018 Climate change and drought: a perspective on drought indices. *Current Climate Change Reports* **4**, 145–163.

Najafi, B., Faizollahzadeh Ardabili, S., Shamshirband, S., Chau, K. W. & Rabczuk, T. 2018 Application of ANNs, ANFIS and RSM to estimating and optimizing the parameters that affect the yield and cost of biodiesel production. *Engineering Applications of Computational Fluid Mechanics* **12** (1), 611–624.

Nicholson, S. E., Davenport, M. L. & Malo, A. R. 1990 A comparison of the vegetation response to rainfall in the Sahel and East Africa, using normalized difference vegetation index from NOAA AVHRR. *Climatic Change* **17** (2–3), 209–241.

Nischitha, V., Ahmed, S., Varikoden, H. & Revadekar, J. 2014 The impact of seasonal rainfall variability on NDVI in the Tunga and Bhadra river basins, Karnataka, India. *International Journal of Remote Sensing* **35** (23), 8025–8043.

Roushangar, K. & Alizadeh, F. 2018 Entropy-based analysis and regionalization of annual precipitation variation in Iran during 1960–2010 using ensemble empirical mode decomposition. *Hydroinformatics* **2** (2), 468–485.

Roushangar, K., Nourani, V. & Alizadeh, F. 2019 Corrigendum to a multiscale time-space approach to analyze and categorize the precipitation fluctuation based on the wavelet transform and information theory concept. *Hydrology Research* **49** (3), 724–743.

Roushangar, K., Chamani, M., Ghasempour, R., Azamathulla, H. M. & Alizadeh, F. 2021 A comparative study of wavelet-and empirical mode decomposition-based GPR models for river discharge relationship modeling at consecutive hydrometric stations. *Water Supply*. doi.org/10.2166/ws.2021.073.

Sahoo, A. K., Sheffield, J., Pan, M. & Wood, E. F. 2015 Evaluation of the tropical rainfall measuring mission multi-satellite precipitation analysis (TMPA) for assessment of large-scale meteorological drought. *Remote Sensing of Environment* **159**, 181–193.

Sakizadeh, M., Mohamed, M. M. & Klammler, H. 2019 Trend analysis and spatial prediction of groundwater levels using time series forecasting and a novel spatio-temporal method. *Water Resources Management* **33** (4), 1425–1437.

Smola, A. J. 1996 *Regression Estimation with Support Vector Learning Machines*. Master's Thesis, Technische Universität München, Germany.

Tayfur, G. 2012 *Soft Computing Methods in Water Resources Engineering: Artificial Neural Networks, Fuzzy Logic, and Genetic Algorithm*. WIT Press, Southampton, UK.

Tsakiris, G. & Vangelis, H. 2004 Towards a drought watch system based on spatial SPI. *Water Resources Management* **18** (1), 1–12.

Tucker, C. J. 1979 Red and photographic infrared linear combinations for monitoring vegetation. *Remote Sensing of Environment* **8** (2), 127–150.

Von Hardenberg, J., Meron, E., Shachak, M. & Zarmi, Y. 2001 Diversity of vegetation patterns and desertification. *Physical Review Letters* **87** (19), 198101–198104.

Wang, W. C., Chau, K. W., Qiu, L. & Chen, Y. B. 2015 Improving forecasting accuracy of medium and long-term runoff using artificial neural network based on EEMD decomposition. *Environmental Research* **139**, 46–54.

Wu, Z. H. & Huang, N. E. 2009 Ensemble empirical mode decomposition: a noise assisted data analysis method. *Advances in Adaptive Data Analysis* **1**, 1–41.

Wu, L., Peng, Y., Fan, J. & Wang, Y. 2019 Machine learning models for the estimation of monthly mean daily reference evapotranspiration based on cross-station and synthetic data. *Hydrology Research* **50** (6), 1730–1750.

Xie, P. & Arkin, P. A. 1996 Analyses of global monthly precipitation using gauge observations, satellite estimates, and numerical model predictions. *Journal of Climate* **9** (4), 840–858.

Xu, L., Chen, N., Zhang, X. & Chen, Z. 2018 An evaluation of statistical, NMME and hybrid models for drought prediction in China. *Hydrology* **566**, 235–249.

Yilmaz, M. T., Hunt, E. R. & Jackson, T. J. 2008 Remote sensing of vegetation water content from equivalent water thickness using satellite imagery. *Remote Sensing of Environment* **112** (5), 2514–2522.

Yin, Z., Wen, X., Feng, Q., He, Z., Zou, S. & Yang, L. 2017 Integrating genetic algorithm and support vector machine for modeling daily reference evapotranspiration in a semi-arid mountain area. *Hydrology Research* **48** (5), 1177–1191.

Zhang, R., Chen, Z. Y., Xu, L. J. & Ou, C. Q. 2019 Meteorological drought forecasting based on a statistical model with machine learning techniques in Shaanxi province, China. *Science of the Total Environment* **665**, 338–346.

Zhou, Q., Leng, G. & Peng, J. 2018 Recent changes in the occurrences and damages of floods and droughts in the United States. *Water* **10** (9), 1109.

First received 24 March 2021; accepted in revised form 21 May 2021. Available online 18 June 2021

doi: 10.2166/nh.2021.030

Stochastic modeling of artificial neural networks for real-time hydrological forecasts based on uncertainties in transfer functions and ANN weights

Shiang-Jen Wu[a,*], Chih-Tsung Hsu[b] and Che-Hao Chang[c]

[a] Department of Civil and Disaster Prevention Engineering, National United University, Miaoli 36003, Taiwan
[b] National Center for High-Performance Computing, Hsinchu 30076, Taiwan
[c] Department of Civil Engineering, National Taipei University of Technology, Taipei 10608, Taiwan
*Corresponding author. E-mail: sjwu@nuu.edu.tw

ABSTRACT

This study proposes a stochastic artificial neural network (named ANN_GA-SA_MTF), in which the parameters of the multiple transfer functions considered are calibrated by the modified genetic algorithm (GA-SA), to effectively provide the real-time forecasts of hydrological variates and the associated reliabilities under the observation and predictions given (model inputs); also, the resulting forecasts can be adjusted through the real-time forecast-error correction method (RTEC_TS&KF) based on difference between real-time observations and forecasts. The observed 10-days rainfall depths and water levels (i.e., hydrological estimates) from 2008 to 2018 recorded within the Shangping sub-basin in northern Taiwan are adopted as the study data and their stochastic properties are quantified for simulating 1,000 sets of rainfall and water levels at 36 10-days periods as the training datasets. The results from the model verification indicate that the observed 10-days rainfall depths and water levels are obviously located at the prediction interval (i.e., 95% confidence interval), revealing that the proposed ANN_GA-SA_MTF model can capture the temporal behavior of 10-days rainfall depths and water levels within the study area. In spite of the resulting forecasts with an acceptable difference from the observation, their real-time corrections have evident agreement with the observations, namely, the resulting adjusted forecasts with high accuracy.

Key words: artificial neural network, genetic algorithm, Monte Carlo simulation, real-time error correction, transfer function

HIGHLIGHTS

- This study presents a stochastic ANN model.
- Use GA-SA method to calibrate the ANN weights.
- A large training dataset is accomplished via a non-nomral multivariate Monte Carlo simulation approach.
- Real-time correction method is used to adjust the hydrological forecast.

1. INTRODUCTION

Recently, physically and statistically-based hydrological models are frequently applied to predict the rainfall-induced hydrological variates in response to the effect of extreme rainfall on runoff dynamics under conditions of the observed and forecasted hydrological data given (e.g., precipitation and water levels) (e.g., Bertoni *et al.* 1992; Dawson & Wilby 2001; Xia 2002; Sun & Kim 2014; Chang *et al.* 2018; Sun *et al.* 2020; Yang *et al.* 2020). The resulting hydrological forecasts with high accuracy and efficiency play an important role in the flooding early-warning/risk-management and water resources management (e.g., Wang *et al.* 2009; Sayers *et al.* 2014; Sung *et al.* 2017; Sankaranarayanan *et al.* 2019; Yang *et al.* 2020). Generally speaking, the aforementioned relevant hydrological-analysis modeling can be grouped into three types: data-driven model, conceptual model and distributed model in describing the hydrological cycle process, such as the rainfall-runoff process and rainfall-induced inundation (Kan *et al.* 2016). Of three types of hydrological models, the reliability and accuracy of the resulting hydrological forecasts might be influenced by uncertainties in the perceptual model, the measurement of input data and parameter calibration (Melching 1995; Gupta *et al.* 2005; Wagener & Gupta 2005). Furthermore, the hydrological modeling with high resolution in time and space, such as the watershed hydrological models which focus on simulating the entire hydrological cycle process, might be established through numerical coding, especially for the real-time forecasting of hydrological variates (e.g. precipitation, water level/inundation depth); accordingly, this is possibly executed with long computation time in order calibrating the optimal values of model parameters and in achieving the hydrology forecasts (e.g., Paschalis *et al.* 2013: Kan *et al.* 2019; Sun *et al.* 2020). That is to say, desipte the hydrological modeling can be advantageous in emulating the rainfall-runoff-water level characteristic by means of numerical coding based on a deterministic and conceptual relationship given in advance, its performance and effectiveness are probably impacted due to the uncertainties in the complicated model structure and a number of associated parameters which represent the spatial scale differing more than the observation or which have no direct measurements regrading physical signification (Wagener & Gupta 2005; Melsen *et al.* 2016).

To overcome the above disadvantage, the artificial intelligent (AI) modeling is widely applied in the simulations and predictions in regard to the flood-related hydrological variates (e.g., precipitation, discharge and water level) (e.g., Campolo *et al.* 1999; Imrie *et al.* 2000; Cigizoglu & Kisi 2005; Firat 2008; Wang *et al.* 2009; Shamseldin 2010; Maca *et al.* 2014; Khan *et al.* 2016; Sung *et al.* 2017; Malik *et al.* 2019). For example, Campolo *et al.* (1999) utilized the logistic function as the transfer function, namely, the activation function, to train an ANN model which describes the spatial relationship between rainfall and water levels to issue forecasting information on the distributed water levels; Shamseldin (2010) proposed an ANN-derived rainfall-runoff model based on the structure of the multi-layer perceptron with a specific transfer function (i.e., the logistic/sigmoid function) to provide the river-runoff forecasts using the weighted average of rainfall and expectation of the rainfall index and the observed discharge as model inputs. Numerous types of the neural-network (NN) models have been proposed to forecast the hydrological variables and simulate the water resource process, such as the rainfall-runoff-water levels in the watershed and urban area, including the adaptive neuro fuzzy inference system (ANFIS), artificial neural network methods (ANN), generalized regression neural networks (GRNN), feed forward networks (FFNN), support vector machine (SVM) and genetic programing (Firat 2008; Wang *et al.* 2009). It is well known that although the ANFIS model is generally superior to the ANN model in forecasting hydrological estimates, the ANN model can be efficiently and usefully applied in modeling difficult and complicated phenomena described in terms of nonlinear mathematic relationship by constructing the linear multi-layers network using all possible predictor variables through the multiple training algorithm (Tu 1996; Khan *et al.* 2016).

Since the ANN-derived models are formatted by establishing the linear relationships between the neurons at input-hidden-output layers, the number of hidden layers and associated number of neurons should be given in advance. Sheela & Deepa (2013) summarized a variety of formulae for estimating the number of connection neurons required in an ANN model under a given number of hidden layers; the above coefficients of the aforementioned linear relationships between neurons (named ANN weights) are commonly determined by means of the optimization method with a known objective function (i.e. loss function) under a specific network structure. Among the well-known optimization approaches, the back-propagation (BP) algorithm integrated with the gradient descent method, is frequently adopted to calibrate ANN weights (e.g., Campolo *et al.* 1999; Cigizoglu & Kisi 2005; Sung *et al.* 2017; Chen *et al.* 2019). In detail, with the BP algorithm used in training the ANN model, the adjusted ANN weights and bias are calculated by the gradient method with a value of the loss function (e.g., root mean square error, RMSE) under given initial value of neurons weights and bias; as the minimum RMSE can be obtained, the corresponding adjusted ANN weights and bias are viewed as the optimal parameters regarding the ANN model. However, the resulting ANN weights fail to be the local optimal values with high likelihood attributed to their given inappropriate initial values; also, the optimal ANN weights are obtained with difficulty due to the unreasonable learning rate which causes the problem of oscillation; this possibly slows down the convergence speed of the BP neural network (Wang *et al.* 2015). Additionally, although the reliability of the estimates provided by the ANN model is propionate to the number of hidden layers and neurons, but the calibration of their optimal parameters are accomplished with greater difficulty, especially for the large number of model inputs (Xu *et al.* 2015). To solve the above problem, a genetic algorithm is employed to optimize the parameters of the ANN model, including the connection weights and network structure (i.e., number of hidden layers) (e.g., Xu *et al.* 2015; Tahboub *et al.* 2016; Chen *et al.* 2019). Nevertheless, the accuracy of the resulting optimal model parameters is possibly affected by the uncertainties in the observations and the types of objective functions considered (Wu *et al.* 2012a, 2012b). The resulting neuron weights from a traditional GA possibly also fall into the local optimal values under consideration of the complex neural network (Chen *et al.* 2019). To avoid obtaining the local optimizations for a complex model structure, Wu *et al.* (2012a, 2012b) modified a GA based on the sensitivity of model outputs to the model parameters, named GA-SA, in which the chromosome and genes are selected in the crossover and mutation based on the sensitivity of model parameters. Thus, in this study, the GA-SA algorithm with a specific loss function is employed to training the ANN model in order to determine the connection weights (i.e., the ANN weights) and bias for a specific transfer function under consideration of a specific network structure.

Apart from the forecast-related investigations, the effect of the uncertainty factors on the hydrological estimates is supposed to be evaluated through the numerical hydrological/hydraulic models integrated with the uncertainty/risk analysis (e.g., Tung & Yen 2005; Wu *et al.* 2012a, 2012b, 2015); the results from the uncertainty/risk analysis enable to assess the reliability of the hydrological estimates. Therefore, stochastic modeling for the ANN model could be presented by integrating the statistical analysis with the results from the ANN models (e.g., Wegman & Habib 1990; Yetilmezsoy & Saral 2007; Malik *et al.* 2019). In general, the above stochastic ANN model can be classified into two types: one is to quantify the uncertainty of model input by adding a resulting bias from a prior probability function (Wegman & Habib 1990) or using the large number of generated model inputs (Yetilezsoy and Saral); the other is to quantify the stochastic properties by carrying out the statistical analysis for the forecasts provided by the ANN model (Malik *et al.* 2019). As a result, to efficiently provide the real-time hydrological forecasts with high reliability and accuracy, stochastic modeling can be achieved by quantifying the statistical features of results from the ANN models with a number of appropriate parameters calibrated by the SA-GA approach. This can take into account the uncertainties in ANN parameters on the hydrological estimates.

In addition to the uncertainties in the calibrated ANN weights attributed to a limited range of available observations (Imrie *et al.* 2000), the selection of the transfer function (i.e. activation functions) should be regarded as an uncertainty factor in the development of the ANN model which possibly influences the generation of the neural network (Imrie *et al.* 2000; Maca *et al.* 2014), To quantify the effect of selecting the modules/approaches used in the estimation of hydrological variables, Wu *et al.* (2011) presented a weighted-average estimator applied in the estimation of rainfall amount at ungauged grids via the Kriging method with a variety of semivariogram functions. In detail, by evaluating the performance of optimizing the model parameters in comparison to the observations, the rainfall amount at ungauged grids can be obtained by calculating the weighted average of the estimated gridded rainfall through the different semivariogram functions of interest. Note that the aforementioned weight regarding the specific semivariogram functions is defined as the ratio of the inverse values of its objective-function values associated with the optimal parameters to the sum of the inverse value of its objective-function values obtained from all semivariogram functions. The weighted-average estimator could effectively respond to the effect of uncertainties in the transfer/activation functions selected on the estimations of hydrological variables by means of the ANN model.

Recently, the real-time observations from the internet of thing (IoT) are comprehensively applied in the flooding early system in order to facilitate the accuracy of hydrological forecasts (Sankaranarayanan *et al.* 2019; Wu *et al.* 2020; Yang & Chang 2020). For illustration, Sankaranarayanan *et al.* (2019) employed the deep neural network to predict the flood occurrence using the temperature and rainfall intensity collected through IoT. Wu *et al.* (2020) and Yang & Chang (2020) carried out the real-time correction of inundation forecasts, provided by the machine learning techniques and 2-D hydraulic dynamic numerical modeling, respectively, in accordance with the observed inundation depth received through the IoT-derived roadside water-level sensors. In particular, Wu's method (i.e., RTEC_TS&KF), directly corrects the inundations-depth forecasts using the time series approach and Kalman filtering technique with real-time observed inundation depths recorded at IoT sensors without adjusting the parameters of hydrological/hydraulic numerical model; its advantage is not to influence the computation time and convergence speed of hydraulic numerical modeling. As a result, in this study, more accurate hydrological forecasts would be offered by adjusting the results from the proposed ANN model through Wu's method with the real-time observations.

It is well-known that in training the ANN model, the training hydrological-dataset size, such as the precipitation, river flow and water level, might impact the performance of optimizing the associated parameters (Foody & McCulloch 1995). Furthermore, there are frequently high correlations and resolution among hydrological variables in time and space (Wu *et al.* 2006; Haas *et al.* 2018). Accordingly, applying a big dataset in successfully training ANN model is a progressive and significant task in order to avoid the model overfitting or underfitting of the model associated with numerous parameters (Chiroma *et al.* 2018; Chalumuri *et al.* 2020). Therefore, this study focuses on taking into account the effect of uncertainties in the parameters of ANN model for a network structure set up in advance, the associated neuron weights, and selections of transfer functions, on the reliability and accuracy of the resulting hydrological forecasts to propose a stochastic ANN model (called ANN_GA-SA_MTF model). Eventually, it is expected that the proposed ANN_GA-SA_MTF model would not only provide hydrological forecasts with acceptable accuracy, but also quantify their reliability in terms of confidence intervals of a specific significant level.

2. METHODOLOGY

2.1. Model concept

Generally speaking, in training the ANN model, a large training dataset can effectively facilitate the performance of optimizing the associated parameters (Foody & McCulloch 1995), Thus, in this study, reproducing a large training dataset is the first task of the development of the proposed stochastically-based ANN model. Since the hydrological data have correlations in time and space, especially for the precipitation and runoff (e.g., Nandakumar & Mein 1997; Wu *et al.* 2006; McMillan *et al.* 2018), they are commonly regarded as the non-normal correlated variates. Thus, a correlated multivariate Monte Carlo simulation approach is adopted to generate the massive hydrological data used in the training of the proposed stochastically-based model.

According to Section 1 Introduction, this study proposes a stochastically-based ANN model in which a variety of the transfer functions (i.e. activation functions) are adopted under a given multiple network structure. However, it is generally difficult to determine a number of the connection weights between the neural nodes at different layers (i.e., ANN weights) under a multiple network structure using the back-propagation (BP) algorithm attributed to the vanishing and exploding gradient (Hochreiter 1998; Manchev & Spratling 2020). Therefore, in this study, the ANN weights are calibrated by means of a modified genetic algorithm based on sensitivity to the parameters of the complex hydrological models (GA-SA) (Wu *et al.* 2012a, 2012b). As a result, the proposed stochastically-based ANN model is developed by means of the modified genetic algorithm GA-SA method with a variety of transfer functions to obtain the numerous sets of appropriate parameters, including the ANN weights, bias and adjusting factors, named the ANN_GA-SA_MTF model. Accordingly, a number of the hydrological estimates can be accomplished by the proposed ANN_GA-SA_MTF model with numerous sets of calibrated appropriated parameters for the given hydrological factors defined as the model inputs. Eventually, the resulting hydrological forecasts are calculated using a number of the hydrological estimates with the corresponding objective functions associated with the appropriate parameters regarding various transfer functions. Also, the stochastic properties of the resulting hydrological estimates can be quantified through the statistical approach (i.e., weighted-average estimator) with the numerous sets of hydrological estimates, including the first four statistical moments and the 95% of confidence intervals. Note that the quantified 95% of confidence interval is treated as the prediction interval in which the 95% of hydrological estimates can be found. By doing so, in this study, through the above prediction interval, the likelihood of the forecasts approaching the observations can be briefly quantified and evaluated by comparing the upper and lower bounds of the prediction interval with the observations.

Since the proposed ANN_GA-SA_MTF model is expected to be applied in forecasting the flood-related hydrological variates, such as the inundation depths, water levels, and runoff, the results from the proposed ANN_GA-SA_MTF model is represented in terms of the forecasts. However, in reality, a variety of uncertainties are inherent in the observations (e.g., Nandakumar & Mein 1997; Chen & Wang 2018; McMillan *et al.* 2018); the resulting responses from the hydrological models might contain variations which affect their accuracy and reliability (Wu *et al.* 2012a, 2012b, 1015). Therefore, to enhance the performance of forecasting the hydrological variates through the proposed ANN_GA-SA_MTF model, the resulting hydrological forecasts be corrected by incorporating the real-time correction error method RTEC_TS&KF (Wu *et al.* 2020) in comparison to the real-time observations received through IoT which is the major difference from the well-known stochastic ANN models.

In summary, the development of the proposed ANN_GA-SA_MTF model comprises the five parts: (1) the generation of hydrological data as the training dataset; (2) the calibration of the model parameters; (3) the calculation of the weighted average regarding the resulting estimates; (4) the quantification of the stochastic properties of the resulting estimates; and (5) the real-time correlation of the resulting forecasts. The above concepts and methods mentioned can be introduced as follows.

2.2. Introduction to conventional ANN

An artificial neural network (ANN) is an empirical data-driven model which utilizes a parallel computing system with interconnections comparable to a biological neural network. It well-known that the ANN model is advantageous to the situations in which it is difficult to mathematically establish the relationship of the dependent variables with the independent ones for the physical phenomena (Khan *et al.* 2016). Furthermore, ANN models are capable of simulating nonlinear complicated system without the prior assumption of the relationship between model outputs and inputs (ASCE 2000a, 2000b).

The popular type of the neural network extensively applied in the hydrological modeling is the multi-layer perceptron (see Figure 1) in which neurons are mainly set up in input/output-layer and one hidden layer inter-connected by their own connection strengths (Shamseldin 2010; Sung *et al.* 2017); the three-layers neural network is comprehensively applied in the

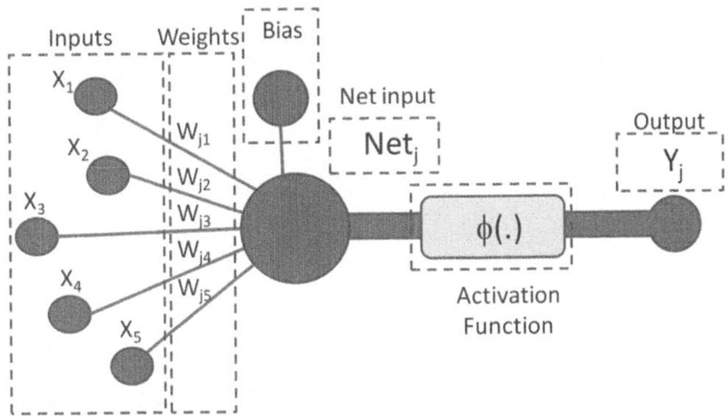

Figure 1 | Graphical structure of artificial neural network (ANN) (Gestal 2011).

hydrological forecasts (e.g., Khan *et al.* 2016; Sung *et al.* 2017), including the input layer, a hidden layer and output layer. Accordingly, the resulting forecasts from the ANN model ($\hat{y}_k$) can be computed by the following linear combination of the transform input variable (x_i):

$$\hat{y}_k = f_0\left[\sum_{j=1}^{N_{\mathrm{IP}}} w_{k,j} f_h\left(\sum_{j=1}^{M_{\mathrm{HN}}} w_{ji}x_i + w_{jb}\right) + w_{kb}\right] \tag{1}$$

in which N_{IP} and M_{HN} account for the number of input and hidden neurons, respectively; x_i and $\hat{y}_k$ serve as the ith observed model input and the kth forecasted model output, respectively; f_0 and f_h denote the transfer (or activation) function for the output and hidden neurons, respectively; $w_{k,j}$ stands for the ANN weights connecting the jth neuron at the hidden layer and the kth neuron at the output layer (called interconnection weight); w_{ji} represents the ANN weight connecting the jth neuron at the input layer and the ith neuron at the hidden layer; and w_{jb} as well as w_{kb} are defined as the ANN bias values regarding the ith hidden neuron and the kth output neuron, respectively.

Specifically, the numbers of neurons at hidden layers are significantly induced due to the performance of training the ANN model: in case of few neurons considered, the corresponding network structure difficulty emulates the underlying function due to the insufficient parameters; on the contrary, using too many neurons in the network structure possibly causes the over-fitting problem (Dawson & Wilby 2001). Therefore, several methods are proposed to estimate the number of neurons as listed in Table 1. Additionally, based on the types of the transfer functions used, the transform input variable (x_i) can be standardized by the following equations:

$$x = \frac{\mathrm{x}_o - X_{o,\min}}{X_{o,\max}}, \quad 0 \ll x \ll 1 \tag{2}$$

$$x = \frac{2(\mathrm{x}_o - \mathrm{x}_{o,\min})}{\mathrm{x}_{o\max}} - 1, \quad -1 \ll x \ll 1 \tag{3}$$

Table 1 | Formula for estimating the number of hidden neurons

No. of formula	Formula	References
1	$N_{\mathrm{HN}} = (\sqrt{1 + 8 \times N_{\mathrm{IP}}} - 1)/2$	Li *et al.* (1995)
2	$N_{\mathrm{HN}} = N_{\mathrm{IP}} - 1$	Tamura & Tateishi (1997)
3	$N_{\mathrm{HN}} = 2 \times N_{\mathrm{IP}}/N_{\mathrm{IP}} + 1$	Zhang *et al.* (2003)
4	$N_{\mathrm{HN}} = \sqrt{N_{\mathrm{IP}} \times N_{\mathrm{OP}}}$	Shibata & Ikeda (2009)
5	$N_{\mathrm{HN}} = 2^{N_{\mathrm{IP}}} - 1$	Hunter *et al.* (2012)
6	$N_{\mathrm{HN}} = [4 \times (N_{\mathrm{IP}})^2 + 3]/[(N_{\mathrm{IP}})^2 - 8]$	Sheela & Deepa (2013)

where x_o and x are the real and normalized values of model inputs, respectively; and $X_{o,min}$ and $x_{o,max}$ separately stand for the minimum and maximum of model inputs x_o.

Generally speaking, the above interconnection weights for two neurons can be determined by a back-propagation algorithm (Rumelthart *et al.* 1986) in association with the gradient descent method with the objective of minimizing the sum of errors of networks E (Supharatid 2003) as:

$$E(w) = \sum_{p=1}^{P} \sum_{k=1}^{n} (y_{pk} - \hat{y}_{pk})^2 \tag{4}$$

in which $E(w)$ represents the error for all input patterns and y_{pk} and $\hat{y}_{pk}$ stand for the target outputs and estimated outputs, respectively. Specifically, in the BP algorithm, the interconnection ANN weights can be updated through the training procedure with the following equation:

$$\Delta w_{ij}(l) = -\varepsilon^* \frac{\partial E(w)}{\partial w_{ij}} + \alpha^* \Delta w_{ij}(l-1) \tag{5}$$

where ε^* and α^* are the learning rate and the momentum, respectively; Δw_{ij} represent the increment of interconnection weights between the ith and j^{th} neuron for the lth and $(l-1)^{th}$ iteration. It is noted that the formula of differential $\partial E(w)/\partial w_{ij}$ for the transfer functions can be referred to in Table 2 in which the suitable normalization equations are suggested. After that, the adjusted interconnection weights should be equal to $w_{ij} + \Delta w_{ij}$. Eventually, the above training process would be stopped to determine the optimal interconnection weights, as the variation of the error $E(w)$ approach to negligible with the process of time or the resulting error $E(w)$ continuously decreasing (Campolo *et al.* 1999).

2.3. Generation of hydrological training dataset

As training the ANN model, a sufficient dataset size is required in order to calibrate reliable ANN weights between neurons at various layers. Since the hydrological data are possibly regarded as the correlated variates in time and space, in this study, the correlated multivariate Monte Carlo simulation approach (name MMCS) (Wu *et al.* 2006), which focuses on the simulation of spatially and temporally corrected non-normal variables based on their uncertainties, would be used to reproduce a large

Table 2 | Transform functions commonly used (Maca *et al.* 2014)

Function		Formula	Derivative	Suggested normalization equation
TF1	Logistic(soft step、sigmoid)	$f(x) = \frac{1}{1+e^{-\propto x}}$	$f'(x) = f(x)(1 - f(x))$	Equation (2)
TF2	Tanh	$f(x) = \tanh(x) = \frac{2}{1+e^{-2\propto x}} - 1$	$f'(x) = 1 - f(x)^2$	Equation (3)
TF3	Arctan	$f(x) = \tan^{-1}(\propto x)$	$f'(x) = \frac{1}{(\propto x)^2 + 1}$	Equation (3)
TF4	Identity	$f(x) = \propto x$	$f\wedge'(x) = \propto$	Equation (3)
TF5	Rectified linear unit (ReLU)	$f(x) = \begin{cases} 0 \text{ for } x < 0 \\ 1 \text{ for } x \geq 0 \end{cases}$	$f'(x) = \begin{cases} 0 \text{ for } x < 0 \\ 1 \text{ for } x \geq 0 \end{cases}$	Equation (2)
TF6	Parameteric rectified linear unit (PReLU、leaky ReLU)	$f(x) = \begin{cases} \propto x \text{ for } x < 0 \\ x \text{ for } x \geq 0 \end{cases}$	$f'(x) = \begin{cases} \propto \text{for } x < 0 \\ 1 \text{ for } x \geq 0 \end{cases}$	Equation (3)
TF7	Exponential linear unit(ELU)	$f(x) = \begin{cases} \propto(e^x - 1) \text{for } x < 0 \\ x \text{ for } x \geq 0 \end{cases}$	$f'(x) = \begin{cases} f(x) + \propto \text{for } x < 0 \\ 1 \text{ for } x \geq 0 \end{cases}$	Equation (3)
TF8	Inverse abs (IA)	$y(x) \frac{x}{1+\lvert\propto x\rvert}$	$y'(a) = \frac{1}{(1+\lvert a \propto x\rvert)^2}$	Equation (3)
TF9	Rootsig (RS)	$y(x) = \frac{\propto x}{1+\sqrt{1+(\propto x)^2}}$	$y'(x) = \frac{1}{\left(1+\sqrt{1+(\propto x)^2}\right)\sqrt{1+a(\propto x)^2}}$	Equation (3)
TF10	Sech function (SF)	$y(x) = \frac{2}{\exp(\propto x) + \exp(-\propto x)}$	$y'(x) = -y(x)\tanh(\propto x)$	Equation (3)

number of hydrological data for the training of the proposed ANN_GA-SA_MTF. The above MMCS method is briefly introduced as follows.

The multivariate Monte Carlo simulation (MMCS) model for the non-normal random variables developed by Chang *et al.* (1997) is adopted herein to simulate the rainfall duration and grid-based rainfalls by three steps: transformation to standard normal space, orthogonal transform and inverse transformation which can be implemented by the Nataf bivariate distribution model (Nataf 1962):

$$\rho_{ij} = \int_{-\infty}^{\infty}\int_{-\infty}^{\infty}\left[\frac{x_i - \mu_i}{\sigma_i}\right]\left[\frac{x_j - \mu_j}{\sigma_j}\right]\emptyset_{ij}\left(\left\langle z_i, z_j|\rho^*_{ij}\right\rangle\right)\mathrm{d}z_i\mathrm{d}z_j \tag{6}$$

where X_i and X_j are the correlated variables at the point i and j with, respectively, the means μ_i and μ_j, the standard deviations σ_i and σ_j, and the correlation coefficient ρ_{ij}; Z_i and Z_j are corresponding bivariate standard normal variables to the variable X_i and X_j with the correlation coefficient ρ^*_{ij} and the joint standard normal density function $\emptyset_{ij}(\cdot)$. Through an extensive numerical experiment, a set of semi-empirical formula has been developed by Liu & Der Kiureghian (1986) who developed a set of semi-empirical formulae from which a transformation factor T_{ij}, depending on the marginal distributions and correlation of X_i and X_j, can be determined to modify ρ_{ij} in the original space to ρ^*_{ij} in the normal space by:

$$\rho^*_{ij} = T_{ij} \times \rho_{ij} \tag{7}$$

Chang *et al.* (1997) derived several formulae for calculating the transform factor T_{ij} between the correlated multivariate standard normal variables and uncorrelated standard normal variables by which random variates can be easily generated. Eventually, upon the generation of multivariate standard normal random variates z^*_i, one can obtain the corresponding random variates in the original space by the following inverse transformation:

$$\mathrm{x}_i = F_i^{-1}[\Phi(z^*_i)] \tag{8}$$

where $F_i[\Phi(z^*_i)]$ is the marginal cumulative distribution function (CDF) of random variable X_i in the original space, and $\Phi(\cdot)$ is the standard normal CDF. The framework of generating the correlated multivariates by means of MMCS method can be seen in Figure 2.

In summary, in the proposed ANN_GA-SA_MTF model, the MMCS method is applied to simultaneously generate the hydrological variables as the training dataset.

2.4. Parameter calibration using the GA-SA method

In this study which intends to consider the effect of uncertainties in the network structure used in the training of the proposed ANN_GA-SA_MTF model, a multi-layer network structure is adopted. As mentioned in Section 2.1, the gradient method is widely used in determining the ANN weights between two neurons at various layers; however, as the gradient shown in Equation (5) $\partial E(w)/\partial w_{ij}$ approaches to zero and 1.0, named the problems of the vanishing and exploding gradient, respectively, it is difficult to optimize the neuron weights (Hochreiter 1998; Manchev & Spratling 2020).

In addition, regarding the artificial intelligence (AI) method, it is necessary to identify appropriate model inputs in order to achieve the goal of successfully training the AI model (Wang *et al.* 2009). Therefore, to figure out the problems with the vanishing and exploding gradients, this study utilizes the modified genetic algorithm (GA-SA) based on the sensitivity of the model parameters to the model outputs which can provide the stochastic model parameters in response to the uncertainty in the hydrological data (Wu *et al.* 2012a, 2012b). Therefore, regarding the proposed ANN_GA-SA_MTF model, the associated appropriate parameters (i.e. ANN weights, bias and adjusting factor) could be calibrated by using the GA-SA method with the numerous transfer functions of interest. Accordingly, the stochastic information on the hydrological estimates, including the statistical moments and 95% of confidence interval (i.e. prediction interval) can be quantified through the proposed ANN_GA-SA_MTF model using a number of calibrated parameters via the GA-SA method with the multiple transfer functions. The aforementioned GA-SA method is introduced as follows.

The parameter-calibration method (GA-SA) for the parameter calibration of the conceptual hydrological model using a real-value coding genetic algorithm (GA) takes into account runoff estimation sensitivity to model parameters (Wu *et al.*

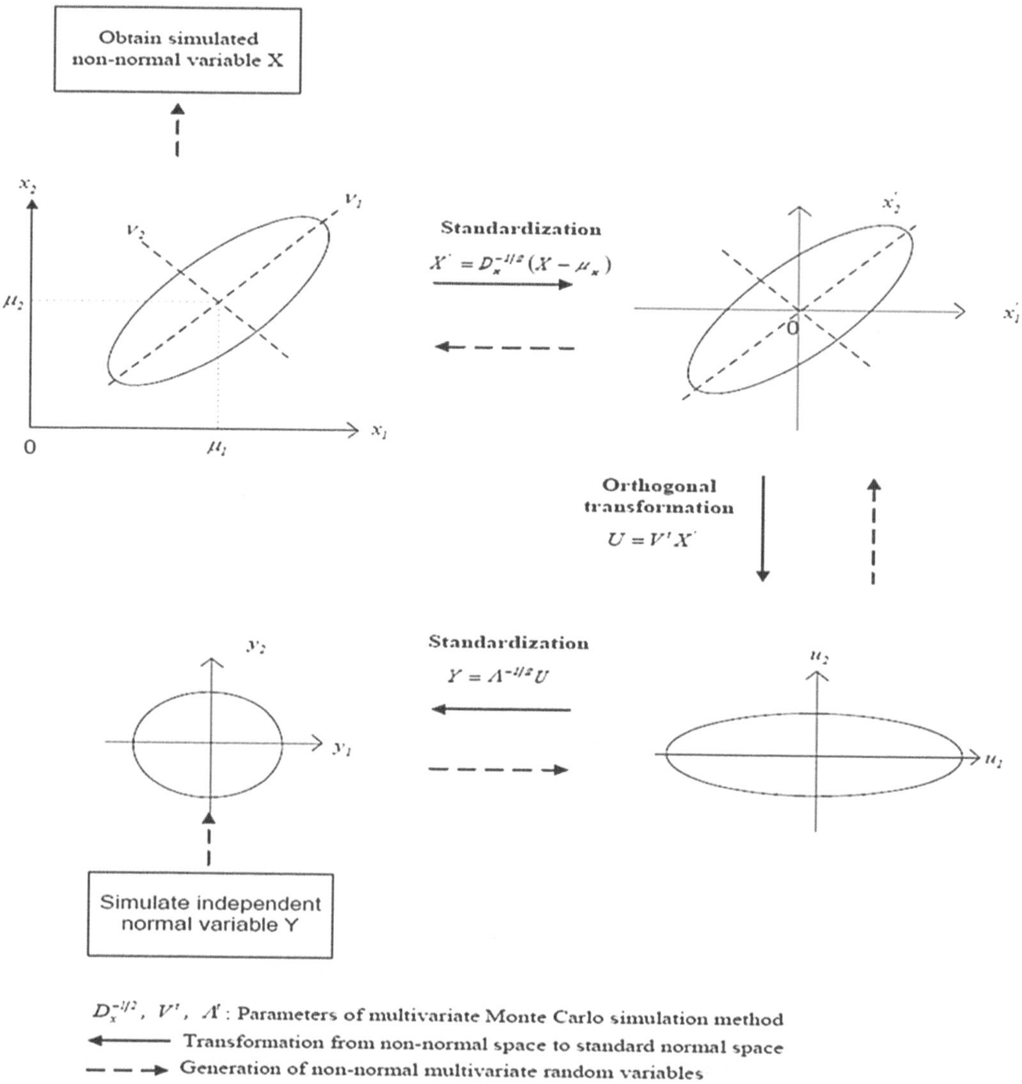

Figure 2 | Flowchart of generating the correlated multivariate by MMCS method (Tung & Yen 2005).

2012a, 2012b). The above sensitivity can be quantified through the standardized regression equation (Speed & Yu 1993) as:

$$\frac{Y - \bar{Y}}{S_Y} = \sum_{i=1}^{n} \beta_i \frac{X_i - \hat{X}_i}{S_{X_i}} \tag{9}$$

where Y and X are the model output and inputs; $\bar{Y}$ and $\hat{X}$ account for the mean of the model output and inputs, respectively; S_Y and S_X separately represent the corresponding standard deviation; and β_i denotes the standard regression coefficients. In the GA_GA method, the standardized values of the model parameters are treated as the real-value code through the following equation:

$$z_{\theta_i} = \frac{\theta_i - \hat{\theta}_i}{S_{\theta_i}} \tag{10}$$

in which z_{θ_i} is the standardized value of the model parameter θ_i ranging from -1 and 1; and $\hat{\theta}_i$ and S_{θ_i} denote the mean and standard deviation of model parameters, respectively.

It is well-known that as carrying out the GA, the chromosomes used in the crossover and mutation should be identified in advance; thus, the GA-SA method selects the chromosomes based on the probability $P_{\text{chrom}}(\theta_i)$ calculated using the resulting objective-function values $f(\theta_i)$ from the difference between the observations and model outputs as:

$$P_{\text{chrom}}(\theta_i) = \frac{f(\theta_i)}{\sum_{i=1}^{n_{\text{chrom}}} f(\theta_i)} \tag{11}$$

in which n_{chrom} stands for the number of chromosomes given in advance; and θ_i represents the calibrated parameters with a minimum value of objective function $f(\theta_i)$ (i.e., the root mean square of error, RMSE) computed using the following equation:

$$f(\theta_i) = \left[\frac{1}{N_{\text{data}}}\sum_{\text{k}=1}^{N_{\text{data}}} [Y_k - \hat{Y}_k(\theta_i)]^2\right]^{0.5} \tag{12}$$

in which N_{data} is the number of observed hydrological estimates; and Y_k and $\hat{Y}_k(\theta_i)$ stand for the observed hydrological data and estimated ones, resepctively. After identifying the desired chromosomes using the multinomial trial process with the probability $P_{\text{chrom}}(\theta_i)$, the genes for the crossover and mutation are also determined by the multinomial trial process with the probabilities P_{gene} in terms of the sensitivity of model outputs to the model parameters, i.e. the coefficients of the standardized regression equation β_i:

$$P_{\text{gene}} = \frac{|\beta_i|}{\sum_{i=1}^{n_{\text{par}}} |\beta_i|} \tag{13}$$

The genes selected for the crossover and mutation can be referred to in Figure 3 (Wu *et al.* 2012a, 2012b).

2.5. Calculation of weighted average of hydrological estimates

In addition to stochastic information on the hydrological estimates, the proposed ANN_GA-SA_MTF model can take into account the combination of the hydrological estimates through the transfer functions considered by means of the weighted-average estimator method as mentioned in Section 1; also, its parameters can be calibrated by the GA-SA method (Wu *et al.* 2012a, 2012b). Thus, in calibrating the parameters of the proposed ANN_GA-SA_MTF model with the ith transfer function (TF) of interest (θ_{TF}^i) conducted by the GA-SA method, the corresponding weighted factors of the hydrological estimates (W_{TF}^i) can be calculated using the following equation:

$$W_{\text{TF}}^i = \frac{\dfrac{1}{E(\theta_{\text{TF}}^i)}}{\sum_{i=1}^{N_{\text{TF}}} \dfrac{1}{E(\theta_{\text{TF}}^i)}} \tag{14}$$

in which N_{TF} represents the number of transfer functions used; and $E(\theta_{\text{TF}}^i)$ is the value of the desired objective function associated with the appropriate parameters of the ith transfer function θ_{TF}^i calculated by the following equation in referring to Equation (4):

$$E(\theta_{\text{TF}}^i) = \sqrt{\frac{1}{N_{\text{data}}}\sum_{\text{k}=1}^{N_{\text{data}}} [Y_k - \hat{Y}_k(\theta_{TF}^i)]^2} \tag{15}$$

where N_{data} is the number of observed hydrological estimates; and Y_k and $\hat{Y}_k(\theta_{\text{TF}}^i)$ stand for the observed hydrological data and estimated ones by the proposed ANN_GA-SA_MTF model with the j_{th} appropriate parameters θ_{TF}^i, respectively.

By means of the weighted-average estimator, the weighted average of hydrological estimates $\hat{Y}_{\text{WA}}$, defined as the resulting hydrological forecast, can be obtained by combining the estimates from the ith transfer functions $\hat{Y}(\theta_{\text{TF}}^i)$ with the

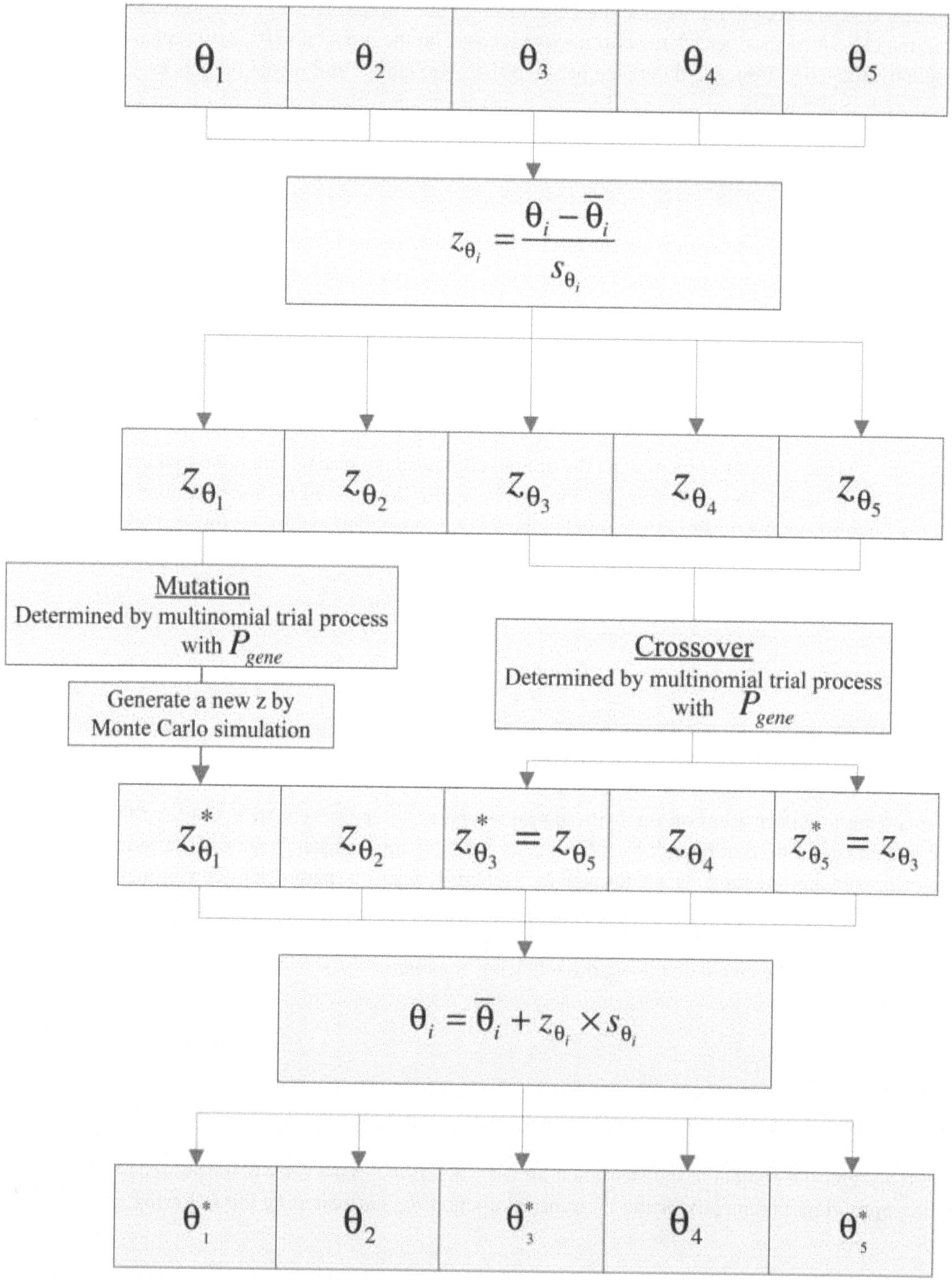

Figure 3 | Graphical illustration of selecting the genes used in the crossover and mutation (Wu *et al.* 2012a, 2012b).

corresponding weighted factor W^i_{TF} as:

$$\hat{Y}_{WA} = \sum_{i=1}^{N_{TF}} (W^i_{TF} \times \hat{Y}(\theta^j_{TF_i})) \qquad (16)$$

2.6. Quantification of stochastic properties of hydrological forecasts

Since the proposed ANN_GA-SA_MTF can provide a number of estimated hydrological variables with numerous sets of appropriate parameters regarding a variety of transfer functions calibrated by the GA-SA method, the quantiles of estimated hydrological variables, i.e. the probability distribution, can be derived from the aforementioned large number of resulting hydrological variables. In general, the best-fit theoretical probability function should be identified in advance via a statistical test method, e.g., the K-S test and Chi-square methods in order to establish the variable's quantile relationship, i.e. the cumulative probability distribution. Nevertheless, it is difficult to identify the best-fit probability function from a large number of candidate distributions and parameter estimation procedures (Haddad & Rahman 2011). To avoid this problem, this study adopts a nonparametric method to calculate the quantiles of resulting hydrological variables from the proposed ANN_GA-SA_MTF method, i.e. the weighted likelihood sample quantile estimator method (Yang & Tung 1996). The weighted likelihood sample quantile estimator method is addressed as follows.

Consider a set of n independent observations, $(X_1, X_2, \ldots, X_n)$, of a random variable from a distribution function, $F(x)$. They can be arranged in ascending order to form a set of n order statistics as $X_{(1)} \leq X_{(2)} \leq \ldots \leq X_{(n)}$ with $X_{(1)}$ being the smallest observation and $X_{(n)}$ the largest. The maximum likelihood estimator of the unknown distribution function is an empirical one, $F_n(x)$, which is defined as:

$$F_n(x) = \frac{1}{n}\sum_{i=1}^{n} I(X_i \leq x) \tag{17}$$

where $I(X_i \leq x)$ is the indicator variable having a value of 1 if $X_i \leq x$ or 0 otherwise. Hence, when $x = X(r)$, then the value of the empirical distribution function is $F_{r:n}[X_{(i)}] = 1/n$ for $r = 1,2,\ldots, n$.

Since the sample quantile estimator hardly satisfies the symmetrical property of the population quantile; it experiences a substantial lack of efficiency due to the variability of the individual order statistics (Sheather & Marron 1990). Hence, to estimate a more stable quantile estimator and smoother distribution, a weighted likelihood estimator is employed herein to estimate the quantiles presented here where the estimator for the pth quantile is defined using the weighted average of the order statistics (Yang & Tung 1996):

$$X_{p,\mathrm{WL}} = \sum_{r\in\omega} W_{r,n,\mathrm{WL}} X_{(r)} \tag{18}$$

where $W_{r,n,\mathrm{WL}} = F_{r,n}(p)/\sum_{s\in\omega} F_{s:n}(p)$ with ω being the band width that contains a set of order statistics that are deemed significant in contributing to the estimation of X_p as a result of the band width being no smaller than 2.0.

In summary, the stochastic properties of the resulting hydrological forecasts in terms of quantiles under different cumulative probabilities from the proposed ANN_GA-SA_MTF model can be quantified by using the weighted likelihood sample quantile estimator method with the hydrological estimates provided by the proposed ANN_GA-SA_MTF model with various transfer functions of interest. Thus, the estimated quantiles can be advantageous to the reliability assessment for the resulting hydrological forecasts and the variance analysis for the effect of climatic change on the observations.

2.7. Real-time correction of estimation using RTEC_TS&KF

According to the aforementioned model concept, the proposed ANN_GA-SA_MTF can quantify the stochastic information of the hydrological estimates $\hat{Y}(\theta^{i}_{\mathrm{TF}})$, including statistical moment and 95% of confidence intervals (i.e., prediction intervals). However, the accuracy of the resulting hydrological forecasts $\hat{Y}_{\mathrm{WA}}$ from Equation (16) are possibly impacted by the uncertainties in observations as the training dataset attributed to the climate change and occurrence of extreme hydrological events. Therefore, this study collaborates the real-time error correction method RTEC_TS&KF (Wu *et al.* 2012a, 2012b), developed using the time-series method with the Kalman filtering approach based on the difference between the observation and forecasts at the previous time steps during an event, to facilitate the accuracy of the resulting hydrological forecasts, i.e. the weighted average of hydrological estimates $\hat{Y}^{k}_{\mathrm{WA}}$ provided by the proposed ANN_GA-SA_MTF model. The brief introduction to RTEC_TS&KF method applied in the correction of hydrological forecasts is addressed as follows.

In the case of the RTEC_TS&KF method, the correcting procedure by means of the time-series method and Kalman filtering approach includes three steps as shown in Figure 4. At the first step, the RTEC_TS&KF method predicts the forecast

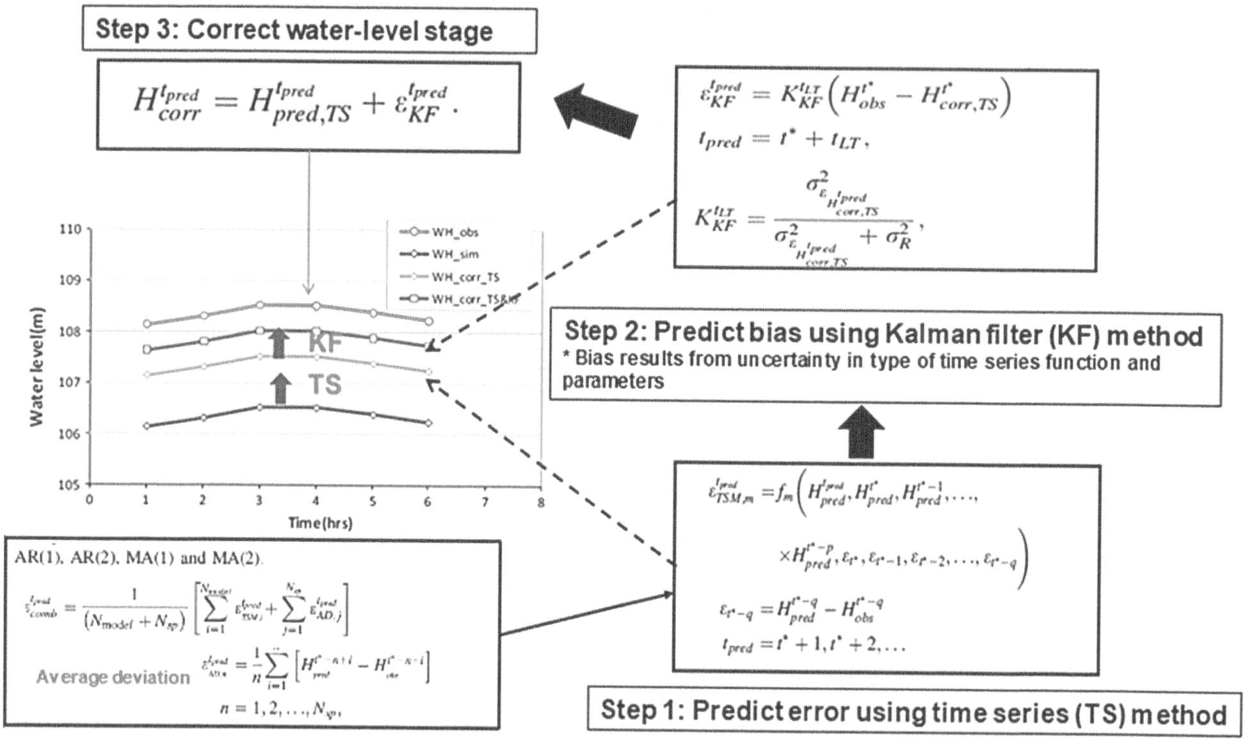

Figure 4 | Brief concept of correcting water-stage forecasts by using RTEC-TS&KF method (Wu *et al.* 2020).

errors via the following equations:

$$\varepsilon_{\text{comb}}^{t_{\text{pred}}} = \frac{1}{N_{\text{model}}} \sum_{i=1}^{N_{\text{model}}} \varepsilon_{\text{TSM},i}^{t_{\text{pred}}} + \frac{1}{N_{sp}} \sum_{j=1}^{N_{sp}} \varepsilon_{\text{AD},i}^{t_{\text{pred}}} \tag{19}$$

$$\varepsilon_{\text{AD},i}^{t_{\text{pred}}} = \frac{1}{n} \sum_{i=1}^{n} [Y_{\text{pred}}^{t^*-n+i} - Y_{\text{obs}}^{t^*-n+i}] \tag{20}$$

In which $\varepsilon_{\text{comb}}^{t_{\text{pred}}}$ denotes the combined forecast errors in terms of $\varepsilon_{\text{TSM},i}^{t_{\text{pred}}}$; $\varepsilon_{\text{AD},i}^{t_{\text{pred}}}$ serves as the forecast error estimated by the ith one of N_{model} time series models and the average deviation model for a specific period, N_{sp},respectively; t^*stands for the current time; t_{LT} and t_{pred} denote the lead time and the prediction time, respectively; and $Y_{\text{pred}}^{t^*-q}$ and $Y_{\text{obs}}^{t^*-q}$ separately represent the predicted and observed hydrological variates at the previous time $t^* - q$. After that, through the time series approach, the corrected forecast $H_{\text{corr,TS}}^{t_{\text{pred}}}$ results from adding the forecast error to the forecasts $Y_{\text{pred}}^{t_{\text{pred}}}$ using the following equation:

$$Y_{\text{corr,Ts}}^{t_{\text{pred}}} = Y_{\text{pred}}^{t_{\text{pred}}} + \varepsilon_{\text{comb}}^{t_{\text{pred}}} \tag{21}$$

Furthermore, to reduce the measurement bias which might affect the accuracy and reliability of correcting the water-level forecast by means of the time series approach, the Kalman filter (KF) algorithm is adopted in Step (2) to improve the corrected forecast $Y_{\text{corr,Ts}}^{t_{\text{pred}}}$ in order to further enhance the correction performance. Finally, the error of the corrected forecast error can be quantified by the KF method based on the measurement bias $\varepsilon_{\text{KF}}^{t_{\text{pred}}}$ as follows:

$$\varepsilon_{\text{KF}}^{t_{\text{pred}}} = K_{\text{KF}}^{t_{\text{LT}}}(Y_{\text{obs}}^{t^*} - Y_{\text{corr,TS}}^{t^*}),\ t_{\text{pred}} = t^* + t_{\text{LT}} \tag{22}$$

$$K_{\text{KF}}^{t_{\text{LT}}} = \frac{\sigma^2_{\varepsilon_{Y_{\text{corr,TS}}^{t_{\text{pred}}}}}}{\sigma^2_{\varepsilon_{Y_{\text{corr,TS}}^{t_{\text{pred}}}}} + \sigma_R^2} \tag{23}$$

where $Y_{\text{obs}}^{t^*}$ *is* the observed hydrological variate of interest at the current time t*; $Y_{\text{corr,TS}}^{t^*}$ represents the corrected hydrological forecast at the current time t^* by the time series approach accomplished at the time step $t^* - t_{\text{LT}}$; $K_{\text{KF}}^{t_{\text{LT}}}$ denotes the Kalman gain for the lead time t_{LT}; $\sigma^2_{\varepsilon_{\text{HY}_{\text{corr,TS}}^{t_{\text{pred}}}}}$ is the variance of the corrected forecast $Y_{\text{corr}}^{t_{\text{pred}}}$, and σ_R^2 represents the variance of measurement bias. Moreover, the $\sigma^2_{\varepsilon_{Y_{\text{corr,TS}}^{t_{\text{pred}}}}}$ is calculated from the differences between the observed hydrological variate $Y_{\text{obs}}^{t^*}$ and corrected forecast $Y_{\text{corr, TS}}^{t^*}$ at the past prediction time ($t_{\text{pre}} < t^*$).

Eventually, at the third step of RTEC_TS&KF method, the final corrected forecasts $Y_{\text{corr}}^{t_{\text{pred}}}$ can be obtained by:

$$Y_{\text{corr}}^{t_{\text{pred}}} = Y_{\text{pred,TS}}^{t_{\text{pred}}} + \varepsilon_{\text{KF}}^{t_{\text{pred}}} \tag{24}$$

Therefore, in this study, the correction of the resulting hydrological forecast from the proposed ANN_GA-SA_MTF model could be carried out through the updating process within RTEC_TS&KF method as shown in Figure 4.

2.8. Model framework

To sum up the above introduction to the concepts of interest, the model framework for the proposed ANN_GA-SA_MTF model could be classified into two parts, model training and model application. Regarding the model development, the proposed ANN_GA-SA_MTF is developed on the basis of the conventional ANN model whose the parameters, including the ANN weights and bias at neurons within the input-hidden-output layers for the transfer functions of interest, are calibrated through the GA-SA method with a large number of simulated hydrological data as the training dataset by the non-normal correlated multivariate Monte Carlo approach (Wu *et al.* 2006). With respect to the model application, for each transfer function, the hydrological variables are estimated by means of ANN_GA-SA_MTF model under the given observations/predictions. The stochastic properties of resulting hydrological forecasts are then quantified by the weighted likelihood sample quantile estimator method; and their weighted averages are computed as model output, i.e. the hydrological forecast. Eventually, the resulting model outputs are corrected using the RTEC_TS&KF method in accordance with the difference between observations and forecasts at the previous time steps. The development and application framework of the proposed ANN_GA-SA_MTF method can be summarized as.

2.8.1. Model development

Step [1] Generate the hydrological data as the training dataset using the non-normal correlated multivariate Monte Carlo simulation approach with the statistical features of the hydrological variables.

Step [2] Identify the appropriate hydrological variates as the model inputs through the sensitivity analysis with the simulated training dataset.

Step [3] Standardize the desired model inputs and model outputs as the dimensionless variates listed at (0,1) or (−1,1) using Equations (1) and (2) in order to reduce the effect of the scales in time and space on the hydrological data.

Step [4] Specify the number of neuron nodes at hidden layers through the equations as shown in Table 1.

Step [5] Calibrate the parameters of the proposed ANN_GA-SA_MTF model, i.e. the ANN weights, and bias at each hidden layers at different layers as well as the adjusting factor, using GA-SA approach with the following objective or loss function Equation (12).

Step [6] Calculate the weighted factors of transfer functions used in the calculation of weighted average of hydrological estimates using Equation (14).

2.8.2. Model application

Step [1] Estimate the hydrological variates and the corresponding 95% of confidence intervals (i.e., prediction interval) from the ANN_GA-SA_MTF model with numerous sets of calibrated parameters for a variety of transfer functions of interest.

Step [2] Calculate the weighted average of the resulting hydrological estimates, defined as the resulting hydrological forecasts, from the proposed ANN_GA-SA_MTF model via Equation (16).

Step [3] Carry out the real-time correction of weighed weighted estimations of the resulting hydrological forecasts by the RTEC_TS&KF method via Equations (19)–(24). The introduced model framework can be referred to in Figure 5.

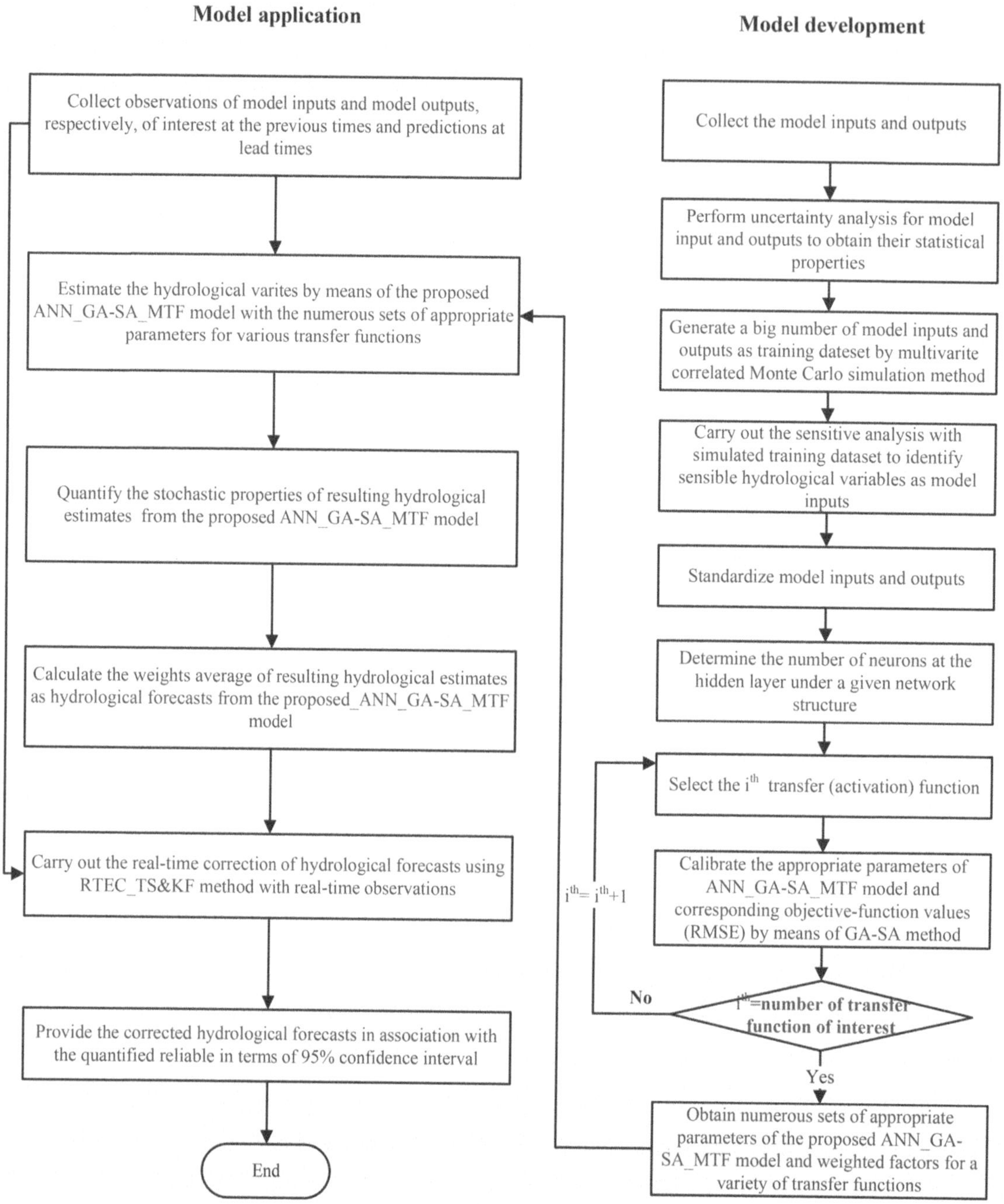

Figure 5 | Graphic framework of developing and applying the proposed ANN_GA-SA_MTF model.

3. RESULTS AND DISCUSSION

To demonstrate the reliability and accuracy of the resulting hydrological forecasts from the proposed ANN_GA-SA_MTF model, 10-days rainfall-induced water levels are regarded as the hydrological variates of interest used in the model development and verification. The results and relevant discussion are described as follows.

3.1. Study area and data

In this study, Shanping Gauge located within Shanping River, a major tributary to Touquian River in northern Taiwan, is selected as the study area as shown in Figure 6. The observations regarding 10-days rainfall depths and water levels are adopted as the study data. Note that the 10-days rainfall depths is the areal average of rainfall amount of 10 days calculated

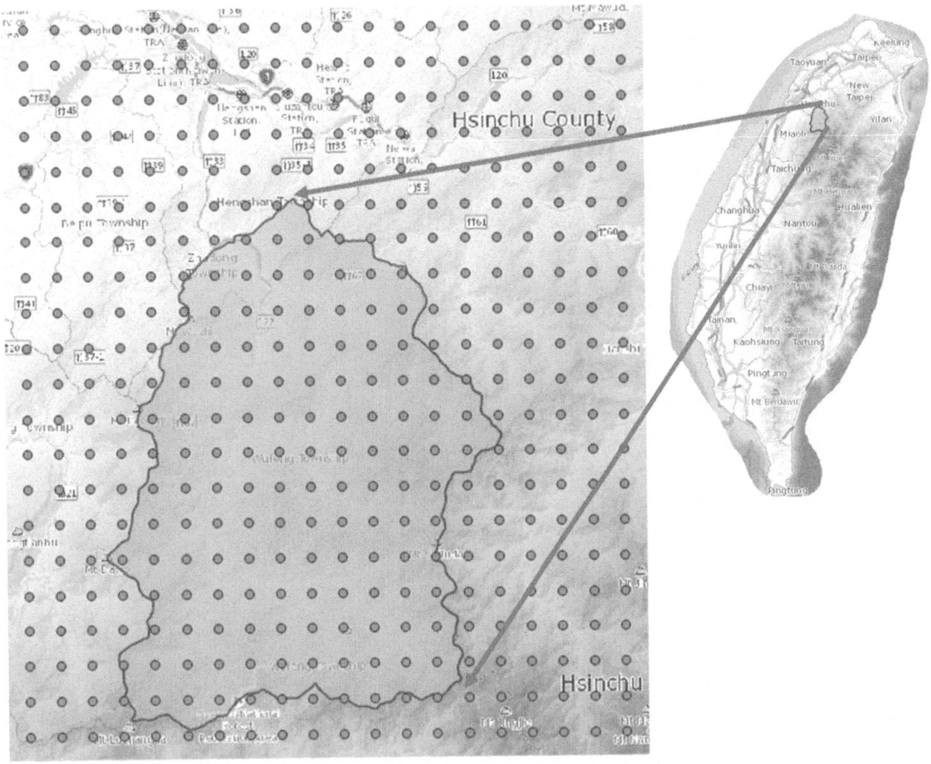

Figure 6 | Location and boundary of study area (note: blue circle stands for the QPE grids). Please refer to the online version of this paper to see this figure in color: http://dx.doi.10.2166/nh.2021.030.

from the quantitation precipitation estimations (QPE) at grids within the study area provided by the Taiwan Central Weather Bureau from 2005 to 2018 as shown in Figure 7. Also, the corresponding the water level observations at Shanping Gauge (see Figure 6) are received from Taiwan Water Resources Agency. Since this study focuses on the rainfall and induced water levels every 10 days, the observed 10-days rainfall depths and water levels at 36 10-days periods are used in the model development and validation. Accordingly, uncertainty analysis for the 10-days rainfall depths and water levels are carried out to quantify the variation in the hydrological data within the study area. The aforementioned results from the uncertainty are applied in the generation of a large number of 10-days rainfall depths and water levels as the training and validating datasets for the proposed ANN_GA-SA_MTF model.

3.2. Quantification of uncertainties in 10-days rainfall depth and water level

As introduced in Section 2.7 Model framework, uncertainty analysis for the hydrological estimates (i.e., 10-days rainfall depths and water levels) should be conducted in advance to obtain their statistical properties, including the first statistical moments, 95% of confidence intervals and correlation coefficients as shown in Figure 8. By observing Figure 8, 10-days water level averagely ranges between 209.37 and 209.84 m. Of the statistics derived from the 10-days water level, although the average of 10-days water levels merely approximates 209.6 m, the corresponding statistics, the standard deviation of 0.47 m and the 95% of confidence interval (209.31 m–210.9 m), are significantly greater than those in the remaining 10-days periods; this result means that more variation possibly exists in the 10-days rainfall depths in the 26th 10-days period. With respect to the investigation of statistics of 10-days rainfall depths, the corresponding average ranges from 65 to 435 mm; the standard deviations and the 95% of confidence interval from the 20th to 26th (i.e., rainy season in Taiwan) obviously exceeds those in the remaining 10-days period; namely, similar to the 10-days water level, the variation in the 10-days rainfall depths varies with the time and more oscillation can be found at the 20th–26th 10-days periods. The above results reveal that the variation of the 10-days rainfall depth and water level markedly changes with time (10-days periods); thus, it possibly impacts the performance of calibrating the parameters (i.e., ANN weights and bias as well as adjusting factors) of the proposed ANN_GA-SA_MTF model. As a result, a large training dataset regarding the generated 10-days rainfall depths and water levels in response to the uncertainty in the observations is needed in the training of the proposed ANN_GA-SA_MTF model.

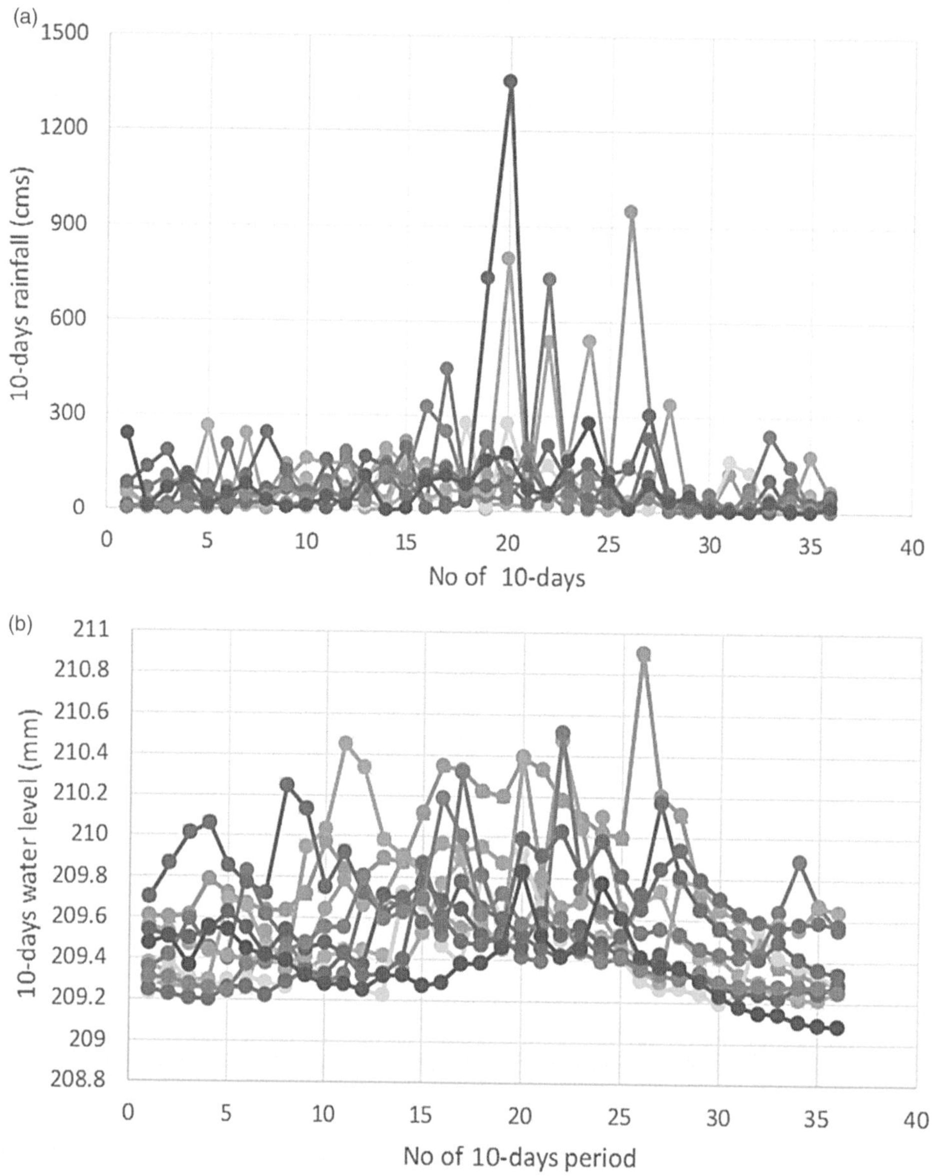

Figure 7 | Observation of hydrological data of interest recorded at the study area Shanping Gauge (2008–2018). (a) 10-days rainfall depth. (b) 10-days water level.

Apart from the statistical moments and 95% of confidence intervals, the dependence among 10-days rainfall depths and water levels in various 10-days periods, respectively, are represented in terms of the correlation coefficients as shown in Figure 9. Note the above correlation coefficients are separately calculated using the 10-days rainfall depth and water level at a specific time step (i.e. 10-days period) with those at the associated forward time steps. In the case of the dependence among the 10-days rainfall depths as shown in Figure 9(a), it can be seen that the correlation coefficient of the 10-days rainfall depth decreases from the forward 1st time step (0.33) to the 3rd time step (0.12) and then stays at a constant at the forward 4th to 5th time step (0.12). After that, the change in the correlation coefficient of 10-days rainfall depth between the forward 6th time step and 9th time step resembles the results from the correlation coefficients among the forward 1st and 5th time steps; that is, the correlation coefficient decreases from 0.12 to −0.03. It concludes that the 10-days rainfall depth in the current time steps is evidently and highly related to those at the continuous three time steps (10-days period); that is to say, every four 10-days rainfall depths can be regarded as the temporally correlated variables with a similar varying trend in time.

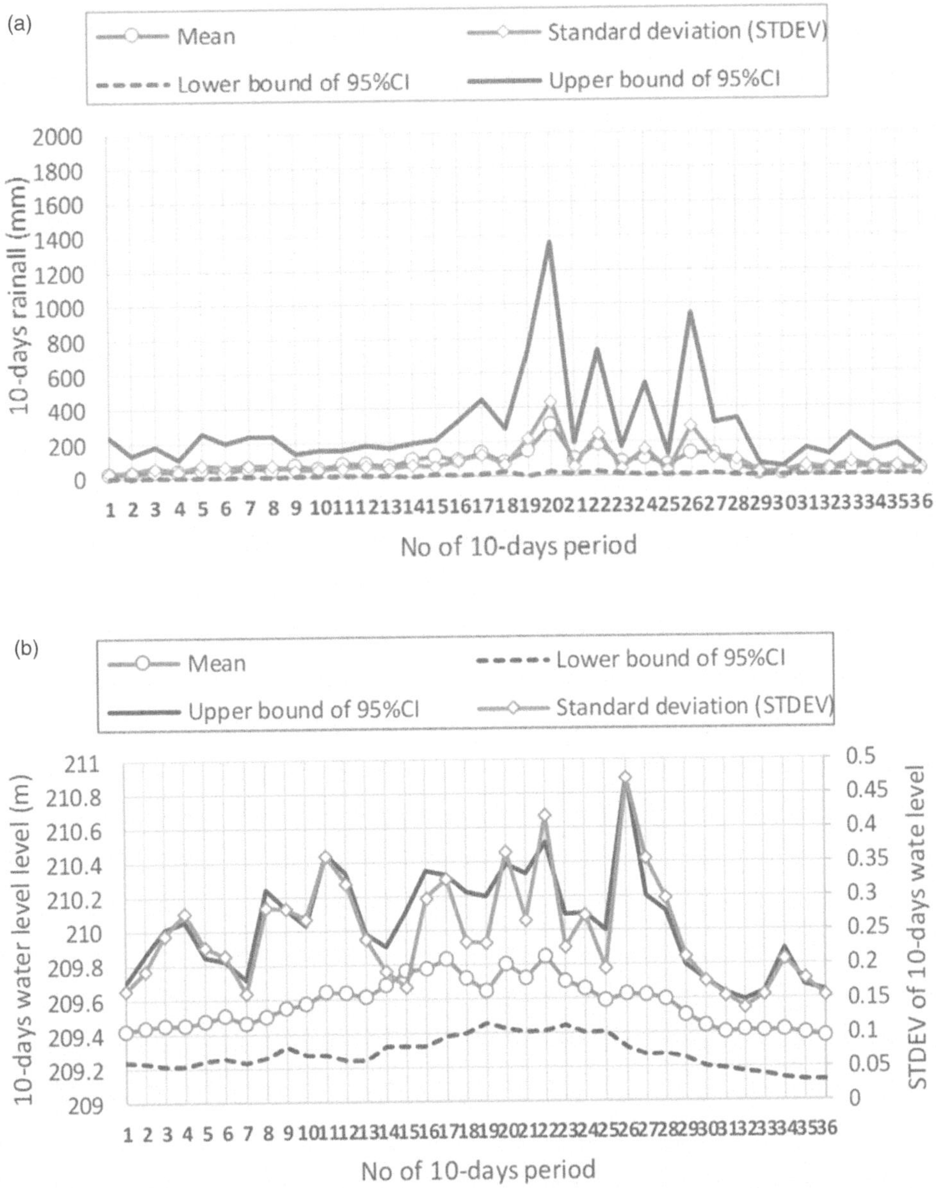

Figure 8 | Summary for the results from the uncertainty analysis for 10-days rainfall depths and water levels in 36 10-days period within the study area. (a) 10-days rainfall depth. (b) 10-days water level.

Also, From Figure 9(b), it can be observed that the correlation coefficient for the 10-days water level declines with the forward time steps; clearly, the correlation coefficient of the 10-day water level decreases from 0.79 (the forward 1 10-day period) to 0.31 (the forward 14 10-days periods). In particular, the correlation coefficient of 10-days water level for the first three forward time steps, on average, reaches 0.7; this implies that the 10-days water levels at the present time step is strongly related to those in the first three 10-days periods.

The above results summarize that the uncertainties of different degree can be found among the 10-days rainfall depths and water levels; furthermore, the 10-days rainfall depths and water levels among the continuous 10-days periods can be classified according to their high correlation. By doing so, the aforementioned results from the uncertainty analysis could be applied in the simulation of a large number of simulated 10-days rainfall depths and water levels and used in the model training and validating.

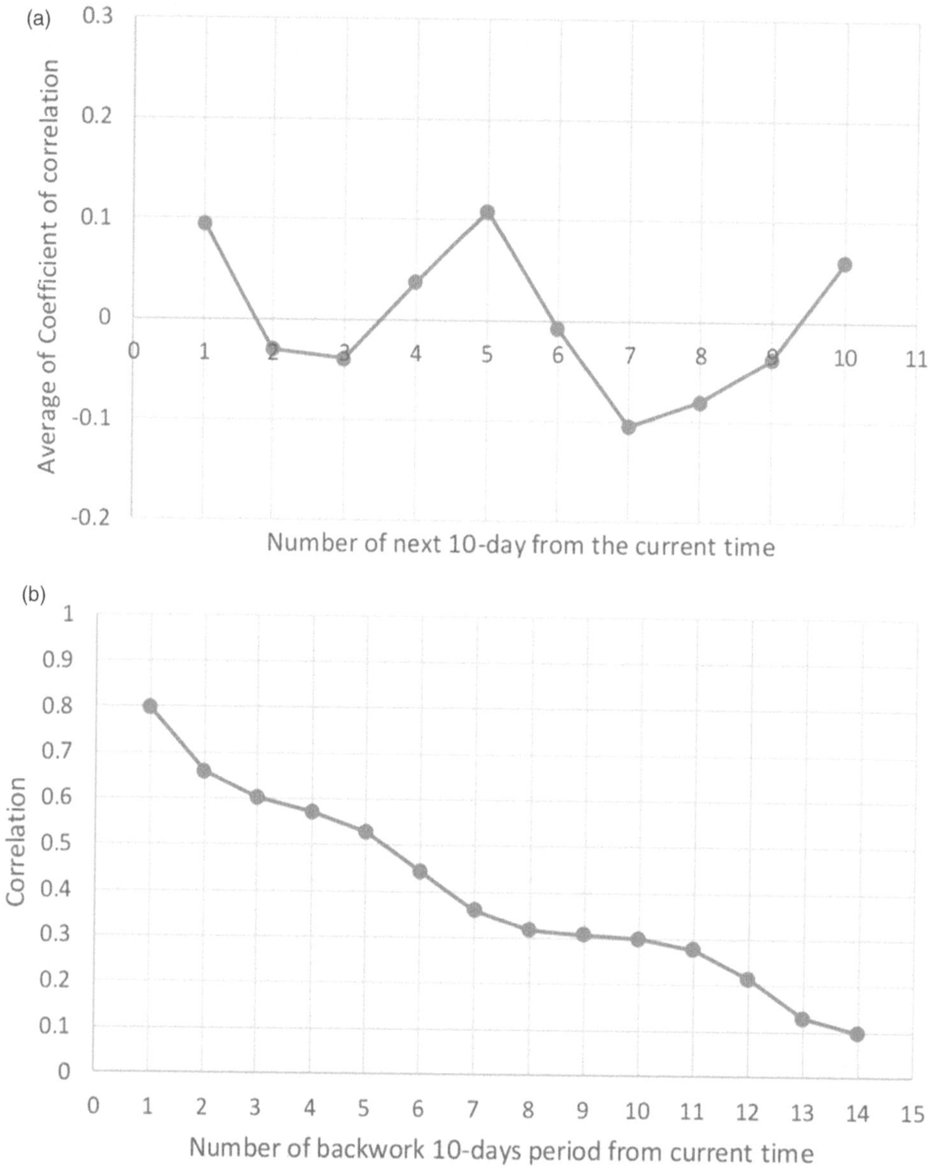

Figure 9 | Correlation coefficients among the 10-days rainfall depths and water levels. (a) 10-days rainfall depth. (b) 10-days water level.

3.3. Simulation of 10-days rainfall depths and water levels

Since a large training dataset is expected in training the ANN model in order to provide more reliable estimates, the multivariate Monte Carlo simulation (MMCS) (Wu *et al.* 2006) method in which a number of non-normal correlated variables can be simulated based on the statistical moments and correlation structure, including the water resource, would be widely applied in reproducing data based on the results from the uncertainty analysis. Hence, using the MMCS method with the statistical properties as shown at Figure 8, 1,000 simulations of 10-day rainfall and water levels are achieved as the training dataset (see Figure 10). Note that not only the dependence among the 10-days rainfall depths and water levels are taken into account, but also the correlations between the 10-days rainfall depths and water levels in 36 10-days periods are considered in the simulations of 10-days rainfall depths and water levels. Accordingly, the parameters of the proposed ANN_GA-SA_MTF model for forecasting the 10-days rainfall depths and water levels are calibrated by adopting the resulting 1,000 simulations of the 10-days rainfall depths and water levels.

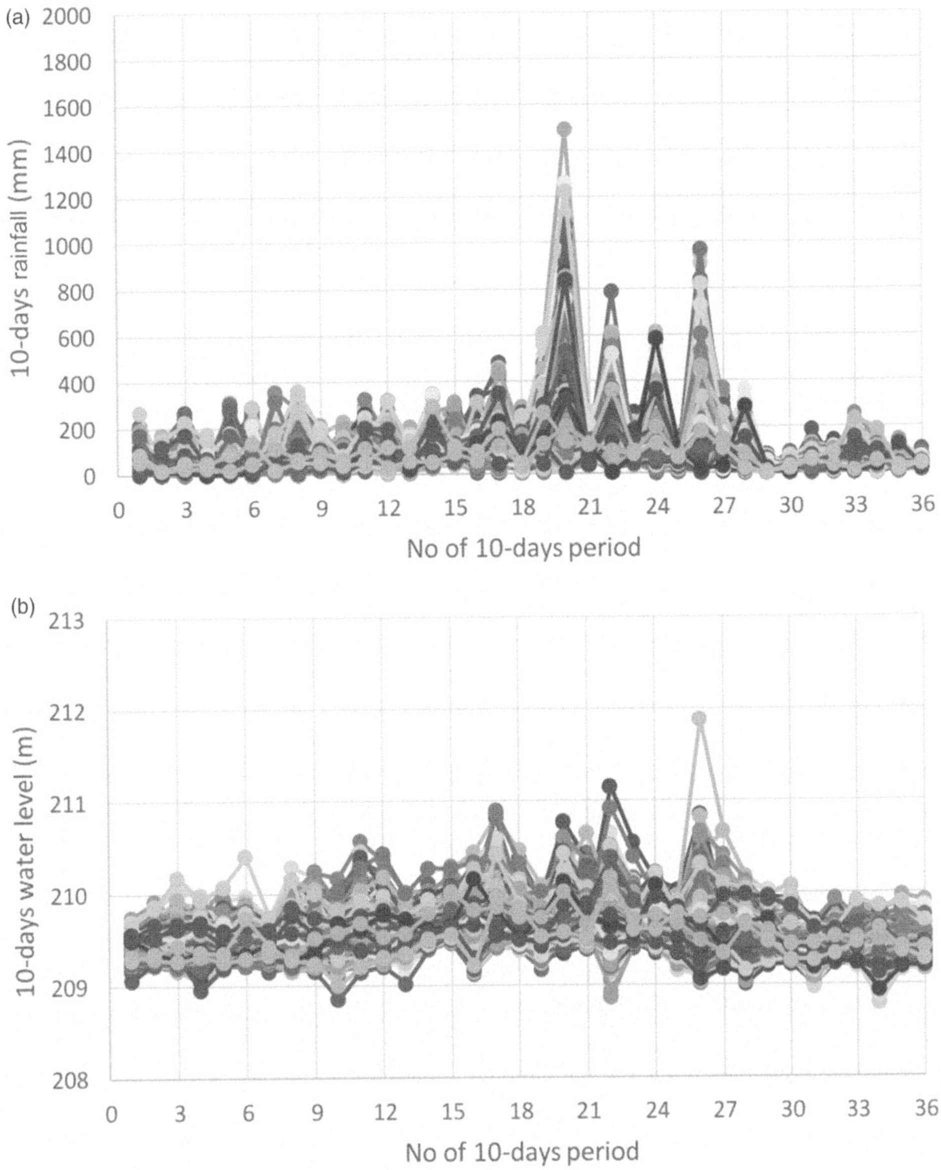

Figure 10 | 1,000 simulations of 10-days rainfall depths and water levels in 36 10-days periods within the study area. (a) 10-days rainfall depth. (b) 10-days water level.

3.4. Sensitivity analysis for correlations between the rainfall and water levels in time

According to results from Section 3.2, the 10-days rainfall depths and water levels are separately correlated to those in the forward three 10-days periods. However, rainfall is a key hydrological factor significantly inducing the discharge and affecting the river stages. Thus, this study utilizes the standardized regression equation introduced in Section 2.4 to evaluate the impact of the rainfall depth in current and forward three 10-days periods on the water-level forecasts in the specific 10-days periods using 1,000 simulations. Figure 11 shows the regression coefficients of rainfall at the forward 6th 10-days periods regarding the water levels in the specific 10-day period, approximately ranging from −0.36 to 0.25. This can be observed that the regression coefficients of the 10-day rainfall change at the various forward 10-days periods; however, on average, the regression coefficient declines with the 10-days period. In detail, the regression coefficient decrease from 0.08 at the current 10-days period to 0.02 at the forward 1st 10-days period; it then reaches a constant (about −0.008) from the forward 2rd to 6th

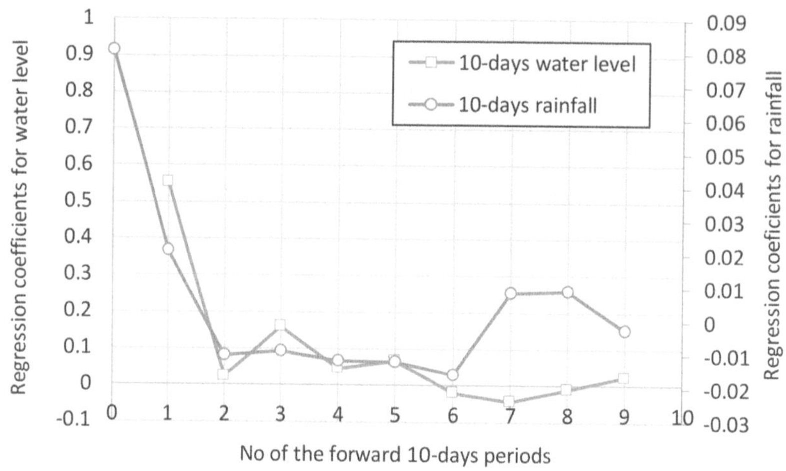

Figure 11 | Averaged regression coefficients of the rainfall depths and water levels in the various forward 10-days period.

10-days period. After that, the regression coefficient slightly rises to 0.008, indicating that the water levels in the specific 10-days period is sensible to the rainfall at the current and forward 1st 10-days periods.

In referring to the results from uncertainty and sensitivity analysis for the 10-days rainfall depth and water level, the rainfall and water levels in the specific 10-days periods are related to the associated rainfall in the forward three 10-days periods, respectively. With respect to the 10-days water levels, in addition to the water levels in the forward three 10-days periods, the rainfall in the current 10-days period should be treated as the dominant factor for estimating the 10-days water level; and it should be taken as reference regarding the establishment of the proposed ANN_GA-SA_MTF model for forecasting the 10-days rainfall depth and water level in the period of interest.

3.5. Establishment of ANN_GA-SA_MTF model

To sum up the results from Sections 3.3 and 3.4, the water level in the specific 10-days period should be the function of the rainfall depth in the current 10-days period and the water levels in the forward four 10-days periods, respectively. Accordingly, the proposed ANN_GA-SA_MTF model for the estimation of the water level (H^D_{fest}) can be written as:

$$H^D_{\text{fest}} = f_{\text{ANN-GA-MTF}}(R^D_{\text{est}}, H^{D-1}_{\text{obs}}, H^{D-2}_{\text{obs}}, H^{D-3}_{\text{obs}}) \tag{25}$$

where $R^{D-1}_{\text{obs}}, R^{D-2}_{\text{obs}}, R^{D-M3}_{\text{obs}}$ represent the observed rainfall in the forward 1st, 2nd, and 3rd 10-days periods, and $H^{D-1}_{\text{obs}}, H^{D-2}_{\text{obs}}, H^{D-3}_{\text{obs}}$ stand for the measured water levels in the forward 1st, 2nd, and 3rd 10-days period.

In particular, using the Equation (25) to forecast the 10-days rainfall depth, the predicted rainfall depth at the lead time of the 10-days period should be given in advance. In accordance with the results from the Sections 3.3 and 3.4, the 10-days rainfall depth in the current time step is related to those in the forward three 10-days periods. Thus, the proposed ANN_GA-SA_MTF model for estimating the 10-days rainfall depth can be addressed as the following equation:

$$R^D_{\text{est}} = f_{\text{ANN-GA-MTF}}(R^{D-1}_{\text{obs}}, R^{D-2}_{\text{obs}}, R^{D-3}_{\text{obs}}) \tag{26}$$

where $R^{D-1}_{\text{obs}}, R^{D-2}_{\text{obs}}, R^{D-3}_{\text{obs}}$ are the rainfall depths in the forward 1st, 2nd, and 3rd 10-days period, respectively.

In summary, as forecasting the water level in the specific 10-days period through Equation (25), the associated 10-days rainfall depth R^D_{est} should be estimated through Equation (26) in advance. Moreover, the parameters of the proposed ANN_GA-SA_MTF model and corresponding objective function (i.e., loss function) values can be determined via the GA-SA method with the training dataset in terms of 1,000 simulations of rainfall depths and water-levels in 36 10-days periods. Specifically, Equations (25) and (26) used in the proposed ANN_GA-SA_MTF model are individually derived for the 36 10-days periods corresponding to their own parameters calibrated.

3.5.1. Parameter calibration

It is well-known that in training the ANN model, i.e., calibrating the parameters of the ANN model, the number of hidden layers and the associated ANN weights are supposed to be defined in advance. Nevertheless, the neural network structures change in the model complexity and efficiency and the performance of the neural network can be obviously improved with the number of connections at various layers (Hunter *et al.* 2012). Generally speaking, regarding the relevant hydrological/hydraulic analysis, a three-layer network structure, i.e., the input/output layer and one hidden layer is comprehensively applied in training the ANN model in which the adjusting neuron weights could be effectively estimated via the back-propagation (BP) algorithm (e.g. Dawson & Wilby 2001; Khan *et al.* 2016; Sung *et al.* 2017); nevertheless, a large number of hidden layers, possibly exhibiting long-term dependencies, should be a challenge as the training process sets up a complicated network structure (Bengio 1991; Dawson & Wilby 2001). In this study, in order to the proposed ANN_GA-SA_MTF model in response to the nonlinear characteristics among hydrological variables, a four-layer network structure, consisting of the input and output layers as well as two hidden layers are used. Additionally, regarding the number of the hidden neurons defined in the network structures of the ANN model, several methods (see Table 1) can be used to estimate the number of hidden neurons as shown in Figure 12. In view of Figure 12, the number of hidden neurons used in the proposed ANN_GA-SA_MTF for 10-days rainfall-depth and water-level forecasts are located between 2–39 (10-days rainfall depth) and 2–17 (10-days water level), respectively; on average, the number of hidden neurons can be assigned as 10 (10-days rainfall depth) and 6 (10-days water level). To reduce the uncertainties in the selection of transfer functions in training the ANN model, a variety of the transfer functions as shown in Table 2 are adopted in this study. Additionally, this study utilizes a stochastic GA-based optimization method (GA-SA) to calibrate the parameter of the proposed ANN_GA-SA_MTF model, including the ANN weights, bias and adjusting factor; so that, the prior statistics of the calibrated parameters, including the mean and standard deviation, are summarized in Table 3.

Note that in this study, the 650 simulations randomly extracted from the training dataset (named development set), comprising the 1,000 simulations of 10-days rainfall depths and water levels (i.e., training sets), are used in training the proposed ANN_GA-SA_MTF model; the remaining 350 simulations (i.e., the 35% of the training dataset) are treated as the validation set. Thus, the training of the proposed ANN_GA-SA_MTF model for forecasting the 10-days rainfall depth and water level can be accomplished to obtain numerous sets of appropriate associated parameters regarding 10 transfer functions in 36 10-days periods. Tables 4 and 5 illustrates the calibrated ANN_GA-SA_MTF parameters of the for 10-days rainfall depth and water level in the 20th 10-days period under consideration of the transfer function TF1. In addition to appropriate parameters, Figure 13 shows the weighted factors of transfer functions of interest obtained via Equation (14) for quantifying the corresponding weighed average of hydrological estimates in the 20th 10-days period; it can be seen that the weighted factors regarding the ten transfer function for the 10-days rainfall depth resemble those for the 10-days water level. In comparison to the resulting weighted factors, TF1 (Sigmoid function) and TF5 (ReLU) make more significant contributions to predict the hydrological estimates as a result of the largest weighted factors

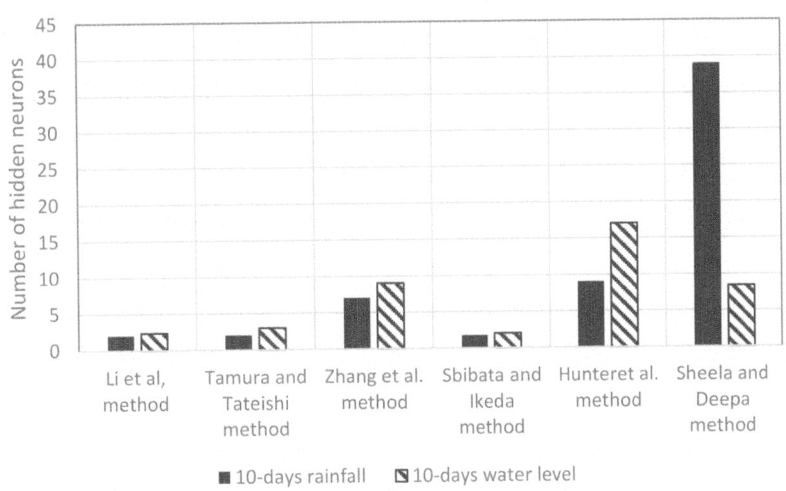

Figure 12 | Estimated number of hidden neurons for 10-days rainfall depths and water levels.

Table 3 | Definition of parameters used in the proposed ANN-GA-SA_MTF model

Parameters	Definition			
Transfer functions used	TF1-TF10			
Input factors	10-days rainfall		R_{obs}^{D-3}, R_{obs}^{D-2}, R_{obs}^{D-1}	
	10-days water level		R_{est}^{D}, H_{obs}^{D-3}, H_{obs}^{D-2}, H_{obs}^{D-1}	
Output factor	10-days rainfall		R_{est}^{D}, $R_{est,95\%CI}^{D}$,	
	10-days water level		H_{est}^{D}, $RH_{est,95\%CI}^{D}$,	
Number of hidden levels	2			
Number of hidden neurons	10-days rainfall	10	1st hidden layer	5
			2nd hidden layer	5
	10-days water level	6	1st hidden layer	3
			2nd hidden layer	3
Calibration of parameters of transfer function	Number of optimizations		1,000	
	Weights of neurons (ω_{HL})		Mean	1
			Standard deviation	5
	Bias of function (θ_{TF})		Mean	0
			Standard deviation	1
	Adjusting factor ($\propto_{TF}$)		Mean	0.5
			Standard deviation	0.35

Table 4 | Summary for the calibrated parameters of the proposed ANN_GA-SA_MTF model for 10-days rainfall depths regarding the 1st 10-days period

Transfer function	No of optimization	Adjust factor ($\propto_{TF}$)	0.43034							
TF1	OPT1	Weights of neurons ω_{HL}	The 1st Hidden Layer		Input factors					
					1	2	3	Bias		
			Neuron	1	3.55675	1.22755	−2.27802	−6.76864		
				2	7.84666	2.81362	−1.71866	1.09987		
				3	3.31515	1.08956	−5.00773	4.12584		
				4	−7.8968	−4.79188	2.54334	1.74716		
				5	−6.33654	1.69217	−4.92362	−0.64359		
			The 2nd Hidden Layer		The 1st Hidden Layer					
					1	2	3	4	5	Bias (θ_{TF})
			Neuron	1	9.45802	−1.28606	8.31515	2.74016	1.78777	−0.36351
				2	1.28778	−1.65618	2.11505	2.38406	−0.93583	−0.29953
				3	−1.51024	3.78917	4.79595	−1.02425	−3.07127	−1.90227
				4	3.47	1.7702	6.90849	0.77034	1.04623	1.02147
				5	−1.95808	−1.22736	9.85076	−0.89409	−0.0028	−1.56067
			Output layer		The 2nd Hidden Layer					
					1	2	3	4	5	Bias (θ_{TF})
			Output factor	1	−0.53822	−1.1665	−0.73957	0.39918	0.70249	−0.39683

(approximately 0.18); on the contrary, TF8 (Inverse absolute function) make few contributions to the hydrological forecasts in association with the worst factor. This reveals that estimated hydrological variates by the ANN models are significantly affected by the transfer functions selected and relevant parameters with high likelihood. Consequently, the effect of

Table 5 | Summary for the calibrated parameters of the proposed ANN_GA-SA_MTF model for 10-days water level in the 1st 10-days period

Transfer function	No of optimization	Adjust factor ($\propto_{TF}$)	.99824						
TF1	OPT1	Weights of neurons ω_{HL}	1st Hidden Layer		Input factors				
					1	2	3	4	Bias (θ_{TF})
			Neuron	1	1.38743	1.16656	5.79542	2.14252	−6.80367
				2	2.24727	−0.17489	5.19199	2.52	−3.89677
				3	9.2057	−3.59509	−5.04945	0.89079	7.97452
			The 2nd Hidden Layer		The 1st Hidden Layer				
					1	2	3	Bias (θ_{TF})	
			Neuron	1	4.96698	10.03204	6.52164	−0.47488	
				2	0.91084	−1.69658	6.81181	0.54777	
				3	0.34674	−6.11597	1.55485	−0.30177	
			Output layer		The 2nd Hidden Layer				
					1	2	3	Bias (θ_{TF})	
			Output factor	1	1.17024	0.09756	−1.07135	0.40452	

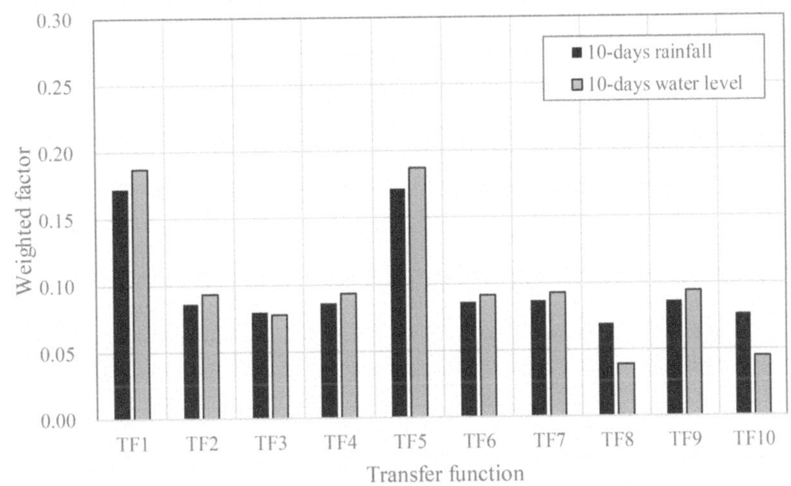

Figure 13 | Summary for weighted factors of transfer functions for calculating the weighted average of hydrological estimates in the 20th 10-days period.

uncertainty in the selection of transfer functions on the estimation of hydrological variable through the ANN model is supposed to be considered in training the ANN model.

3.5.2. Model validation

As the ANN_GA-SA_MTF parameters regarding 36 10-days period are determined, the model validation can be made by comparing the resulting estimations of 10-days rainfall depths and water levels (defined as the estimation sets) and the corresponding quantiles from the proposed ANN_GA-SA_MTF model using model inputs of 350 validation sets with the associated model outputs. The above quantiles of estimated 10-days rainfall depths and water levels are quantified by means of the weighted likelihood sample quantile estimator method as shown in Figures 14 and 15. It can be observed that the change in the quantiles of 10-days rainfall depths (about 10–700 mm) is significantly greater than those of 10-days water levels, approximately 208.8 and 210.8 m, indicating that the reliable 10-days rainfall depths might be obtained with more difficulty than the 10-days water levels. Additionally, Figures 16 and 17 show the comparisons among the validation sets, estimation sets and the associated prediction intervals regarding the 10-days rainfall depth and water level. It is noted that the prediction intervals of the 10-days rainfall depth and water level can be computed according to the corresponding

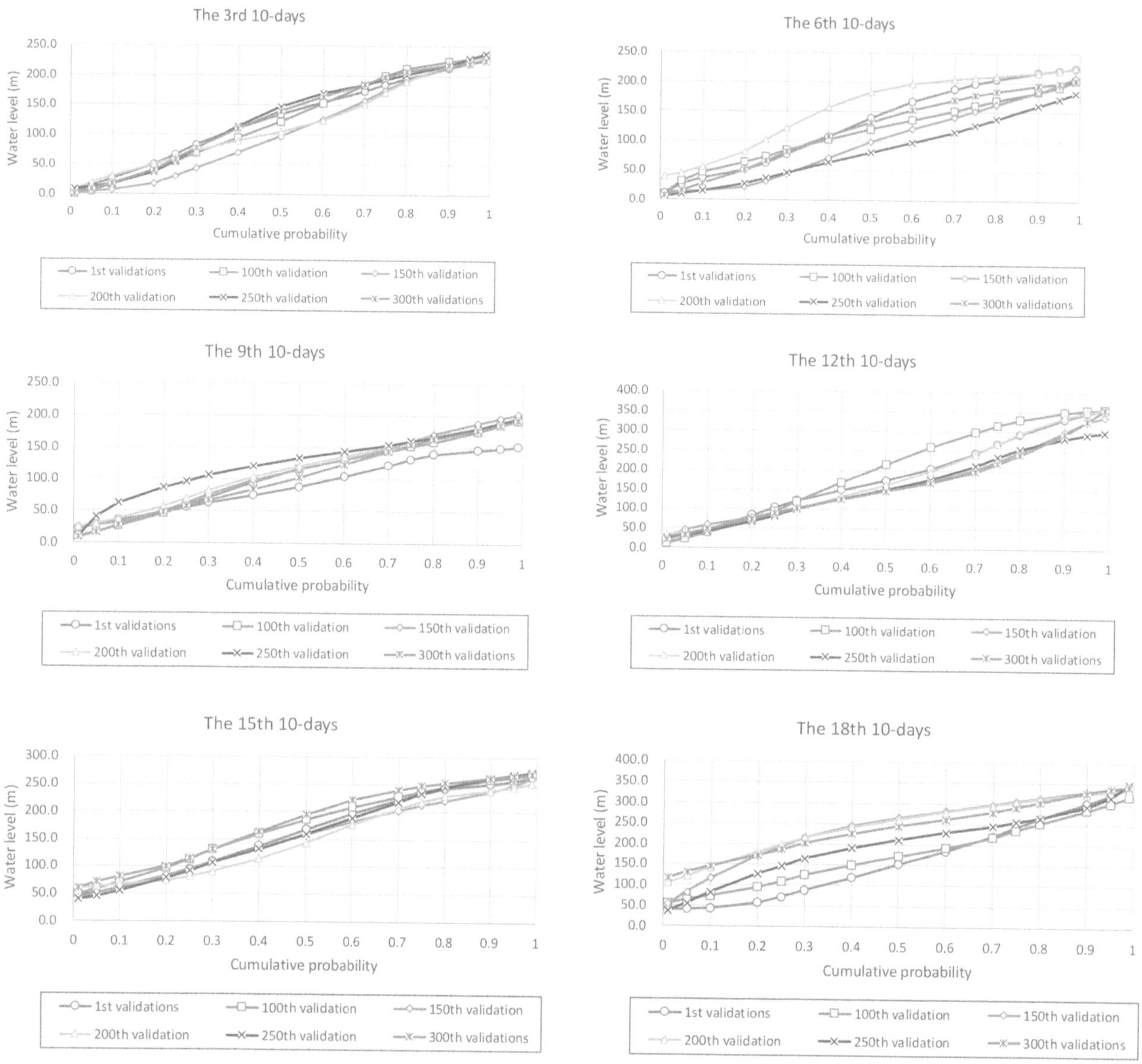

Figure 14 | Rainfall-depth quantiles calculated by the proposed ANN_GA-SA_MTF method for numerous 10-days periods. *(continiued.)*

quantile as shown in Figures 14 and 15. Also, Figure 18 shows the performance indices, i.e., the root mean square error (RMSE) and Geometric reliability index (KG) in various 10-day periods. The geometric reliability index KG can be calculated using the following equation, on average, approximates 1.28.

$$\mathrm{KG} = \frac{1 + \sqrt{\dfrac{1}{N_Y}\displaystyle\sum_{i=1}^{N_Y}\left[\dfrac{1-\left(\dfrac{\hat{Y}_i}{Y_i}\right)}{1+\left(\dfrac{\hat{Y}_i}{Y_i}\right)}\right]^2}}{1 - \sqrt{\dfrac{1}{N_Y}\displaystyle\sum_{i=1}^{N_Y}\left[\dfrac{1-\left(\dfrac{\hat{Y}_i}{Y_i}\right)}{1+\left(\dfrac{\hat{Y}_i}{Y_i}\right)}\right]^2}} \tag{27}$$

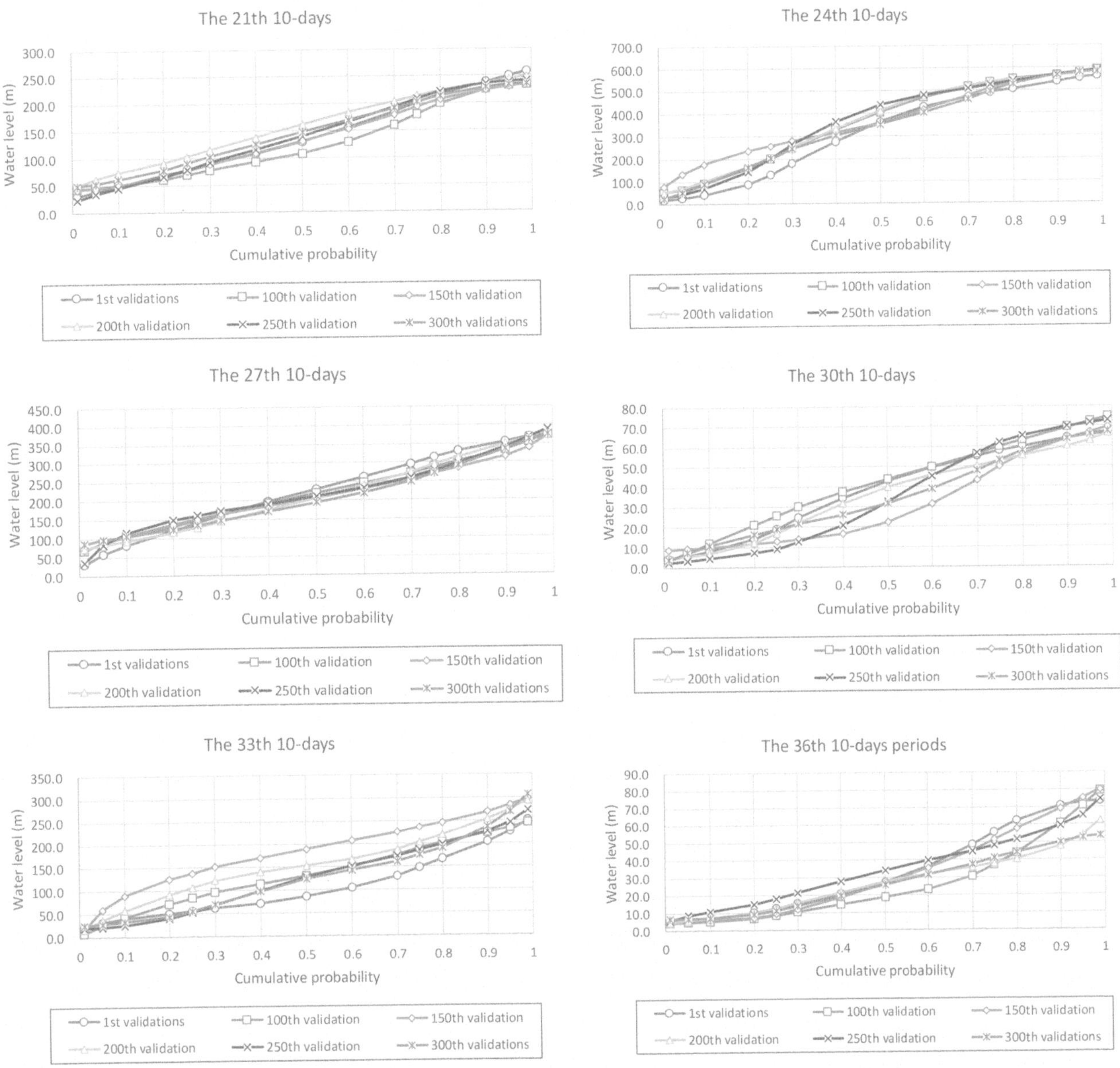

Figure 14 | Continued.

in which Y and $\hat{Y}$ serve as the validation and the estimation of the hydrological variate; N_Y denotes the number of hydrological data used in the performance evaluation. Note that when KG approaching one indicates that the varying trend of the estimated hydrological variates in time resembles that regarding validated ones.

(1) 10-days rainfall depth

From Figure 16, it can be observed that the validated rainfall depths in 36 10-days periods obviously exceed the resulting estimations from the proposed ANN_GA-SA_MTF model; also, Figure 18(1) indicate the root mean square error (RMSE) drastically varies between 25 and 450 mm (on average 128 mm) and the KG value approximate 1.28, meaning that the temporal change in the estimated 10-days rainfall depth is somewhat unlike that regarding the validated ones. These results reveal that the proposed ANN_GA-SA_MTF model possibly overestimates the 10-days rainfall depths which depart from the validation sets in 10-days periods possibly attributed to high variation in observed 10-days rainfall depth (see Figure 18(1)).

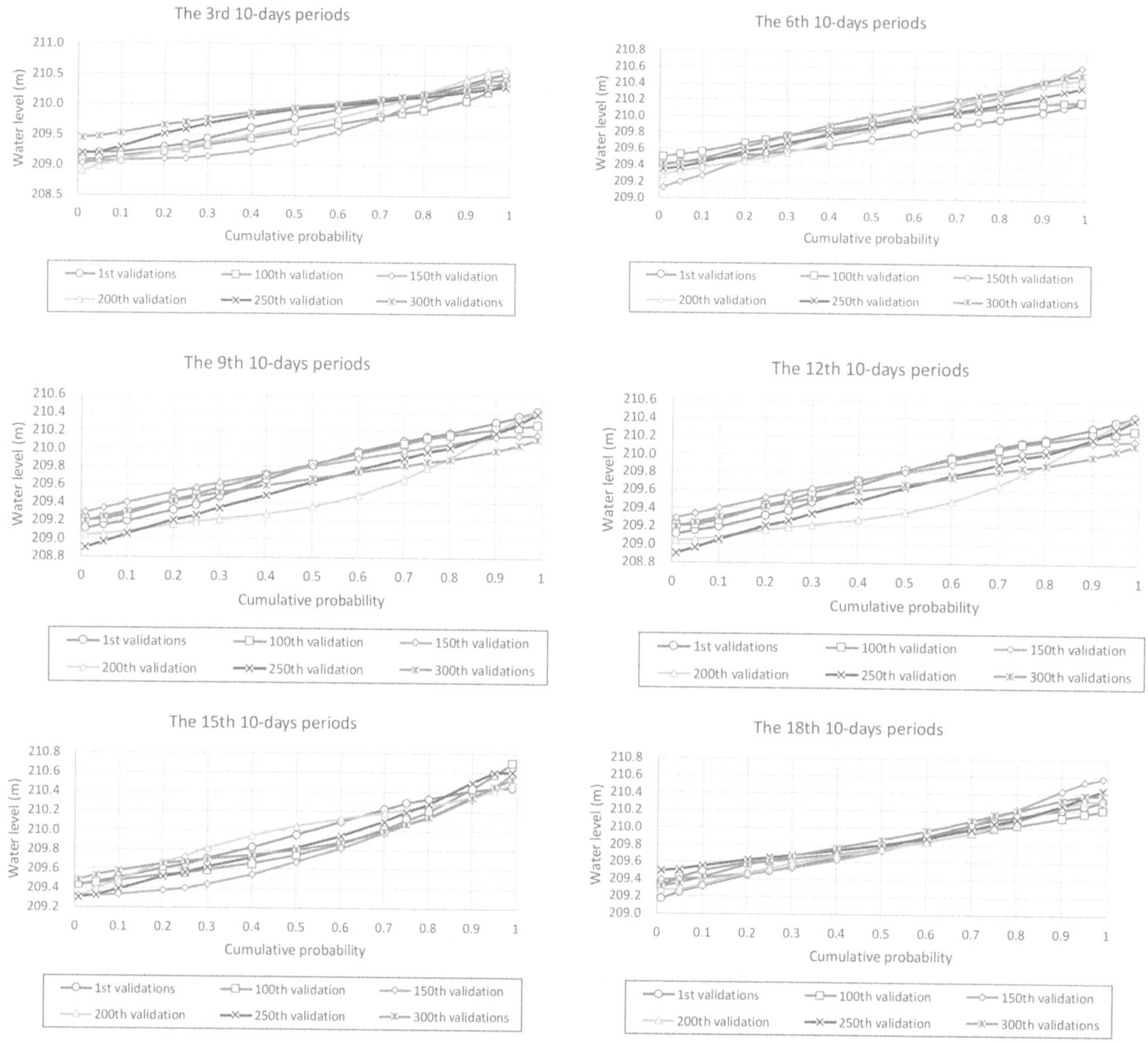

Figure 15 | Water-level quantiles calculated by the proposed ANN_GA-SA_MTF method for numerous 10-days periods. *(continued)*

In spite of overestimating the 10-days rainfall depths, the resulting prediction intervals of 10-days rainfall-depth estimates from the proposed ANN_GA-SA_MTF covers the validated ones in various 10-days periods. Actually, the validations of the 10-day rainfall depth approaches the lower bound of the prediction interval, implying that the 10-days rainfall depth estimates can capture the validation data with high likelihood. That is to say, the proposed ANN_GA-SA_MTF model can provide reliable 10-days rainfall-depth estimates associated with the acceptable bias.

(2) 10-days water level

With respect to the 10-days water-level estimates, the resulting estimation and quantified statistical properties of the water levels in various 10-days periods are compared to 350 sets of validation data (i.e., validation sets) (see Figure 17). It can be seen that as similar to the evaluation of the 10-days rainfall-depth estimates, the validation sets are randomly located between the prediction intervals in 36 10-days periods, unlike the validation sets regarding the 10-days rainfall depth markedly being close to the lower bound. In detail, Figure 18(2) indicates that the RMSE of 10-days water-level estimates changes from 0.2 to 0.7 m with an excellent agreement with the validated data, in which the associated KG is about 1.0, except at the 21th 10-days period with the bias of 0.7 m; the remaining forecasts approach the validations with a small error of 0.4 m. In addition, the

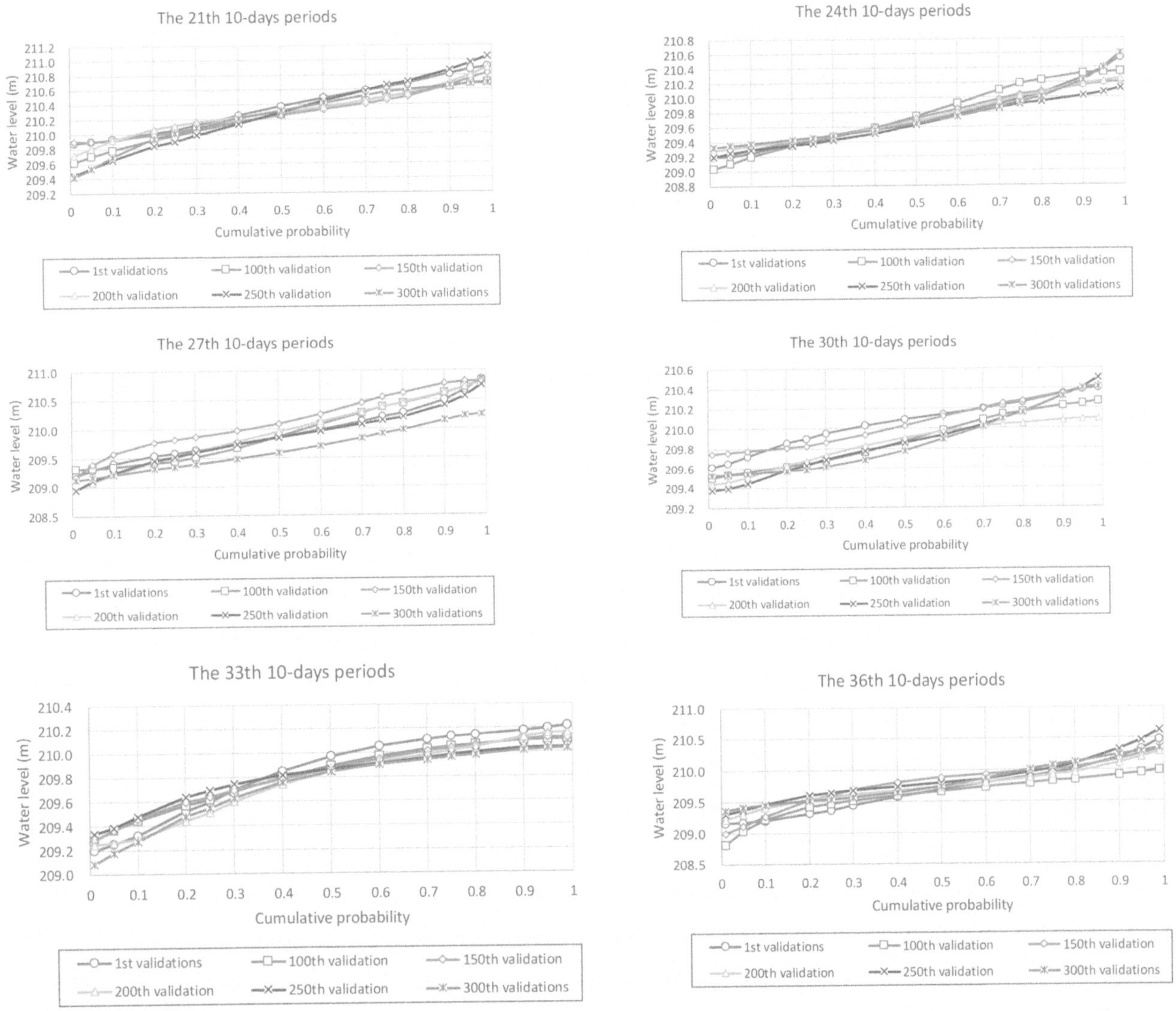

Figure 15 | Continued.

validated data are located within the 95% confidence interval, revealing that the proposed ANN_GA-SA_MTF is capable of providing the reliable 10-days water-level estimates.

(3) Summary

In summary, the results (i.e. weighted averages) of the 10-days rainfall depths and water levels separately provided by the proposed ANN_GA-SA_MTF model are possibly impacted by the uncertainties in the calibrations of appropriate parameters attributed to the variations in the training dataset; accordingly, their accuracy evidently changes with the 10-days periods. Furthermore, the wider ranges of the prediction intervals (i.e., 95% of confidence intervals) regarding 10-days rainfall depths and water levels exist at various 10-days periods. This reveals that even using the optimal ANN weights for the transfer functions, the results exhibit a significant difference among the corresponding estimations using a variety of transfer functions. Nevertheless, the proposed ANN_GA-SA_MTF model can provide the reliable results which can capture the observation with high opportunity in response to the variation in the training dataset and relevant induced uncertainties in relation to the model parameters.

3.6. Real-time correction of resulting forecasts

Since the climate change and occurrence of extreme events probably trigger the uncertainty in hydrological data (e.g., Wu *et al.* 2012a, 2012b), the accuracy of the resulting 10-days rainfall depths and water levels (defined as forecasts) from the

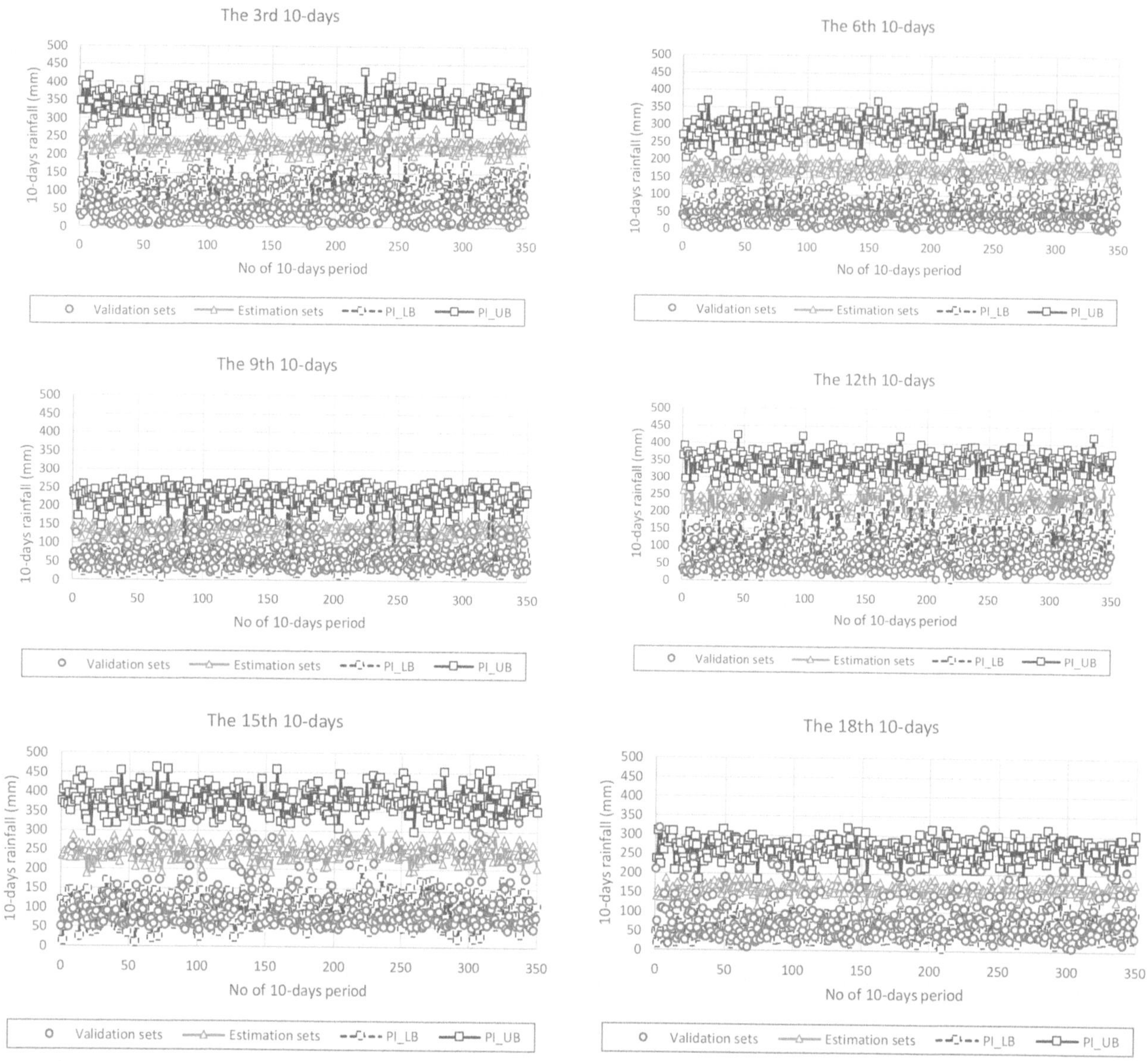

Figure 16 | The forecasts and associated statistical properties of resulting 10-days rainfall-depth estimates from the proposed ANN_GA-SA_MTF model as compared with the validation sets. *(continued)*

proposed ANN_GA-SA_MTF model might be affected by the above data variation. Thus, to provide the 10-day rainfall-depth and water-level forecasts with high accuracy via the proposed ANN_GA-SA_MTF model, the real-time correction method based on the difference between observation and forecasts in the past time steps, i.e., the RTEC_TS&KF developed by Wu *et al.* (2012a, 2012b), is utilized in simultaneously adjusting the results from the proposed ANN_GA-S_MTF model.

This study illustrates the process of correcting the forecasted 10-days rainfall depths and water levels using the observations of 10-days rainfall depths and water levels at the Shanping gauge during the 4th–15th 10-days periods in 2018 (see Figure 19). Figure 19 presents that the observed 10-days rainfall depths are positively related to the 10-days water levels, in which the corresponding correlation coefficient approximates 0.72; also, they reach the maximum values, 209.67 m (10-days water level) and 128.6 mm (10-days rainfall depth), respectively, in the 15th and 7th 10-days periods.

Figure 20 shows the comparison among the observed, forecasted and corrected 10-days rainfall depths and water levels, respectively. Note that the rainfall and water level in the 4th–10th 10-days periods are regarded as the observations, whereas the observations in the remaining 10-days periods are regarded as the validations. In detail, the rainfall depths and water levels in the 11th–15th 10-days period are forecasted and corrected by the proposed ANN_GA-SA_MTF model and

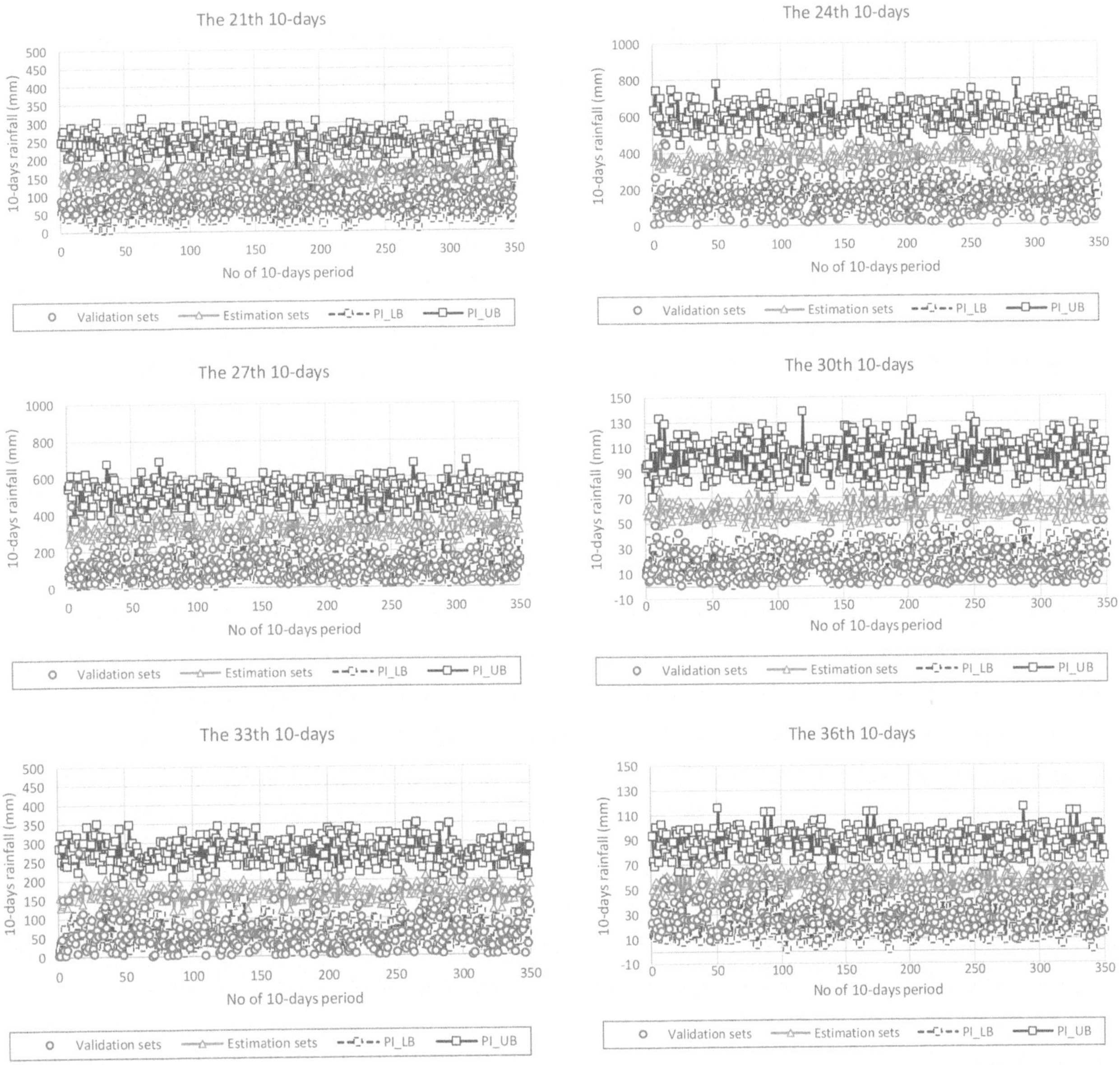

Figure 16 | Continued.

RTEC_TS&KF method, respectively, model with the observation at the 4th–10th 10-days periods. Additionally, the corresponding performance indices (root mean square error RMSE and correction coefficient) are also calculated as shown in Table 6. By referring to Figure 19 and Table 6, it can be seen that the 10-days rainfall-depth forecasts has good match regarding the varying trend in time with the observations, but with an overestimated bias (from 165.7 to −0.73 mm), of which the average approximates 46.64 mm. Moreover, the rainfall forecasts at the 8th and 9th 10-days periods slightly lie outside the preditions; However, after the real-time correction implemented by the RTEC_TS&KF method, the corresponding corrections are effectively improved to be closer to the observations and the corresponding performance indices RMSE and the correlation coefficient significantly declines from 46.5 to 15.43 mm and rises from 0.67 to 0.98, respectively.

In the case of the 10-days water level, similar to the 10-days rainfall depth, the observed 10-days water at the 8th–12th periods in 2019 level are slightly lower than or close to the bound of the prediction interval. Thus, the estimations significantly depart from the observations with a large overestimated bias (i.e. RMSE) being 0.295 mm, but with an acceptable temporal varying trend (correlation coefficient = 0.67). Although the proposed ANN_GA-SA_MTF model overestimates the 10-days

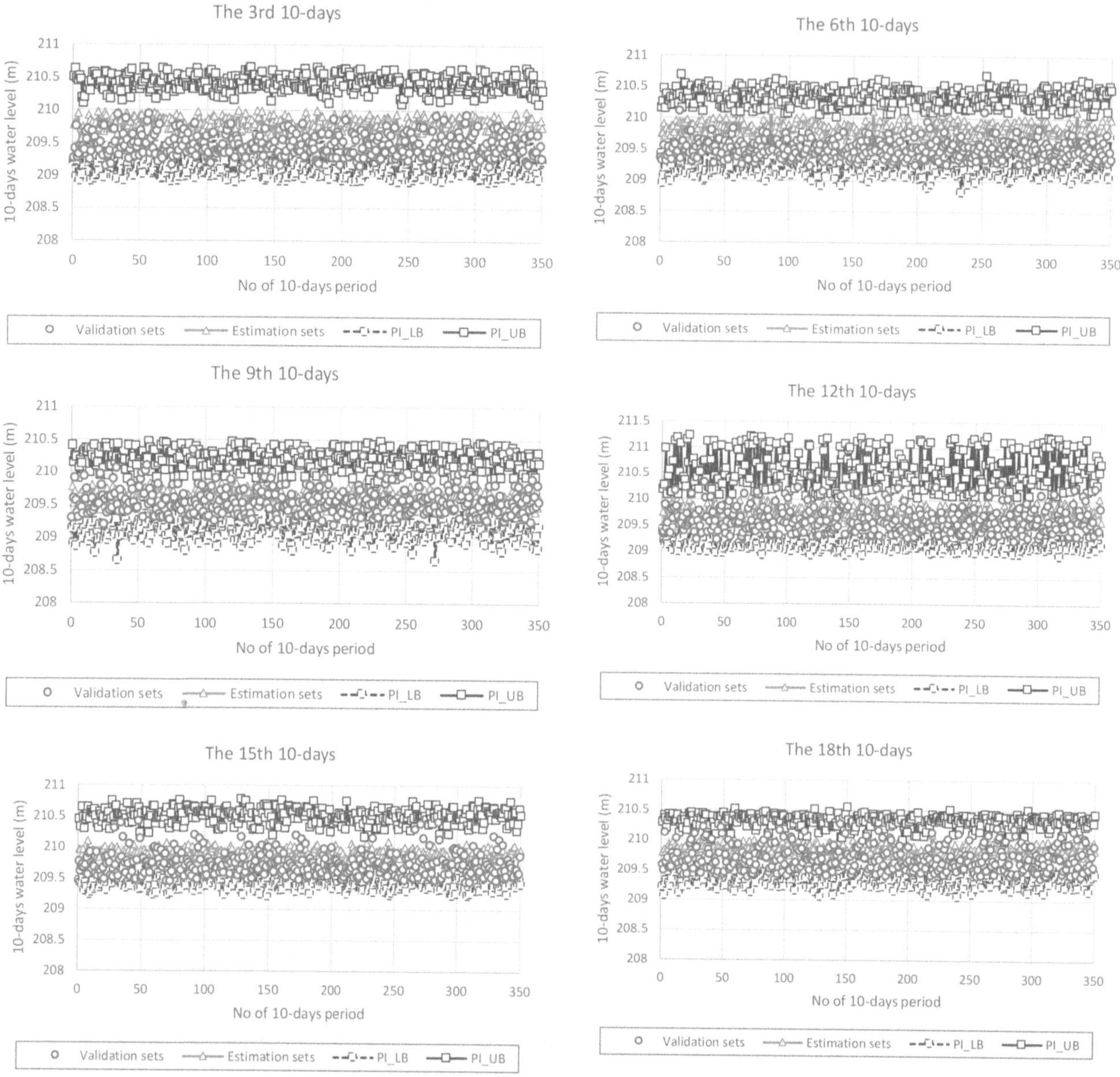

Figure 17 | The forecasts and statistical properties of resulting 10-days water-level estiamtes from the proposed ANN_GA-SA_MTF model as compared with the validation sets. *(continued)*

water level and the associated prediction interval, the RTEC_TS&KF can predict the errors in the future 10-days periods so as to correct the corresponding forecasts; namely, the 10-days water-level forecasts are subtracted by a predicted error changing with time, on average, 0.62 m. Accordingly, the performance indices can be effectively amended, that is, the RMSE values are from 0.298 m down to 0.1 m, and the correction coefficient rises from 0.67 to the 0.84.

To sum up the above results, the observed 10-days rainfall depths and water levels locate within or near the lower bound of the prediction interval quantified by the proposed ANN_GA-SA_MTF model. Moreover, the temporal change in the resulting forecasts from the proposed ANN_GA-SA_MTF model resembles that in the observations, revealing that the proposed ANN_GA-SA_MTF model can produce the 10-days rainfall depth and water-level forecasts with a highly comparable varying trend in time. Despite the resulting forecasts possibly evincing an acceptable bias from the proposed ANN_GA-SA_MTF model, by collaborating the RTEC_TS&KF method, the accuracy of the 10-days rainfall depth and water-level forecasts can be effectively improved based on the difference between the observation and forecasts in the previous 10-days period.

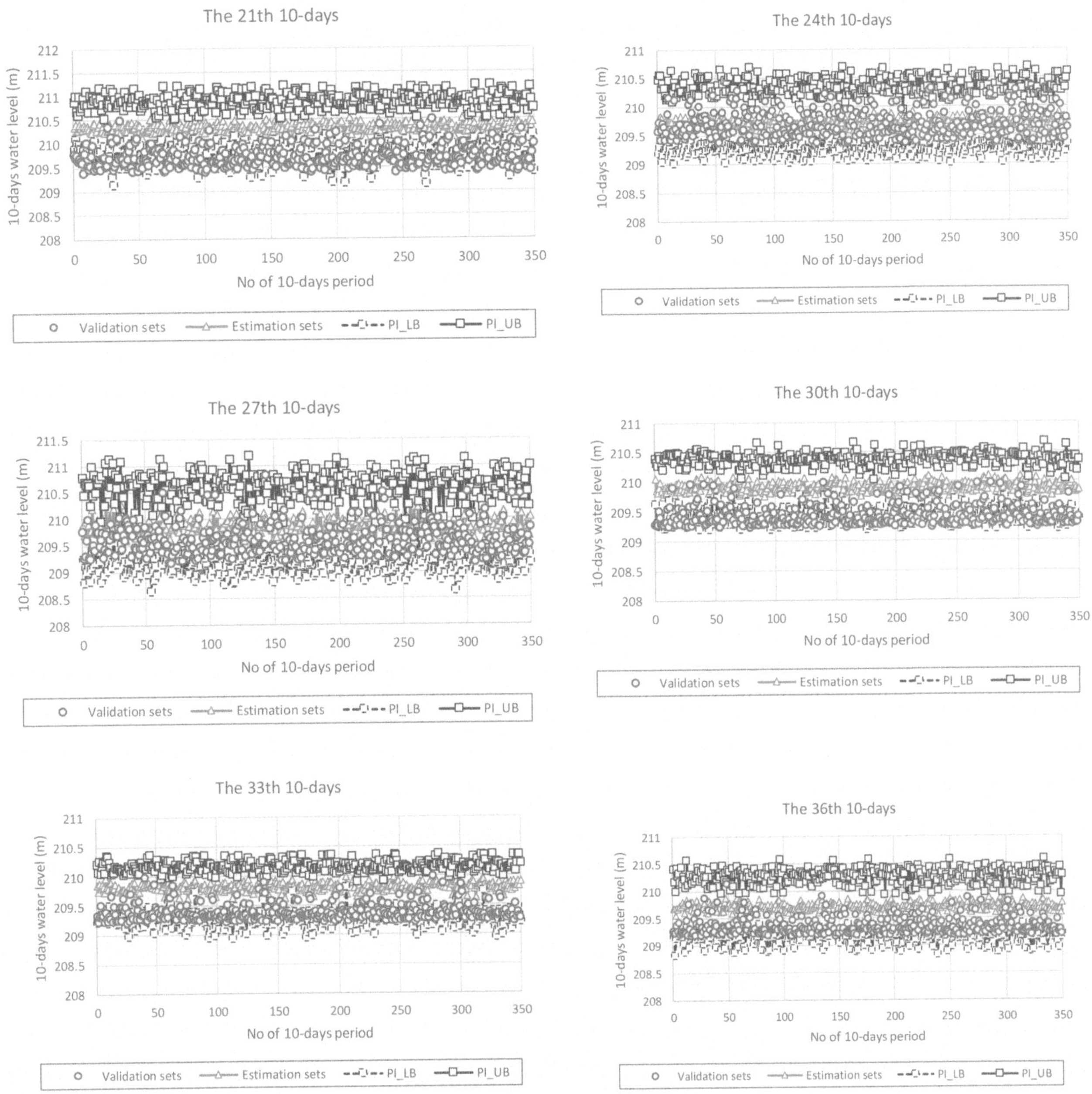

Figure 17 | Continued.

4. CONCLUSION

This study intends to develop a stochastic artificial neural network (ANN_GA-SA_MTF) model for forecasting the hydrological estimates in association with the stochastic properties, including the statistical moments and probabilities. To train the proposed ANN_GA-SA model, the multivariate Monte Carlo simulation (Wu *et al.* 2006) is adopted to generate a large number of hydrological data as the training dataset; they are then used to identify the sensible model inputs via the sensitivity analysis and to calibrate the associated parameters using the modified genetic (GA-SA) algorithm based on the sensitivity to uncertain parameters (Wu *et al.* 2012a, 2012b). Moreover, within the proposed ANN_GA-SA_MTF model, the multi-transfer functions are utilized in the combination of standardized variates at the input layer, hidden layers and output layers. Through the proposed ANN_GA-SA_MTF model, the reliability of the hydrological forecasts, defined as the likelihood of the forecasts approaching the observations, could be evaluated through the prediction interval (i.e., 95% confidence interval) by comparing

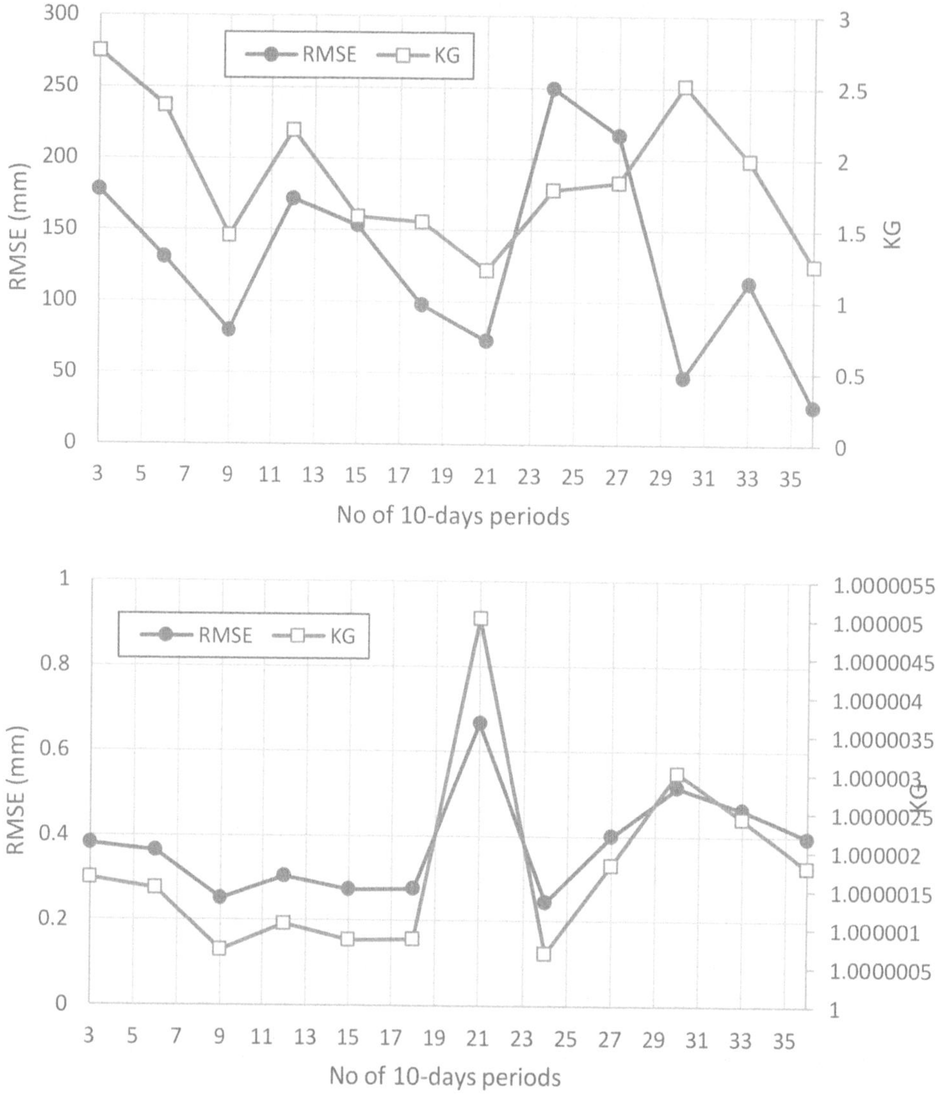

Figure 18 | Performance index of forecasts by the proposed ANN_GA-SA_MTF model. (1) 10-days rainfall depth. (2) 10-days water level.

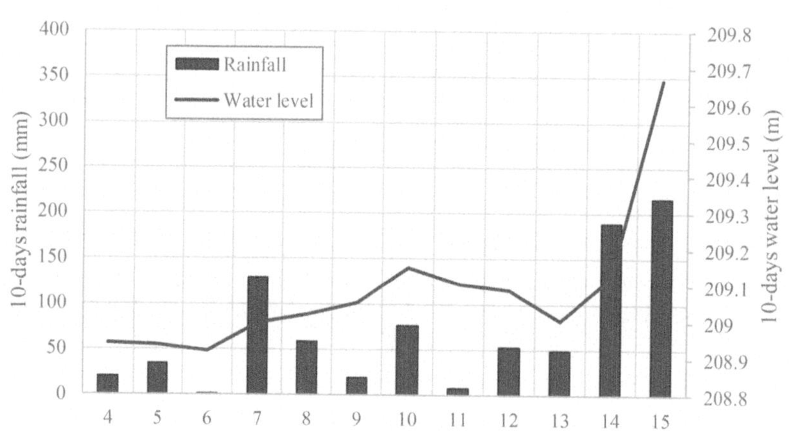

Figure 19 | Time series of observed 10-days rainfall depths and water levels used to assess the performance of the proposed ANN_GA-SA_MTF in real-time forecast.

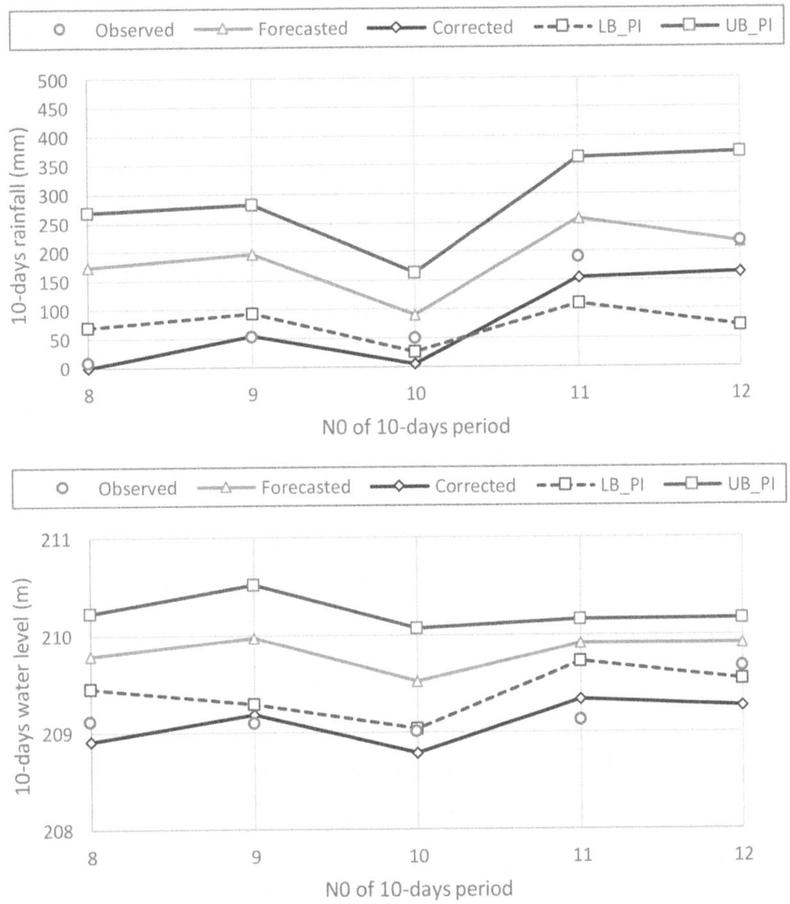

Figure 20 | Comparison among observed, estimated and corrected rainfall depths and water levels as well as the lower and upper bounds of prediction interval (LB_PI and UB_PI) from the 8th 10-days period to the 12th 10-day period.

Table 6 | Comparison in performance index of forecasted and corrected 10-days rainfall and water level

	10-days rainfall depth (mm)		10-days water level (m)	
Performance index	**Forecasted**	**Corrected**	**Forecasted**	**Corrected**
Root mean error square (RMSE)	46.642	15.430	0.295	0.109
Correlation coefficient	0.665	0.976	0.670	0.845

the upper and lower bounds of the prediction interval with the observations under the uncertainties in the selection of transfer functions and associated parameters. Eventually, the hydrological forecasts in terms of the weighted average of results from the proposed ANN_GA-SA_MTF model can be corrected by the real-time error correction method (RTEC_TS&KF) using the time-series and Kalman filtering methods (Wu *et al.* 2012a, 2012b) as to enhance forecast accuracy.

To demonstrate the performance of the proposed ANN_GA-SA_MTF model in real-time forecasts of hydrological variates, the observations of the 10-days rainfall depths sand water levels recorded from 2008 to 2017 at Shanping Gauge in Shinging Creek, a tributary of Toquian River, are selected as study data within the study area. Using 1,000 simulations of 10-days rainfall depths and water levels as the training dataset, the numerous sets of appropriate parameters of the transfer functions used within the proposed ANN_GA-SA_MTF model for 36 10-days periods are calibrated and validated. The results from the model training and validation indicate that the simulated rainfall depths and water levels in 36 10-days periods mostly locate between their prediction intervals as quantified by the proposed ANN_GA-SA_MTF model; this implies that the proposed ANN_GA-SA_MTF is supposed to estimate the 10-days rainfall depths and water levels which approach the observations with high likelihood. In spite of the forecasted 10-days rainfall depth and water-levels by the proposed ANN_GA-SA_MTF model with the difference of varying degree from observations, the aforementioned forecasts is capable of being immediately corrected by means of the RTEC_TS&KF method with real-time observations received through IoT to enhance their accuracy.

The proposed ANN_GA-SA_MTF model can quantify the uncertainties in the forecasted 10-day rainfall depths and water levels with an acceptable bias under consideration of uncertainties in the types of transfer functions and associated connection weights and bias at various layers, but without taking into account network structures (i.e., number of hidden layers). Therefore, to evaluate the effect of network structures on the resulting hydrological forecasts from the proposed ANN_GA-SA_MTF model, the various network structures could be applied in the model training, and the corresponding results can be compared with those from a given number of hidden layers. Furthermore, since the coverage probability of a confidence interval in statistics is defined as the ratio of the true values being located within interval, the quantified 95% of confidence intervals by the proposed ANN_GA-SA_MTF model can be applied in the well-known hydrological/hydraulic modeling for evaluating the performance of the resulting estimations/forecasts under consideration of uncertainties in the observations. Eventually, the proposed model concept and framework of training the proposed ANN_GA-SA_MTF model, including simulating of a large number of training datasets using the multivariate Monte Carlo simulation, calibrating parameters through the GA-SA algorithm and adjusting model outputs carried out by the real-time error correction method RTEC_TS&KF, can be employed in the stochastic modeling the artificial intelligence (AI) methods (e.g., CNN, LSTM) and be advantageous to facilitate the forecast's reliability and accuracy.

REFERENCES

ASCE 2000a Artificial neural networks in hydrology I: preliminary concepts. *Journal of Hydrologic Engineering* **5** (2), 115–123.

ASCE 2000b Artificial neural networks in hydrology II: hydrologic appliction. *Journal of Hydrologic Engineering* **5** (2), 124–137.

Bengio, Y. 1991 *Artificial Neural Networks and Their Application to Sequence Recognition*. PhD thesis, McGill University, Montreal, Quebec, Canada.

Bertoni, J. C., Tucci, C. E. & Clarke, T. 1992 Rainfall-based real-time flood forecasting. *Journal of Hydrology* **131**, 313–339.

Campolo, M., Andreussi, P. & Soldati, A. 1999 River flood forecasting with a neural network model. *Water Resources Research* **35** (4), 1191–1197.

Chalumuri, A., Kune, R. & Manoj, B. S. 2020 Training an artificial neural network using qubits as artificial neurons: a quantum computing approach. *Procedia Computer Science* **171**, 568–575.

Chang, C. H., Yang, J. C. & Tung, Y. K. 1997 Incorporate marginal distributions in point estimate methods for uncertainty analysis. *Journal of Hydraulic Engineering* **123** (3), 244–251.

Chang, C. H., Chung, M. K., Yang, S. Y., Chih-Tsung Hsu, C. T. & Wu, S. J. 2018 A case study for the application of an operational two-dimensional real-time flooding forecasting system and smart gauges on roads in Tainan City, Taiwan. *Water* **10** (5), 574–589.

Chen, L. & Wang, L. 2018 Recent advance in earth observation big data for hydrology. *Big Earth Data* **2** (1), 86–107.

Chen, N., Xiong, C., Du, W., Wang, C. & Lin, X. 2019 An improved genetic algorithm coupling a back-propagation neural network model (IGA-BPNN) for water-level predictions. *Water* **11** (9), 1975–1992.

Chiroma, H., Abdullahi, U. A., Abdulhamid, S. M., Alarood, A. A., Gabralla, L. A., Rana, N., Shuib, L., Hashem, A. T., Gbenga, D. E., Abubakar, A. I., Zeki, A. M. & Herawan, T. 2018 Progress on artificial neural networks for big data analytics: a survey. *IEEE Access* **7**, 70535–70551.

Cigizoglu, H. K. & Kisi, Q. 2005 Flow prediction by three back propagation techniques using k-fold partitioning of neural network training data. *Nordic Hydrology* **33** (1), 49–64.

Dawson, C. W. & Wilby, R. L. 2001 Hydrological modelling using artificial neural networks. *Progress in Physical Geography* **25** (1), 80–108.

Firat, M. 2008 Comparison of artificial intelligence technigques for river flow forecasting. *Hydrology and Earth System Science* **12**, 123–139.

Foody, G. M. & McCulloch, M. B. 1995 The effect of training set size and composition on artificial neural network classification. *Internationa Journal of Remote Sensing* **16** (9), 1707–1723.

Gestal, M. 2011 *Introducción a las Redes de Neuronas Artificiales*. Universidade da Coruña. Available from: http://sabia.tic.udc.es/mgestal/cv/RNAtutorial/TutorialRNA.pdf.

Gupta, H. V., Beven, K. J. & Wagener, T. 2005 Model calibration and uncertainty estimation. In: *Encyclopedia of Hydrologic Sciences* (Anderson, M. G., ed.). Wiley, Chichester (in press).

Haas, J. C., Switanek, M. & Birk, S. 2018 Analysis of hydrological data time correlation matrices: technical implementation and possible applications. *Environmental Earth Sciences* **77**, 310–321.

Haddad, K. & Rahman, A. 2011 Selection of the best fit flood frequency distribution and parameter estimation procedure: a case study for Tasmania in Australia. *Stochastic Environmental Research and Risk Assessment* **25** (3), 415–428.

Hochreiter, S. 1998 The vanishing gradient problem during learning recurrent neural nets and problem solutions. *International Journal of Uncertainty Fuzziness and Knowledge Based System* **6** (2), 107–116.

Hunter, D., Yu, H., Pukish III, M. S., Kolbusz, J. & Wilamowski, B. M. 2012 Selection of proper neural network sizes and architectures: a comparative study. *IEEE Transactions on Indus-Trial Informatics* **8** (2), 228–240.

Imrie, C., Durucan, S. & Korre, A. 2000 River flow prediction using artificial neural networks: generalization beyond the calibration range. *Journal of Hydrology* **233**, 138–153.

Kan, G. Y., Li, J., Zhang, X. N., Ding, L. Q., He, X. Y., Liang, K., Jiang, M. L., Wang, F., Zhang, Z. B. & Hu, Y. B. 2016 A new hybrid data-driven model for event-based rainfall–runoff simulation. *Neural Computing and Applications* **28** (9) doi.10.1007/s00521-016-2200-4.

Kan, G., He, X., Li, J., Ding, L. & Hong, Y. 2019 Computer aided numerical methods for hydrological model calibration: an overview and recent development. *Archives of Computational Methods in Engineering* **26**, 35–59.

Khan, M. Y. A., Hasan, F., Panwar, S. & Chakrapani, G. J. 2016 Neural network model for discharge and water-level prediction for Ramganga River catachment of Ganga Basin. *Hydrological Sciences Journal* **61** (11), 2084–2095.

Li, J. Y., Chow, T. W. S. & Yu, Y. L. 1995 Estimation theory and optimization algorithm for the number of hidden units in the higher-order feedforward neural network. In: *Proceedings of the IEEE International Conference on Neural Networks*, December 1995, Vol. 3, pp. 1229–1233.

Liu, P. L. & Der Kiureghian, A. 1986 Multivariate distribution models with prescribed marginals covariances. *Probabilistic Engineering Mechanics* **1** (2), 105–112.

Maca, P., Pech, P. & Pavlasek, J. 2014 Comparing the selected transfer functions and local optimization methods for neural network flood runoff forecast. *Mathematical Problems in Engineering.* doi:10.1155/2014/782351.

Malik, S., Jafri, Y. Z., Marri, A., Yasmeen, S., Khanum, Z. & Jafri, H. 2019 Stochastic models and neural networks with prediction equations: a comparative study using weather data of Quetta, Pakistan. *International Journal of Statistics and Application* **95**, 143–152.

Manchev, N. & Spratling, M. 2020 Target propagation in recurrent neural network. *Jounral of Machine Learning Research* **21**, 1–33.

McMillan, H. K., Westerberg, I. K. & Krueger, T. 2018 Hydrological data uncertainty and its implications. *Wiley Interdisciplinary Reviews: Water* **5** (4), 1–14.

Melching, C. S. 1995 Reliability estimation. In: *Computer Models of Watershed Hydrology* (Singh, V. P., ed.). Water Resources Publications, Colorado, pp. 69–118.

Melsen, L., Teuling, A., Torfs, P., Zappa, M., Mizukami, N., Clark, M. & Uijlenhoet, R. 2016 Representation of spatial and temporal variability in large-domain hydrological models: case study for a mesoscale pre-Alpine basin. *Hydrology Earth System and Science* **29**, 2207–2226.

Nandakumar, N. & Mein, R. G. 1997 Uncertainty in rainfall-runoff model simulations and the implications for predicting the hydrological effects of land-use change. *Journal of Hydrology* **192**, 211–231.

Nataf, A. 1962 Determination des distributions don't les marges sontdonnees. *Comptes Rendus de L'Académie des Sciences* **225**, 42–43.

Paschalis, A., Molnar, P., Fatichi, S. & Burlando, P. 2013 A stochastic model for high-resolution space-time precipitation simulation. *Water Resources Research* **49**, 8400–8417.

Rumelhart, D. E., McClelland, J. & P. D. P. & Research Group, 1986 *Parallel Distributed Processing: Explorations in the Microstructure of Cognition*, vol. 1. MIT Press, Cambridge, MA.

Sankaranarayanan, S., Prabhakar, M., Satish, S., Jain, P., Ramprasad, A. & Krishnan, A. 2019 Flood prediction based on weather parameters using deep learning. *Journal of Water and Climate Change* **11** (4), 1766–1783.

Sayers, W., Savic, D., Kapelan, Z. & Kellagher, R. 2014 Artificial intelligence techniques for flood risk management in urban environments. *Procedia Engineering* **70**, 1505–1512.

Shamseldin, A. Y. 2010 Artificial neural network model for river flow forecasting in a developing country. *Journal of Hydroinformatics* **12** (1), 22–35.

Sheather, S. J. & Marron, J. S. 1990 Kernel quantile estimator. *Journal of American Statistical Association* **85** (410), 410–416.

Sheela, K. & Deepa, S. N. 2013 Review on methods to fix number of hidden neurons in neural networks. *Mathematical Problems in Engineering.* doi:10.1155/2013/425740.

Shibata, K. & Ikeda, Y. 2009 Effect of number of hidden neurons on learning in large-scale layered neural networks. In: *Proceedings oftheICROS-SICE International Joint Conference (ICCAS-SICE '09)*, pp. 5008–5013.

Speed, T. P. & Yu, B. 1993 Model selection and prediction: normal regression. *Annals of the Institute of Statistical Mathematics* **45** (1), 33–54.

Sun, M. & Kim, G. 2014 Real-time analysis and forecasting of multisite river flow using a distributed hydrological model. *Mathematical Problem in Engineering* **2014**, 1–9.

Sun, M., Li, Z., Yao, C., Liu, Z., Wang, J., Hou, A., Zhang, K., Huo, W. & Liu, M. 2020 Evaluation of flood prediction capability of the WRF-Hydro Model based on Multiple forcing scenarios. *Water* **17**, 874–893.

Sung, J. Y., Lee, J., Chung, I. W. & Heo, J. H. 2017 Hourly water level forecasting at tributary affected by main river condition. *Water* **6**, 644–660.

Supharatid, S. 2003 Application of a neural network model in establishing a stage-discharge relationship for a tidal river. *Hydrological Processes* **17**, 3085–3099.

Tahboub, K. K., Barghash, M., Arafeh, M. & Ghazal, O. 2016 An ANN-GA Framework for optimal engine modeling. *Mathematic Problems in Engineering* **2016**, 1–8.

Tamura, S. & Tateishi, M. 1997 Capabilities of a four-layered feedforward neural network: four layers versus three. *IEEE Transactions on Neural Networks* **8** (2), 251–255.

Tu, J. V. 1996 Advantages and disadvantages of using artificial neural networks versus logistic regression for predicting medical outcomes. *Journal of Clinical Epidemiology* **49** (11), 1225–1231.

Tung, Y. K. & Yen, B. C. 2005 *Hydrosystems Engineering Uncertainty Analysis (ISBN: 0071451595, 9780071451598).* McGraw-Hill Book Company, New York.

Wagener, T. & Gupta, H. V. 2005 Model identification for hydrological forecasting under uncertainty. *Stochastic Environment Research and Risk Assessment* **19**, 378–387.

Wang, W. C., Chau, K. W., Cheng, C. T. & Qiu, L. 2009 A comparison of performance of several artificial intelligence methods for forecasting monthly discharge time series. *Journal of Hydrology* **374**, 294–306.

Wang, L., Zeng, Y. & Chen, T. 2015 Back propagation neural network with adaptive differential evolution algorithm for time series forecasting. *Expert Systems with Applications* **42**, 855–863.

Wegman, E. J. & Habib, M. K. 1990 Stochastic methods of neural systems. *Journal of Statistical Planning and Inference* **33**, 5–23.

Wu, S. J., Tung, Y. K. & Yang, J. C. 2006 Stochastic generation of hourly rainstorm events. *Stochastic Environment Research and Risk Assessment* **21**, 195–212.

Wu, S. J., Chen, P. H. & Yang, J. C. 2011 Application of Weighted Semivariogram Model (WSVM) based on fitness to experimental semivariogram on estimation of rainfall amount. *Hydrology and Earth System Science Discussion* **8** (2), 4229–4259.

Wu, S. J., Lien, H. C. & Chang, C. H. 2012a Calibration of a conceptual rainfall-runoff model using a genetic algorithm integrated with runoff estimation sensitivity to parameters. *Journal of Hydroinformatics* **14** (2), 497–511.

Wu, S. J., Lien, H. C., Chang, C. H. & Shen, J. C. 2012b Real-time correction of water stage forecast during rainstorm events using combination of forecast errors. *Stochastic Environment Research and Risk Assessment* **26** (4), 519–531.

Wu, S. J., Lien, H. C., Hsu, C. T., Chang, C. H. & Shen, J. C. 2015 Modeling probabilistic radar rainfall estimation at ungauged locations based on spatiotemporal errors which correspond to gauged data. *Hydrology Research* **46** (1), 39–59.

Wu, S. J., Chang, C. H. & Hsu, C. T. 2020 Real-time error correction of two-dimensional flood-inundation simulations during rainstorm events. *Stochastic Environmental Research and Risk Assessment*. doi:10.1007/s00477-020-01792-2.

Xia, J. 2002 A system approach to real-time hydrologic forecast in watersheds. *Water International* **27** (1), 91–97.

Xu, B., Zhang, H., Wang, Z., Wang, H. & Zhang, Y. 2015 Model and algorithm of BP neural network based on expanded multichain quantum optimization. *Mathematical Problems in Engineering* **2015**, Article ID 362150, 11. https://doi.org/10.1155/2015/362150.

Yang, S. N. & Chang, L. C. 2020 Regional inundation forecasting using machine learning techniques with the internet of things. *Water* **12**, 1578–1594.

Yang, J. C. & Tung, Y. K. 1996 *Establishment of Flow-Duration Curve and the Assessment of its Certainty.* Final Report of Environment Protection Agency, Taiwan.

Yang, S. C., Yang, T. H., Chang, Y. C., Chen, C. H., Lin, M. Y., Ho, K. J. & Lee, K. T. 2020 Development of a hydrological ensemble prediction system to assist with decision-making for floods during typhoons. *Sustainability* **12**, 4258–4277.

Yetilmezsoy, K. & Saral, A. 2007 Stochastic modeling approaches based on neural network and linear-nonlinear regression techniques for the determination of single droplet collection efficiency of countercurrent spray towers. *Environmental Modeling and Assessment* **12**, 13–36.

Zhang, Z., Ma, X. & Yang, Y. 2003 Bounds on the number of hidden neurons in three-layer binary neural networks. *Neural Networks* **16** (7), 995–1002.

First received 5 April 2021; accepted in revised form 14 July 2021. Available online 17 August 2021

doi: 10.2166/nh.2021.021

Application of temporal convolutional network for flood forecasting

Yuanhao Xu[a,b], Caihong Hu[a,b,*], Qiang Wu[a,b], Zhichao Li[a,b], Shengqi Jian[a,b] and Youqian Chen[a,b]

[a] School of Water Conservancy Science and Engineering, Zhengzhou University, Zhengzhou 450000, China
[b] Yellow River Institute for Ecological Protection & Regional Coordinated Development, Zhengzhou 450001, China
*Corresponding author. E-mail: hucaihong@zzu.edu.cn

ABSTRACT

Rainfall–runoff modeling is a complex nonlinear time-series problem in the field of hydrology. Various methods, such as physical-driven and data-driven models, have been developed to study the highly random rainfall–runoff process. In the past 2 years, with the advancement of computing hardware resources and algorithms, deep-learning methods, such as temporal convolutional network (TCN), have been shown to be good prospects in time-series prediction tasks. The aim of this study is to develop a prediction model based on TCN structure to simulate the hourly rainfall–runoff relationship. We use two datasets in the Jingle and Kuye watersheds to test the model under different network structures and compare with the other four models. The results show that the TCN model outperforms the Excess Infiltration and Excess Storage Model (EIESM), artificial neural network, and long short-term memory and improves the flood forecasting accuracy at different foreseeable periods. It is shown that the TCN has a faster convergence rate and is an effective method for hydrological forecasting.

Key words: deep learning, flood forecasting, rainfall–runoff, temporal convolutional network

HIGHLIGHTS

- Propose a TCN model for flood simulation and forecast.
- Explore the optimal hyperparameter combination of TCN at different forecast periods in flood forecast.
- Compare the performance of the models including physical and neural network models.
- Apply the TCN model in different watersheds to evaluate the practicality.

ABBREVIATIONS

TCN	Temporal convolutional network
LSTM	Long short-term memory
ANN	Artificial neural network
NSE	Nash–Sutcliffe efficiency
RMSE	Root-mean-square error

INTRODUCTION

Floods endanger human lives, hinder sustainable socioeconomic development, and cause inestimable damage to densely populated areas in floodplains or downstream from major rivers (Hu *et al.* 2018). Precise flood forecasting can better reduce the risk of flooding and provide timely and efficient environmental information for management decisions (Le *et al.* 2019; Sahoo *et al.* 2019).

In the past decades, multiple flood forecasting models have been proposed. According to the difference in principle, models for flood forecasting can be divided into process-driven models based on physical mechanisms and data-driven models based on machine learning methods (Douglas-Mankin *et al.* 2010; Qin *et al.* 2018; Yuan *et al.* 2018). Commonly used process-driven models, such as the Xinanjiang model, have been regarded as common techniques for flood process simulation and forecasting (Beven *et al.* 1984; Zhao 1992; Wang *et al.* 2012). They usually require complex mathematical formulas, a large amount of hydrological and meteorological data, and an accurate understanding of runoff mechanisms. However, with the diversification of hydrological data and in-depth research on the mechanism of runoff generation and convergence, these process-driven models are restricted by many factors (Kratzert *et al.* 2018; Tian *et al.* 2018).

The data-driven models focus on the statistical relationship between input and output data and do not consider the physical mechanism of the hydrological process, but rather establish a mathematical analysis of the time series and use the given sample to discover the statistical or causal relationship between the hydrological variables (Adamowski & Sun 2010; Yunpeng *et al.* 2017). The data-driven method has unique advantages for solving numerical prediction problems, reconstructing highly nonlinear functions, and analyzing time series (Miao *et al.* 2019).

With the rapid development of computer technology, data-driven models have gained more applications in the field of hydrological forecasting (Bafitlhile & Li 2019). The artificial neural networks (ANNs), a common data-driven model based on artificial intelligence, were applied in flood forecasting at the end of the last century already (Hsu *et al.* 1995; Shamseldin 1997; Abdellatif *et al.* 2013).

Recent breakthroughs in the field of computational science have created growing interest in learning methods in academic and applied scientific circles (Chang *et al.* 2002; Kisi 2011). One of the most active research points in deep learning is modeling sequential data through a recurrent neural network (RNN), which is particularly suitable for hydrological prediction and gives a precise and timely prediction of time series in systems (Chang *et al.* 2002; Kumar *et al.* 2004).

Since the late 1990s, more modern RNN architectures have been proposed, particularly the long short-term memory (LSTM), which has been successfully applied to image recognition, the Internet of Things, text translation, and stock prediction (Hochreiter & Schmidhuber 1997; Donahue *et al.* 2015; Crivellari & Beinat 2020; Wu *et al.* 2020). While the LSTM neural network is designed to overcome the problem of gradient disappearance in long-term dependent time series of simple RNN neural network, each LSTM unit contains four affine transformations and each time step needs to be run once, which can easily use up the memory available.

However, the development of deep learning is changing with each passing day, and there are still some networks worthy of applying for hydrological modeling and flood forecasting (Gao *et al.* 2020; Lin *et al.* 2020). Although CNNs were originally designed for computer vision tasks, they are also suitable for time-series data since they can extract high-level features from data with a grid topology. Temporal convolutional network (TCN) is a recently proposed convolutional neural network, which combines the 1-dimensional fully convolutional network (1D FCN) and causal convolutions (Bai *et al.* 2018). The 1D FCN keeps the network producing an output of the same length as the input. The causal convolution guarantees that future information will not affect the past. The dilated convolutions and residual modules in TCN can obtain a large receptive field by a few convolutional network layers and allows the network to transmit information in a cross-layer manner (Bai *et al.* 2018). Recently, TCN has been proved superior for time-series data modeling tasks to LSTM (Chen *et al.* 2020). Several works have already successfully used TCNs for time-series forecasting tasks such as the stock trend prediction, anomaly detection, and recognition of sepsis (Deng *et al.* 2019; Yujie *et al.* 2019; Lara-Benitez *et al.* 2020). In the field of hydrology, especially the prediction of rainfall–runoff relationship contains a large number of time-series data, resulting in an urgent demand for modeling. Up to now, people have not applied this new technology to flood forecasting, and more advanced deep-learning technologies such as TCN can provide support for more efficient and accurate forecasting. Therefore, TCN is introduced first in flood forecasting based on deep learning.

The objective of this study is to explore the ability and stability of TCN for flood forecasting with multi-step lead times. To achieve this goal, we process the rainfall, flow, and NDVI data of the Jingle watershed to a size between (0, 1) through normalization to facilitate neural network analysis. Next, TCN is used to predict the future 1–12 h flow process. Additionally, we use the Nash–Sutcliffe efficiency (NSE), root-mean-square error (RMSE), and bias to evaluate the performance of the prediction model and to compare the results with those of EIESM, ANN, and LSTM to experimentally verify the model effectiveness (Shiri *et al.* 2015; Wen *et al.* 2020). In addition, in order to explore the influence of NDVI data as input data on the TCN model, TCN without NDVI is also used as a comparison model. Finally, we also conducted experiments on these models in another basin to prove the stability of TCN in dealing with flood forecasting tasks.

The following section introduces the TCN model structure, model design, settings and parameterization, comparison models, and performance evaluation criteria. The case study, hydrological data, and the data preprocessing method are given in the 'Case study' section. The experimental results and thorough discussion are presented in the 'Result and discussion' section. The conclusions of this study provide a summary of the work and recommendations for future research in the 'Conclusions' section.

METHOD

Temporal convolutional neural network

TCN, a type of convolutional neural network, is applied in many time-series forecasting tasks (Liu *et al.* 2019). It is deliberately kept simple, combining some of the best practices of modern convolutional architectures (Bai *et al.* 2018). TCN has two specific designs: 1D FCN architecture is used in TCN to keep the network output the same length as the input sequence (Long *et al.* 2015); The outputs are only influenced by information of the present and past inputs in each layer by using causal convolutions. Causal convolution is different from standard convolution in that the output at time t is not convolved with future values (Wan *et al.* 2019).

However, simple causal convolution still has the problem of traditional convolutional neural networks, and the modeling length of time is limited by the size of the convolution kernel (van den Oord *et al.* 2016). In this case, if you want to learn longer dependencies between data, you need to stack many layers linearly. To solve this problem, TCN uses one-dimensional expansion convolution. The difference between expansion convolution and traditional convolution is that it allows the input of convolution to have interval sampling. Without the pooling operation, this convolution increases the acceptable range of the network, so there is no loss of resolution.

The complete dilated causal convolution operation F over consecutive layers for a 1-D sequence of a given hydrology input $q = (q_0^1, \ldots, q_t^1)$ and a filter f:{0, ..., $k-1$}, on element s of the sequence, is defined by the following equation:

$$F(s) = (q *_d f)(s) = \sum_{i=0}^{k-1} f(i) \cdot q_{s-d \cdot i} \tag{1}$$

where d is the dilation factor, k is the filter size, and $s - d \cdot i$ accounts for the direction of the past.

As shown in Figure 1, the sampling standard is selected according to d, the first layer $d = 1$ means that each number is selected, and the second layer $d = 2$ means that the data are selected at intervals (Lara-Benitez *et al.* 2020). Choosing a larger filter size k or increasing the expansion factor d can obtain a larger network receptive field. Similar to the common dilated convolutions, d increases exponentially with the depth of the network layers, which allows the network to use a larger effective history while receiving each input. Another common method to further increase the network acceptance domain is to connect several TCN blocks. To avoid the deeper network structure from complicating the learning process,

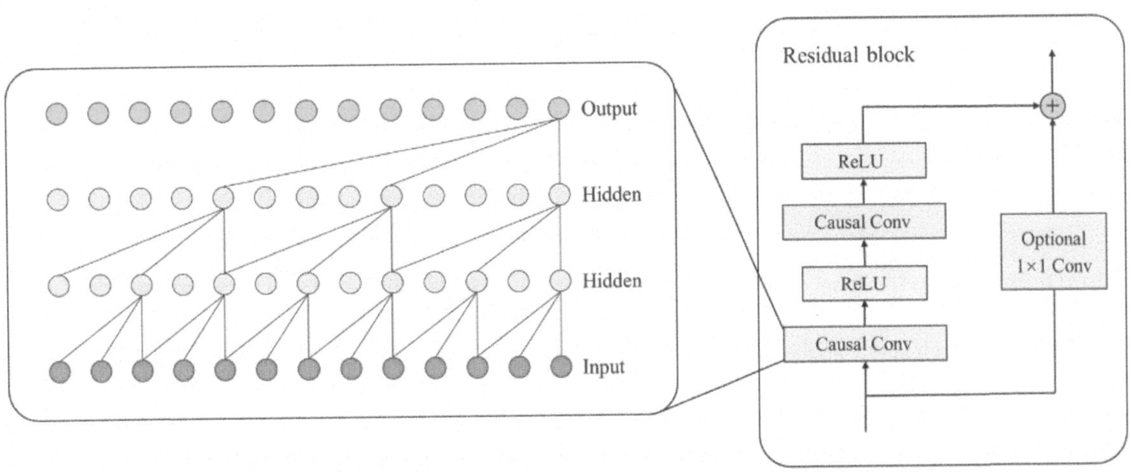

Figure 1 | TCN architectural elements, including a dilated causal convolution with dilation factors of $d = 1$, 2, and 4 and filter size of $k = 3$. The receptive field is able to cover all values from the input sequence. And a 1 × 1 convolution (residual block) is added when residual input and output have different dimensions. The sampling standard is selected according to d, the first layer $d = 1$ means that each number is selected, and the second layer $d = 2$ means that the data are selected at intervals (Lara-Benitez *et al.* 2020). Choosing a larger filter size k or increasing the expansion factor d can obtain a larger network receptive field. To avoid the deeper network structure from complicating the learning process, a residual connection is added to the output (He *et al.* 2016). The residual connection uses 1 × 1 convolution to ensure that the addition operation receives the same tensor (Bai *et al.* 2018).

a residual connection is added to the output (He *et al.* 2016). Since the input and output have different widths, the residual connection uses 1×1 convolution to ensure that the addition operation receives the same tensor (Bai *et al.* 2018).

Flood forecasting model based on TCN

TCN is designed for time-series forecasting tasks. The rainfall–runoff process contains a series of time-series data, such as rainfall evapotranspiration and runoff. To explore the application of TCN in flood forecasting, this paper established a multi-step time-series forecasting model based on TCN for flood forecasting in the next 1–12 h. In TCN, a moving window scheme is used to create input and output pairs, which will be fed into the neural network (Bandara *et al.* 2020). In the prediction process at time t, the input data from time $t-n$ to t are used to simulate the value at time $t+m$, where n is the length of the input data, and m is the prediction time in the future. When the window slides to the next moment (time $t+1$), the value at time $t+m+1$ is simulated using the data at time $t-n+1$ to $t+1$.

In this paper, the factors that affect the process of runoff generation and convergence, such as rainfall, evapotranspiration, and runoff data in the watershed, are selected as the input of the model. In addition, since the middle reaches of the Yellow River where the study area is located, the nature of the underlying surface changes significantly within the data time span, which affects the rainfall–runoff process in the basin. As a piece of data that can reflect the characteristics of the underlying surface, NDVI plays an indispensable role in simulating the relationship between rainfall and runoff. At the same time, deep learning requires a great deal of data as support, and only rainfall and flow data for input are not comprehensive. Although the relationship between rainfall and flow is the most obvious and direct, it is difficult for the model to grasp the subtle differences in the flood process. Moreover, predecessors have done similar studies. Therefore, we consider NDVI as another input to enhance the TCN's ability to learn data; flood flow data that needs to be predicted are used as output. These data constitute the dataset for training and testing the model.

Model setting and parameterization

The programming language of choice is Python 3.7, and the libraries used for preprocessing and managing our data are NumPy and pandas. We use the Google Keras deep-learning framework with TensorFlow backend and the NVIDIA RTX 2080Ti GPU to train the models.

First, the normalization operation is performed. Normalization is a method of rescaling the data in the original range so that all values range from 0 to 1. If the machine learning model uses the gradient descent method to find the optimal solution, dimensions are not uniform, and normalization is often necessary to allow convergence. For efficient learning, all input features (meteorological and hydrological variables) and output data (the discharge) are normalized by subtracting the mean value and dividing by the standard deviation. The output of the network is retransformed to obtain the final discharge prediction, which is an inverse data scaling process. The expression is described as follows:

$$x' = \frac{x - \mu}{\sigma} \tag{2}$$

where x' and x represent the normalized result and sample data, respectively; μ and σ are the mean and standard deviation of the sample data, respectively.

The purpose of this research is to establish a TCN-based deep-learning flood forecasting model to fully explore the ability of TCN to perceive hydrological data. Due to the high complexity of deep-learning models, finding the optimal TCN network structure and hyperparameters is a crucial task. Therefore, under the premise of different foreseeable periods, we conducted multiple experiments on the same dataset and combined them into models containing different network structures by changing the value of the TCN hyperparameters. The kernel size, filter, and residual block are selected as {4, 6, 8}, {32, 64, 128}, and {1, 2, 3}, respectively. In addition, the expansion factor is fixed to [1, 2, 4, 8, 16, 32]. And the length of the input data is fixed to six, which means that a total of six time periods of current and past data are used to predict the future value. All these architectures are then tested for all combinations of the following parameters: the batch size and the epochs are selected as {32, 64, 128} and {20, 50, 100}, respectively. In general, we conducted ($3 \times 3 \times 3 \times 3 \times 3 = 243$) experiments on the dataset in the research basin and found the best network structure and hyperparameter values under different forecast periods.

Performance evaluation criteria

In this study, the performance of different models is evaluated by statistical error measures, including the RMSE, NSE, and bias (Nash & Sutcliffe 1970; Hu *et al.* 2018; Kratzert *et al.* 2018). The mathematical expressions of these metrics can be described as follows:

$$\mathrm{NSE} = 1 - \frac{\sum_{i=1}^{n} (Q_0 - Q_c)^2}{\sum_{i=1}^{n} (Q_0 - \overline{Q_c})^2} \tag{3}$$

where Q_0 (m^3/s) and Q_c (m^3/s) represent the discharge of the observed and simulated hydrographs, respectively; $\overline{Q_0}$ is the mean value of the observed discharge, and n is the data point number.

NSE measures the ability of the model to predict variables different from the mean, gives the proportion of the initial variance accounted for by the model, and ranges from 1 (perfect fit) to $-\infty$. Values closer to 1 provide more accurate predictions.

$$\mathrm{RMSE} = \sqrt{\sum_{i=1}^{n} \frac{(Q_0 - Q_c)^2}{n}} \tag{4}$$

The RMSE is very sensitive to the maximum and minimum errors, enabling it to effectively reflect the accuracy of the prediction results. RMSE values closer to 0 provide more accurate predictions.

$$\mathrm{Bias} = \frac{\sum_{i=1}^{n} (Q_0 - Q_c)}{\sum_{i=1}^{n} (Q_0)} \tag{5}$$

The bias can evaluate the accuracy of the overall water balance of the simulation results and range from −100 to 100%. A value close to 0 means more accurate predictions.

Model benchmarks and methods

We used several benchmarks to evaluate the performance of the TCN model. These benchmarks include LSTM, ANN, and EIESM (physical model). In addition, in order to explore the influence of NDVI as an input on the TCN model, we also added the TCN model without NDVI as a benchmark. The forecast period was set to 1–12 h, which is the premise of comparing the performance of different benchmarks. We simultaneously calculate and compare the performance evaluation indicators of all benchmarks.

LSTM is an RNN with a special memory unit structure (Hochreiter & Schmidhuber 1997). In the memory cell unit, the input information passes through the forget gate, input gate, and output gate in turn. Some information is selected to be forgotten, while other information is selected to be added to the memory, which overcomes the disadvantages of gradient explosion and gradient disappearance of traditional recurrent neural networks. Research in recent years has shown that LSTM is widely used in the modeling of time-series forecasting tasks in the hydrological field, such as flow forecasting and water quality forecasting assessment (Kratzert *et al.* 2018). LSTM was used as a benchmark in this study.

ANN is a black-box simplified model used to solve several water resources problems and can be trained with datasets to identify complex nonlinear relationships between inputs and outputs (Tokar & Johnson 1999). The representative feedforward neural network is composed of an input layer, hidden (containing neurons), and output layers. Recent studies have shown that using ANN is one of the most significant methods to simulate hydrological processes (Ahmad & Hussain 2019). ANN was also used as another benchmark in this study.

EIESM is a physical, improved Xinanjiang model (Hu *et al.* 2003). Compared with the Xinanjiang model, the excess infiltration runoff mode of EIESM is based on the infiltration curve and infiltration capacity distribution curve of the watershed, and the storage runoff mode is based on the water storage capacity distribution curve of the watershed. The two types of runoff modes are organically combined. Recent studies have applied the runoff simulated by EIESM within the range of acceptable accuracy, which is reflected by the goodness-of-fit measure (Wen *et al.* 2020). EIESM was used as a benchmark in this study.

CASE STUDY

The Yellow River – the fifth largest in the world – often suffers from flood disasters. In recent years, a large number of water and soil conservation measures have been applied in the middle reaches of the Yellow River. This study selected the representative Jingle and Guxian watersheds. The first TCN model was developed for the Jingle watershed of the Fenhe River in Shanxi Province, a relatively small watershed that covers 2,799 km^2. Jingle hydrological station is the primary stream control station on the upper Fenhe River and is located at 111°55′ east longitude and 38°20′ north latitude. The annual mean precipitation in the Jingle watershed is approximately 538.38 mm. Devastating frequent flooding in the last few decades has been widely researched.

An additional assessment was conducted in the Kuye watershed to discover if the proposed model architecture operates in different watersheds after training. The watershed covers 8,706 km^2, spans Shanxi and Henan provinces, and is narrow with a long concentration time. The Wenjiachuan station is located at 110°45′ east longitude and 38°26′ north latitude. Annual precipitation in the two watersheds varies greatly, and both are severely affected by flooding. Figure 2 shows the locations of the Jingle and Kuye watersheds.

The underlying data for our study of the Jingle watershed include hourly discharge data from the Jingle station and hourly rainfall data from 14 gauges in the area. Complete records of 98 flood events from 1971 to 2013 were obtained. Of these, 78 flood events (1971–2000, 3,986 datasets) were used for calibration, and 20 events (2000–2013, 1,366 datasets) were used for verification. The type of data used in the Kuye watershed is the same as that of Jingle, but the Kuye watershed contains 19 rainfall stations and the data contain 86 events (1973–2016, 3,479 datasets), of which 66 events (1973–2003, 2,764 datasets)

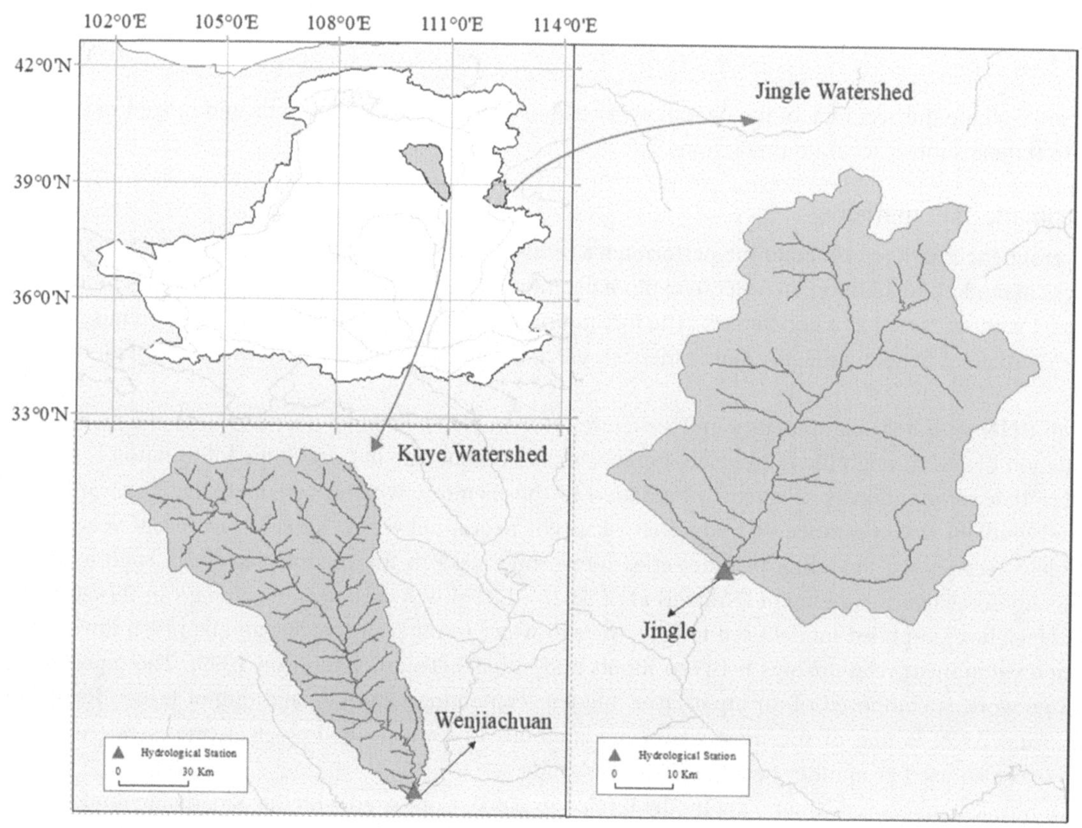

Figure 2 | Two watersheds and their locations in the provinces (top left). The detailed streamflow map of the Kuye watershed (bottom left) and the Jingle watershed (right) with the sub-watershed boundary (red lines) and hydrological stations (red markers). Please refer to the online version of this paper to see this figure in color: http://dx.doi.org/10.2166/nh.2021.021.

are used as the training set and 20 events (2003–2016, 3,479 datasets) as the validation set. In this paper, a typical flood process, with a large volume flow and duration, is selected to verify the performance of the established model.

RESULTS AND DISCUSSION

TCN network hyperparameter experiment

In this study, after experiments with different combinations of TCN models on the Jingle watershed dataset, we finally selected the best TCN model construction scheme based on the loss function (Loss) of the prediction results on the verification set for different forecast periods, including kernel size, filters, residual blocks, batch size, and epochs. Table 1 illustrates the best model construction in each forecast period.

As can be seen from Table 1, there is no obvious difference between the constructions of different forecast periods. The filters and kernel size are mostly selected as 128 and 8, respectively. The number of residual blocks always works best when the value is 2. These results are related to the characteristics of the given input hydrological data. TCN adjusts the receptive field by changing the above three hyperparameters, and the internal learning matches the rainfall–runoff process. The batch size remains unchanged at 128, and the epochs increase with the forecast periods. As the increases of time interval between output data and input data, the convolutional network requires more iterative epochs to capture the relationship in the data.

Understanding TCN in hydrology with model evaluations

In the above research, we discussed the optimal network structure and hyperparameters for TCN to model the rainfall–runoff process. To deeply explore and compare the hydrological process simulation performance of the TCN model, we used the other four models, including TCN (without NDVI input), LSTM, ANN, and EIESM models, for the same experiment and performed a simulation at a forecast period of 1–12 h as TCN. Table 2 illustrates the evaluation index of runoff forecasting at different forecast periods (1–12 h) by the four models. The changes in the three indicators reflect the flood simulation accuracy on the training and testing set. At the same time, in order to more intuitively reflect the performance of the model on the test set, we also calculate the evaluation indicators of all flood events on the test set. Figure 3 shows the Boxplots of different tests. It can be seen that the results of all models are closely related to the forecast periods, and the prediction accuracy decreases with the increase of the forecast period. In the modeling process, the forecast period represents the time interval between input and output data. A longer forecast period increases the difficulty of the prediction of the target value. In terms of flow formation reasons, future flow will be affected by current or earlier rainfall and other factors. The existing data gradually cannot provide an effective reference when it is hoped to obtain more distant future flow.

In the physical model, the results show that the NSEs of EIESM are 0.8848 (training set) and 0.8610 (testing set) at a forecast period of 1 h but fall to 0.7203 (training set) and 0.7106 (testing set) at a forecast period of 12 h. The physical models use the principle of rainfall–runoff to calculate the flow, but the overall accuracy is low due to incomplete consideration and systematic error. The results of ANN show that NSE varies from 0.9 to 0.6 for forecast periods from 1 to 12 h. The simulation effect shows a rapid downward trend with the increase of the foreseeable period. ANN is a relatively simple ANN and uses a backpropagation supervised learning technique for training. As a machine learning model proposed earlier, ANN cannot capture the information in the input required to process sequence data. Compared with the previous two models, LSTM, TCN, and TCN (NDVI) are well simulated and meet the needs of flood forecasting. It is evident that the forecastability of two TCN models is higher than that of LSTM at almost each forecast period, especially for long periods close to 6 h. The performance

Table 1 | The best TCN model settings under different forecast periods

Forecast periods (h)	Kernel size	Residual block	Filters	Batch size	Epochs	Loss
1	8	2	128	128	20	0.0023
2	8	2	64	128	20	0.0035
3	6	2	128	64	50	0.0058
6	8	2	128	128	50	0.0126
9	8	2	64	128	50	0.0249
12	8	2	128	128	50	0.0401

Table 2 | Performance of the five models in the Jingle watershed on the calibration set and validation set

Forecast period (h)	Dataset	Model	NSE	RMSE	Bias (%)
1	Calibration	EIESM	0.8848	46.7094	14.2147
		ANN	0.9536	9.8926	1.3384
		LSTM	0.9836	8.8401	0.6991
		TCN	0.9858	8.4564	0.5341
		TCN (NDVI)	0.9897	7.9532	0.5119
	Validation	EIESM	0.8610	51.9966	14.4069
		ANN	0.9477	11.9942	1.4488
		LSTM	0.9704	9.1252	0.7197
		TCN	0.9825	8.7127	0.5458
		TCN (NDVI)	0.9838	8.1511	0.5176
6	Calibration	EIESM	0.7665	60.4484	26.0965
		ANN	0.8120	53.4280	19.1487
		LSTM	0.8859	45.7676	14.7178
		TCN	0.9061	40.4808	10.6355
		TCN (NDVI)	0.9109	38.8848	8.9400
	Validation	EIESM	0.7588	61.1700	27.1619
		ANN	0.8055	55.5501	20.3971
		LSTM	0.8721	47.7827	14.8497
		TCN	0.8963	43.2517	11.7935
		TCN (NDVI)	0.9031	40.9977	9.9135
12	Calibration	EIESM	0.7203	86.6261	43.0147
		ANN	0.6761	123.2829	52.4860
		LSTM	0.7354	83.3445	40.4083
		TCN	0.7790	75.9966	31.2881
		TCN (NDVI)	0.7983	73.3085	26.1076
	Validation	EIESM	0.7106	91.4497	45.7214
		ANN	0.6800	132.6323	58.1628
		LSTM	0.7185	88.1648	34.6354
		TCN	0.7574	80.0811	29.1310
		TCN (NDVI)	0.7729	77.2392	27.5773

of the three models declined as forecast periods increased. For forecast periods of less than 6 h, the NSE of TCNs is higher than that of LSTM, and RMSE and bias are lower. When the forecast periods exceed 6 h, the prediction accuracy of LSTM sharply decreases, whereas the NSE of TCNs remains above 0.7; RMSE and bias remain below 100 and 40%, respectively. Finally, for the two different input TCN models, the latter model with NDVI input is better simulated, with higher NSE, lower RMSE and bias in all forecast periods. The input including NDVI better reflects the true characteristics of the watershed, so that TCN can learn the relationship between data more fully. At the same time, deep learning requires a large amount of data as support, so adding the NDVI sequence is beneficial for TCN fitting. Among five models, the proposed TCN model shows the highest accuracy compared to normal machine learning and physical models.

To evaluate the ability of the TCN model to forecast the flood process, of which the first three are the large flood events in the verification set, and the last one is the smaller and most common flood during the verification set. We use the previously mentioned model containing the best TCN structure and training hyperparameters to simulate the flood process. Figure 4 shows the observed and estimated hydrographic map of four flood events during the forecast period of 1, 6, and 12 h.

Flood events 1 and 2 were considered abnormal events during the verification set. The rainfall that formed the flood was high in intensity and short in duration, and the peak shape was sharp. At the forecast period of 1 h, the predicted value of the EIESM model for the peak and low tide section is higher than the actual value. ANN fluctuates significantly, especially before the flood peak occurs. However, the forecast runoff curves of LSTM and TCNs both fit the observed runoff curve well. They have strong predictive capabilities of backwater stages in good agreement with the actual process. In contrast, the flood peaks forecasted by TCNs are more realistic than those of LSTM, indicating that TCNs are more sensitive to rainfall and runoff processes. Although there is no obvious difference between TCN and the proposed TCN (NDVI), the latter fits the true value more accurately in the preceding section. At the forecast period of 6 h, the performance of all models has a certain

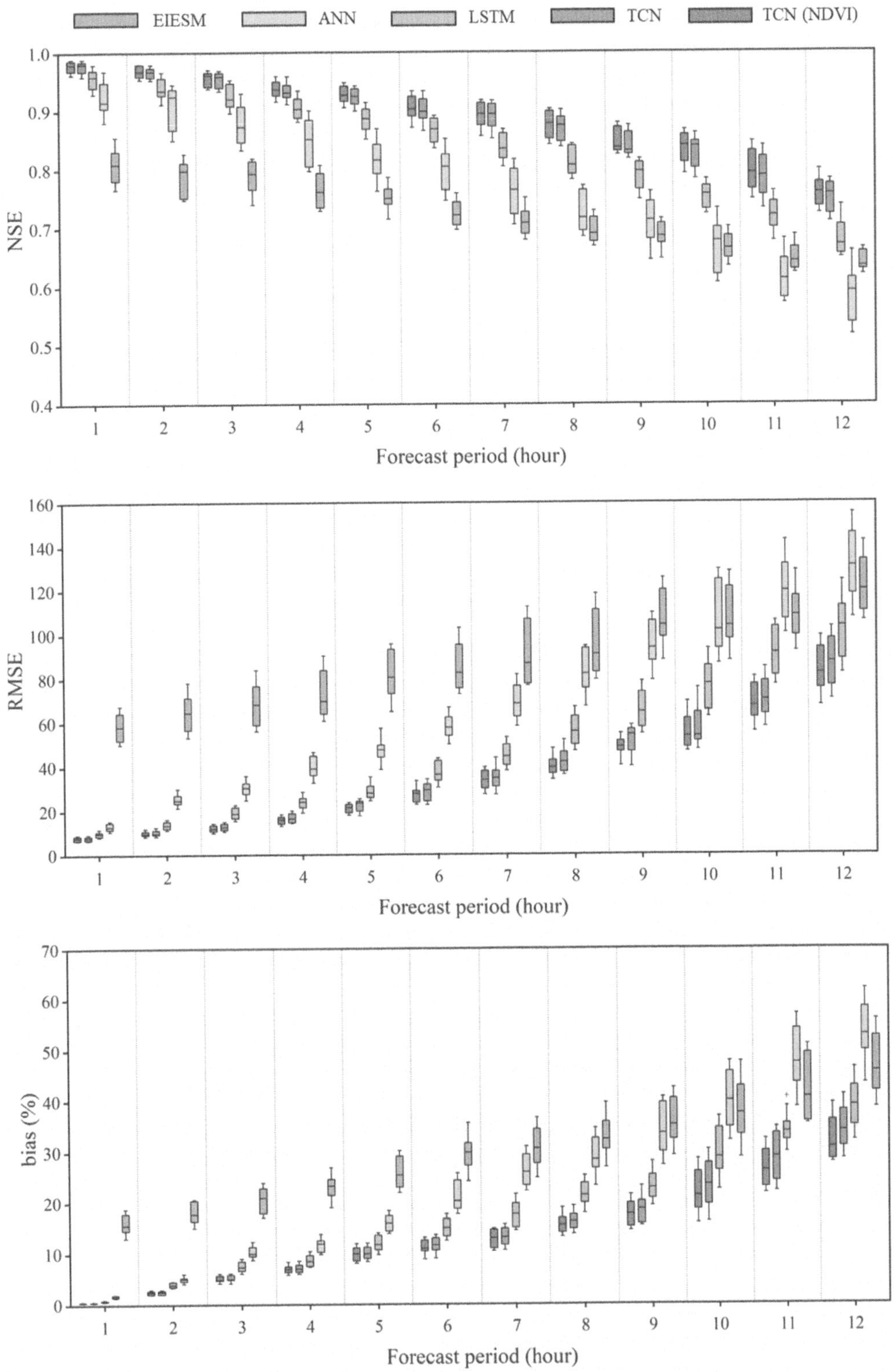

Figure 3 | Boxplots of different tests. Each box is calculated from four realizations of model runs. It shows the average NSE for the prediction with different forecast periods of 1, 6, and 12 h. It can be seen that the results of all models are closely related to the forecast periods, and the prediction accuracy decreases with the increase of the forecast period. Among five models, the proposed TCN model shows the highest accuracy compared to normal machine learning and physical models.

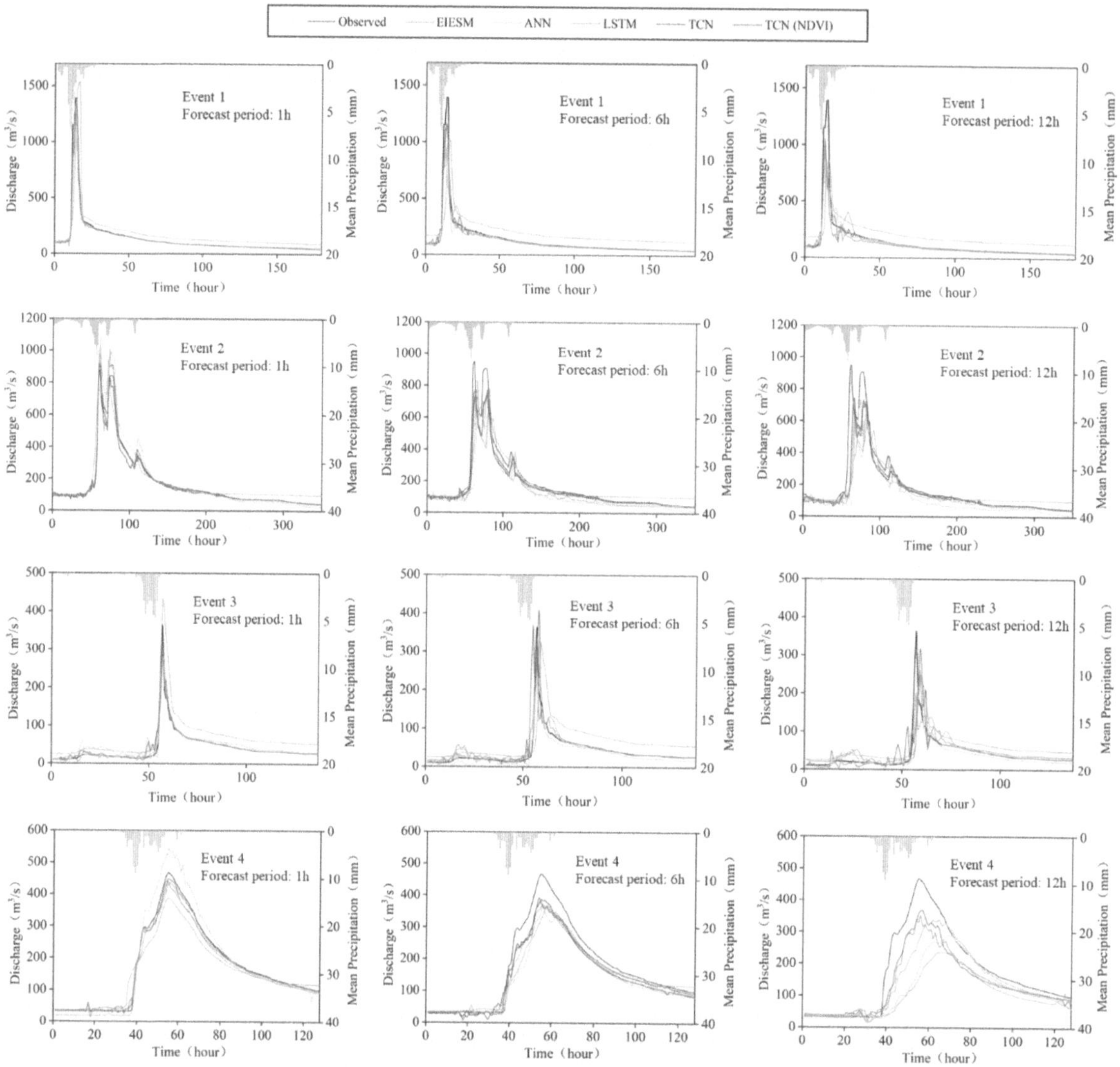

Figure 4 | Observed and estimated hydrographs of the five models at the validation stage (flood events 1, 2, 3, and 4) with different lead times of 1, 6, and 12 h. Flood events 1 and 2 were considered abnormal events during the verification set. And the last two floods have a lower peak flow than the first two events, which are normal flood events. The results show that TCN (NDVI) best predicted the hourly peak flow, whereas the other models were insufficient to predict the values and had lower forecast accuracy.

degree of degradation. Although EIESM expresses the correct flood process trend, the overall accuracy is insufficient. Moreover, the forecasting ability of ANN and LSTM deteriorates significantly, the phenomenon of underestimating flood peaks increase, and the simulated values fluctuate abnormally compared with observed values. ANN has large fluctuations and abnormal values, which may be due to the insufficient ability of ANN to learn long-term data. The flood peak flow forecasted by LSTM is later than the observed flood peak flow, which will seriously affect the flood warning. It is obvious that TCNs better simulate rainfall–runoff and forecast floods well and have higher accuracy than other models. Between the two TCN models, the model with NDVI input has a better grasp of where the flood peak appears and has less fluctuation. When the forecast period exceeds 6 h, due to the lack of hydrological data to form future runoff, the forecast flow curves of all models are much later than future observations. Even so, TCNs are minimally affected and the results are still practical. The results also prove that considering NDVI as an input can effectively improve forecast accuracy.

The last two floods have a lower peak flow than the first two events, which are normal flood events. Both figures show that the overall performance of EIESM is stable and insensitive to forecast period factors. Conversely, changes in the forecast period are likely to cause fluctuations in the neural network models. TCN (NDVI) best predicted the hourly peak flow, whereas the other models were insufficient to predict the values and had lower forecast accuracy.

Model application in a different watershed

To evaluate the practicality of the model structure, we applied the established TCN model to the Kuye watershed and selected the same length data as the previous experiment. The Kuye watershed is larger and narrower than the Jingle watershed and has longer travel time and different topography, soil type, and land use. TCN network modeling used hyperparameter combinations obtained from repeated experiments: the filter, kernel size, and residual blocks are selected as 128, 8, and 2, respectively, and the expansion factor is [1, 2, 4, 8, 16]. Table 3 shows the results based on different forecast periods for four models. In the verification set, NSE of the EIESM, ANN, LSTM, TCN, and TCN (NDVI) models are 0.8416, 0.9463, 0.9787, 0.9798, and 0.9819 at a forecast period of 1 h, respectively. With extended forecast periods, the simulation accuracy of all the models declined to varying degrees. Among them, EIESM and ANN have the most obvious downward trend. Although LSTM maintains a high level of prediction, it is always lower than TCNs. The performance of TCN without NDVI input still lags behind TCN (NDVI). The final TCN (NDVI) model consistently outperformed the other models. These results are the same as those in the Jingle watershed, confirming that the TCN (NDVI) model established in this study responds relatively smoothly to disturbed watershed attributes and can be used to make accurate predictions in multiple watersheds.

CONCLUSIONS

At present, more and more deep neural network methods are applied to rainfall–runoff prediction. On the one hand, with the advancement of technology, the means of obtaining data are more intelligent, and the types of data available are more diverse, such as soil evapotranspiration, wind speed, and pressure in the watershed. Some data that cannot be directly observed can be

Table 3 | Performance of the five models in the Kuye watershed on the calibration set and validation set

Forecast period (h)	Dataset	Model	NSE	RMSE	Bias (%)
1	Calibration	EIESM	0.8667	49.1452	14.9225
		ANN	0.9544	10.3626	1.4078
		LSTM	0.9802	9.2594	0.7345
		TCN	0.9848	8.8991	0.5612
		TCN (NDVI)	0.9874	8.3301	0.5364
	Validation	EIESM	0.8416	54.5497	15.1129
		ANN	0.9463	12.5921	1.5247
		LSTM	0.9766	9.5869	0.7542
		TCN	0.9798	9.1584	0.5716
		TCN (NDVI)	0.9819	8.5426	0.5447
6	Calibration	EIESM	0.7504	63.5939	27.4180
		ANN	0.7929	56.1515	20.0869
		LSTM	0.8665	48.1196	15.4171
		TCN	0.8876	42.4766	11.1822
		TCN (NDVI)	0.8915	40.8133	9.4066
	Validation	EIESM	0.7413	64.2193	27.4510
		ANN	0.7869	56.3258	20.3226
		LSTM	0.8512	48.1237	15.5833
		TCN	0.8771	43.1984	11.3480
		TCN (NDVI)	0.8847	40.9412	9.5563
12	Calibration	EIESM	0.7036	91.0993	45.1899
		ANN	0.6621	129.4762	54.9864
		LSTM	0.7194	87.4816	42.0057
		TCN	0.7590	79.7766	34.7225
		TCN (NDVI)	0.7811	76.8033	27.3611
	Validation	EIESM	0.6964	96.1865	48.0235
		ANN	0.6641	127.6674	52.6108
		LSTM	0.7035	92.4926	39.3347
		TCN	0.7611	83.8810	30.6066
		TCN (NDVI)	0.7649	81.1183	28.9558

obtained by the inversion of remote sensing images. Progress in data requires a framework that can consider multiple factors to interpret the hidden relationships between data. On the other hand, the rainfall–runoff process contains many complicated steps. The runoff mechanism in semi-humid, semi-arid areas is more complicated than in humid areas (Le *et al.* 2019). The physical model is not enough to reflect the complete mechanism. Therefore, the deep neural network has become an effective new way of simulating rainfall and runoff.

As a time-series prediction model based on convolution that has been proposed in the past 2 years, TCN has been successfully applied in some fields. This paper proposes a TCN learning model for rainfall–runoff prediction, which uses one-dimensional convolution operation to process input and output sequences. The TCN model uses the observed rainfall data, evaporation, and NDVI data from rainfall stations in the basin as input, and the outlet section flow as output.

In the process of modeling, some hyperparameters in the structure of TCN neural network need to be considered, such as kernel size, filter size, and residual blocks. Inappropriate hyperparameter combinations will cause the prediction results to deviate from the true value. In each case of the forecast period of 1–12 h, 273 different hyperparameter combinations were tested in the Jingle watershed, and finally, the set of results with the best prediction effect was selected. Several domain characteristics are necessary for a successful application. First, the change of the best combination under different forecast periods is not obvious. The filter and kernel sizes fluctuate between 6, 8 and 64, 128, respectively, and the residual blocks remain at 2. In addition, some of the hyperparameters used for training, such as the batch size, are stable at 128. At the same time, TCN needs more iterations to complete the fitting in response to a longer forecast period. The characteristics of these hyperparameter combinations are considered to be useful for rainfall–runoff simulation.

In this study, we use the TCN network to model two watersheds. The experimental results show that the proposed TCN method can better simulate the rainfall–runoff process, has a better prediction effect than other models, and has less deviation when dealing with long-foreseeing prediction tasks. Traditional physical models and early machine learning models, such as EIESM and ANN, generally have low simulation accuracy. LSTM meets the forecast requirements under the short forecast period, but loses more accuracy with the prolongation of the forecast period, and significantly underestimates flood peaks and abnormal fluctuations. It is worth noting that the TCN model, including NDVI input, has a certain ability to reflect the characteristics of the underlying surface, which leads to better performance in the simulation on the selected watershed. This provides a new idea for the subsequent deep learning for hydrological forecasting tasks: inputting different types of data on the basin that have the potential to affect the flow. At the same time, the simulation in the Kuye watershed showed the stability of the proposed TCN model, and the selected hyperparameter combinations can be applied to a new watershed.

Using TCN to model the rainfall–runoff process can capture the relationship between hydrological data. In TCN, because the filter exists in each layer of the network, convolution can complete the prediction task in parallel, while RNN, such as LSTM, can only process information in one direction, that is, it needs to wait for the forward pass of the previous time step to be completed. The forward pass of the next time step and the deep LSTM network require a larger amount of calculation, resulting in a long time-consuming prediction task. In addition, the adjustable kernel size and expansion coefficient support the receptive field, which enables TCN to handle tasks with varying degrees of complexity. In the experiment, we found that TCN has a faster convergence rate than LSTM and occupies less memory, which is indispensable for prediction. These results also demonstrate the strong potential of applying deep-learning methods to other hydrological problems, specifically other time-series tasks.

The hydrological process contains complex and diverse variables. More variables that affect the target value can be used as input to enhance the deep-learning model's cognition of physical processes in the future, such as soil moisture and wind speed. In addition, deep neural network methods like TCN often have many hyperparameters. At present, we use repeated experiments to determine the best combination of them, which has limitations. If it can be combined with more intelligent parameter optimization methods, it will greatly improve the efficiency of obtaining the desired results.

ACKNOWLEDGEMENTS

The authors acknowledge the financial support received from Projects of National Natural Science Foundation of China (No. 51979250), National Key Research Priorities Program of China (No. 2016YFC040240203), National Key Research Priorities Program of China (No. 2019YFC1510703), Key Projects of National Natural Science Foundation of China (No. 51739009), and Key Research and Promotion Projects (technological development) in Henan Province.

DATA AVAILABILITY STATEMENT

All relevant data are included in the paper or its Supplementary Information.

REFERENCES

Abdellatif, M., Atherton, W. & Alkhaddar, R. 2013 A hybrid generalised linear and Levenberg-Marquardt artificial neural network approach for downscaling future rainfall in North Western England. *Hydrology Research* **44** (6), 1084–1101. doi:10.2166/nh.2013.045.

Adamowski, J. & Sun, K. 2010 Development of a coupled wavelet transform and neural network method for flow forecasting of non-perennial rivers in semi-arid watersheds. *Journal of Hydrology* **390** (1–2), 85–91. doi:10.1016/j.jhydrol.2010.06.033.

Ahmad, S. K. & Hussain, F. 2019 A generic data-driven technique for forecasting of reservoir inflow: application for hydropower maximization. *Environmental Modelling & Software* **119**, 147–165. doi:10.1016/j.envsoft.2019.06.008.

Bafitlhile, T. M. & Li, Z. 2019 Applicability of epsilon-support vector machine and artificial neural network for flood forecasting in humid, semi-humid and semi-arid basins in China. *Water* **11** (1). doi:10.3390/w11010085.

Bai, S., Kolter, J. Z. & Koltun, V. 2018 An empirical evaluation of generic convolutional and recurrent networks for sequence modeling. arXiv:1803.01271.

Bandara, K., Bergmeir, C. & Smyl, S. 2020 Forecasting across time series databases using recurrent neural networks on groups of similar series: a clustering approach. *Expert Systems with Applications* **140**. doi:10.1016/j.eswa.2019.112896.

Beven, K., Kirkby, M., Schofield, N. & Tagg, A. 1984 Testing a physically-based flood forecasting model (TOPMODEL) for three UK catchments. *Journal of Hydrology* **69** (1–4), 119–143.

Chang, F. J., Chang, L. C. & Huang, H. L. 2002 Real-time recurrent learning neural network for stream-flow forecasting. *Hydrological Processes* **16** (13), 2577–2588. doi:10.1002/hyp.1015.

Chen, Y., Kang, Y., Chen, Y. & Wang, Z. 2020 Probabilistic forecasting with temporal convolutional neural network. *Neurocomputing* **399**, 491–501. doi:10.1016/j.neucom.2020.03.011.

Crivellari, A. & Beinat, E. 2020 Trace2trace – a feasibility study on neural machine translation applied to human motion trajectories. *Sensors* **20** (12). doi:10.3390/s20123503.

Deng, S., Zhang, N., Zhang, W., Chen, J., Pan, J. Z. & Chen, H. 2019 Knowledge-driven stock trend prediction and explanation via temporal convolutional network. *Companion of the World Wide Web Conference* 678–685. doi:10.1145/3308560.3317701..

Donahue, J., Hendricks, L. A., Guadarrama, S., Rohrbach, M., Venugopalan, S., Saenko, K. & Trevor, D. 2015 Long-term recurrent convolutional networks for visual recognition and description. In: *IEEE Conference on Computer Vision and Pattern Recognition*. pp. 2625–2634.

Douglas-Mankin, K. R., Srinivasan, R. & Arnold, J. G. 2010 Soil and water assessment tool (SWAT) model: current developments and applications. *Transactions of the ASABE* **53** (5), 1423–1431.

Gao, S., Huang, Y., Zhang, S., Han, J., Wang, G., Zhang, M. & Lin, Q. 2020 Short-term runoff prediction with GRU and LSTM networks without requiring time step optimization during sample generation. *Journal of Hydrology* **589**. doi:10.1016/j.jhydrol.2020.125188.

He, K. M., Zhang, X. Y., Ren, S. Q. & Sun, J. 2016 Deep residual learning for image recognition. In: *IEEE Conference on Computer Vision and Pattern Recognition*. pp. 770–778. doi:10.1109/cvpr.2016.90.

Hochreiter, S. & Schmidhuber, J. 1997 Long short-term memory. *Neural Computation* **9** (8), 1735–1780. doi:10.1162/neco.1997.9.8.1735.

Hsu, K. l., Gupta, H. V. & Sorooshian, S. 1995 Artificial neural network modeling of the rainfall-runoff process. *Water Resources Research* **31** (10), 2517–2530.

Hu, C., Guo, S., Peng, D. & Xiong, L. 2003 Comparative study on hydrological models in semi-dry and semi-humid region. *Engineering Journal of Wuhan University Engineering Edition* **36** (5), 38–42.

Hu, C., Wu, Q., Li, H., Jian, S., Li, N. & Lou, Z. 2018 Deep learning with a long short-term memory networks approach for rainfall-runoff simulation. *Water* **10** (11). doi:10.3390/w10111543.

Kisi, O. 2011 Wavelet regression model as an alternative to neural networks for river stage forecasting. *Water Resources Management* **25** (2), 579–600. doi:10.1007/s11269-010-9715-8.

Kratzert, F., Klotz, D., Brenner, C., Schulz, K. & Herrnegger, M. 2018 Rainfall–runoff modelling using long short-term memory (LSTM) networks. *Hydrology and Earth System Sciences* **22** (11), 6005–6022. doi:10.5194/hess-22-6005-2018.

Kumar, D. N., Raju, K. S. & Sathish, T. 2004 River flow forecasting using recurrent neural networks. *Water Resources Management* **18** (2), 143–161.

Lara-Benitez, P., Carranza-Garcia, M., Luna-Romera, J. M. & Riquelme, J. C. 2020 Temporal convolutional networks applied to energy-related time series forecasting. *Applied Sciences-Basel* **10** (7). doi:10.3390/app10072322.

Le, X.-H., Hung Viet, H., Lee, G. & Jung, S. 2019 Application of long short-term memory (LSTM) neural network for flood forecasting. *Water* **11** (7). doi:10.3390/w11071387.

Lin, K., Sheng, S., Zhou, Y., Liu, F., Li, Z., Chen, H., Xu, C., Chen, J. & Guo, S. 2020 The exploration of a temporal convolutional network combined with encoder-decoder framework for runoff forecasting. *Hydrology Research* **51** (5), 1136–1149. doi:10.2166/nh.2020.100.

Liu, J., Zhu, H., Liu, Y., Wu, H., Lan, Y. & Zhang, X. 2019 Anomaly detection for time series using temporal convolutional networks and Gaussian mixture model. 2018 International Symposium on Power Electronics and Control Engineering. *Journal of Physics Conference Series*. doi:10.1088/1742-6596/1187/4/042111.

Long, J., Shelhamer, E. & Darrell, T. 2015 Fully convolutional networks for semantic segmentation. In: *IEEE Conference on Computer Vision and Pattern Recognition*. pp. 3431–3440. doi:10.1109/cvpr.2015.7298965.

Miao, Q., Pan, B., Wang, H., Hsu, K. & Sorooshian, S. 2019 Improving monsoon precipitation prediction using combined convolutional and long short term memory neural network. *Water* **11** (5). doi:10.3390/w11050977.

Nash, J. E. & Sutcliffe, J. V. 1970 River flow forecasting through conceptual models Part I – a discussion of principles. *Journal of Hydrology* **10** (3), 290.

Qin, J., Liang, J., Chen, T., Lei, X. & Kang, A. 2018 Simulating and predicting of hydrological time series based on TensorFlow deep learning. *Polish Journal of Environmental Studies* **28** (2), 795–802. doi:10.15244/pjoes/81557.

Sahoo, B. B., Jha, R., Singh, A. & Kumar, D. 2019 Long short-term memory (LSTM) recurrent neural network for low-flow hydrological time series forecasting. *Acta Geophysica* **67** (5), 1471–1481. doi:10.1007/s11600-019-00330-1.

Shamseldin, A. Y. 1997 Application of a neural network technique to rainfall-runoff modelling. *Journal of Hydrology* **199** (3–4), 272–294.

Shiri, J., Marti, P., Nazemi, A. H., Sadraddini, A. A., Kisi, O., Landeras, G. & Fard, A. F. 2015 Local vs. external training of neuro-fuzzy and neural networks models for estimating reference evapotranspiration assessed through k-fold testing. *Hydrology Research* **46** (1), 72–88. doi:10.2166/nh.2013.112.

Tian, Y., Xu, Y.-P., Yang, Z., Wang, G. & Zhu, Q. 2018 Integration of a parsimonious hydrological model with recurrent neural networks for improved streamflow forecasting. *Water* **10** (11). doi:10.3390/w10111655.

Tokar, A. S. & Johnson, P. A. 1999 Rainfall-runoff modeling using artificial neural networks. *Journal of Hydrologic Engineering* **4** (3), 232–239. doi:10.1061/(ASCE)1084-0699(1999)4:3(232).

van den Oord, A., Dieleman, S., Zen, H., Simonyan, K., Vinyals, O., Graves, A., Kalchbrenner, N., Senior, A. & Kavukcuoglu, K. 2016 WaveNet: a generative model for raw audio. arXiv: 1609.03499.

Wan, R., Mei, S., Wang, J., Liu, M. & Yang, F. 2019 Multivariate temporal convolutional network: a deep neural networks approach for multivariate time series forecasting. *Electronics* **8** (8). doi:10.3390/electronics8080876.

Wang, G. Q., Zhang, J. Y., Jin, J. L., Pagano, T. C., Calow, R., Bao, Z. X., Liu, C. S., Liu, Y. L. & Yan, X. L. 2012 Assessing water resources in China using PRECIS projections and a VIC model. *Hydrology and Earth System Sciences* **16** (1), 231–240. doi:10.5194/hess-16-231-2012.

Wen, Y., Hu, C., Zhang, G. & Jian, S. 2020 Response of the parameters of excess infiltration and excess storage model to land use cover change. *Journal of Hydrology and Hydromechanics* **68** (2), 99–110. doi:10.2478/johh-2020-0006.

Wu, D., Jiang, Z., Xie, X., Wei, X., Yu, W. & Li, R. 2020 LSTM learning with Bayesian and Gaussian processing for anomaly detection in industrial IoT. *IEEE Transactions on Industrial Informatics* **16** (8), 5244–5253. doi:10.1109/tii.2019.2952917.

Yuan, X., Chen, C., Lei, X., Yuan, Y. & Adnan, R. M. 2018 Monthly runoff forecasting based on LSTM-ALO model. *Stochastic Environmental Research and Risk Assessment* **32** (8), 2199–2212. doi:10.1007/s00477-018-1560-y.

Yujie, L., Hongbin, D., Xingmei, W. & Shuang, H. 2019 Time series prediction based on temporal convolutional network. In: *2019 IEEE/ACIS 18th International Conference on Computer and Information Science*, pp. 300–305.

Yunpeng, L., Di, H., Junpeng, B. & Yong, Q. 2017 Multi-step ahead time series forecasting for different data patterns based on LSTM recurrent neural network. In: *2017 14th Web Information Systems and Applications Conference (WISA)*. pp. 305–310. doi:10.1109/wisa.2017.25.

Zhao, R. 1992 The Xinanjiang model applied in China. *Journal of Hydrology* **135**, 371–381. doi:10.1016/0022-1694(92)90096-E.

First received 10 March 2021; accepted in revised form 23 June 2021. Available online 7 July 2021

IWA Publishing's authorised EU representative for General Product Safety Regulations is Diane D'Arras, 15 rue Duret, 75116 Paris, France, e-mail: safety@iwap.co.uk.

Printed and bound by CPI Group (UK) Ltd, Croydon, CR0 4YY
12/05/2026
02108879-0003